Teaching Geography

Teaching Geography

Phil Gersmehl

THE GUILFORD PRESS
New York London

A Division of Guilford Publications, Inc.
72 Spring Street, New York, NY 10012
www.guilford.com

Printed in the United States of America

This book is printed on acid-free paper.

Last digit is print number: 9 8 7 6 5 4 3 2 1

Library of Congress Cataloging-in-Publication Data

Gersmehl, Philip.
Teaching geography / Phil Gersmehl.
p. cm.
Includes indexes.
ISBN 1-59385-155-3 (hardcover) – ISBN 1-59385-154-5 (pbk.)
1. Geography–Study and teaching (Middle school). 2. Geography–Study and teaching (Secondary). 3. Geography–Study and teaching (Middle school)–Activity programs. 4. Geography–Study and teaching (Secondary)–Activity programs. 5. Creative activities and seat work. I. Title.

G73.G47 2005
10′.71′2–dc22

2005001897

Cartography and research by Althea Melisse Willette and Clay Mering

To Herbert Gross,

who once pointed
to a place on a map
and asked a class
if there were fjords
there
in that part of the world

and then gasped
for dramatic effect
when some students
answered correctly

those who were there
know what it was like
to learn from a master

Acknowledgments

Many people have contributed to this book in many ways. The list of people I am indebted to includes Carolyn Anderson, Eric Anderson, Peter Anthamatten, Clark Archer, Chandra Balachandran, Les and Lois Bayer, Bob Bednarz, Sarah Bednarz, Jerry Benson, Mark Bockenhauer, Elaine Bosowski, Osa Brand, Dwight Brown, Jae-Heon Choi, Hee-Bang Cho, Greg Chu, Bryce Decker, Jon Dicus, Ron Dorn, Charlie Fitzpatrick, Larry Ford, Carol Gersmehl, Glen Gersmehl, Herold Gersmehl, Louise Gersmehl, Paul Hanson, James Harrington III, J. Fraser Hart, A. David Hill, Sidney Jumper, William Kammrath, Kathy Klink, Cary Komoto, Charles Kovacik, Raymond Krishchyunas, Hsiang-te Kung, David Lanegran, Mui Le, Ann Lewandowski, Pierce Lewis, Catherine Lockwood, Gail Ludwig, Jim Marran, Clay Mering, Judith Meyer, Peter Muller, Shuichi Nakayama, Darrell Napton, Alexei Naumov, Harris Payne, Jerry Pitzl, Milton Rafferty, Karl Raitz, Merrill Ridd, Sergei Rogachev, Carlos Ruiz-Rodriguez, Cathy Riggs Salter, Sula Sarkar, Ted Schmudde, Jonathon Schroeder, Leonid Smirnyagin, Everett Smith, Michael Solem, Rod Squires, Joseph Stoltman, John Tichy, Fumio Wada, Fred Walk, Stephen White, Brenda Whitsell, Althea Willette, Jim Young, and Susy Zeigler.

Comments about anything that is in this manual (or should be!) are welcome at any time; please send them to: gersmehl@umn.edu.

Why do we study geography?

To find out why
something we know "for sure,"
here,
is wrong,
over there,

and why
something that works for someone else,
where they are,
won't necessarily work for us,
here.

Contents

Contents

INTRODUCTION

Before We Start

A Los Angeles policeman abruptly stops the car, takes his pistol out of its holster, and checks the safety. He puts the gun in his lap, with the muzzle aimed out through a marked area in the reinforced door. Then, seeing a "more than mildly concerned" expression on the face of his tour passenger (me!), he says:

"You're here to learn about graffiti code, right? how gangs tag their territory? Let's review: what does that asterisk sign on that wall over there mean?"

"It's a warning to a rival gang – don't go here or you could get hurt, maybe even killed."

"Right. Well, look farther to the right. See that number? That's my badge number, with the kill sign on it. It's been there awhile, and it's probably just bluff, but why take chances?"

Like barbed-wire fences in Kansas or mine-perimeter signs in West Virginia, the graffiti on the wall in this part of Los Angeles are **territorial markers**, physical signs on the landscape to warn people that the area behind the sign is different in some way. Usually, there are particular kinds of behavior that are expected (or not allowed) in that marked area.

Territorial marker

Landscape feature that tells others who claims control over an area

A month later, near an entrance to the Blue Line of the Metro in Washington, DC, I asked what some graffiti on the wall meant. The answer: "nothing much; we don't mark territory the same way here as they do in LA."

The subject we call geography emerged because people need ways to organize, teach, and learn what is appropriate in a given place. Geography tells you how to dress – for the climate, the company, or the culture. Like a language guide or a book of etiquette, it can deal with topics that are trivial or extremely important, depending on what specific content is chosen and how it is taught.

About here, some self-styled "educational reformers" would say, "see, this is the 'new' geography, and it's not about states and capitals." But in fact, geography *is* about states and capitals, among other things, because a state is a marked territory, and its capital is one place where people make rules about how to behave when you are in that territory. To do that, citizens and their representatives must know two things about truth, time, and space:

Truth, time, and place

Some truths change (while others stay the same) from one *time* to another. Learning why is the goal of history, and it is important. As an old adage says, people who are ignorant of history are condemned to repeat the mistakes of the past.

Some truths change (while others stay the same) from one *place* to another. Learning why is the goal of geography. People ignorant of this principle may do the wrong thing in the right place, or the right thing in the wrong place.

Of course, people can also study history and geography because they are interesting and fun! Exploratory curiosity is as important as practical applications in justifying a study of geography.

In an age of tight budgets and competing demands, however, it is difficult to argue in favor of teaching a subject unless it appears to have a clear value to students, parents, and school boards. It pays, therefore, to follow a variant of the old scout motto: be prepared, at any time, to explain why it is important to learn geography.

A Few Comments on the Practical Value of the Kind of Geography Described in This Book

We assume you are reading this book because you are interested in teaching geography (or because you have been assigned to teach geography, and you know that you cannot keep students' attention unless you appear interested in the subject).

It's also fair to assume that you occasionally have to deal with students, parents, colleagues, and administrators. Some of these people, unfortunately, may think that geography is a trivia game or a primary-school subject. To them, geography consists primarily (pun intended) of knowing the names of the capitals, rivers, and mountains in each country of the world.

If you have to deal with that kind of audience, your personal collection of convincing rationales might get thin. In that case, you may find a use for some rationales that we have devised (or borrowed) over the years.

In that spirit, here is a local rationale for geography:

> Geography is about the locations of things. Students (present and future business-people, voters, and elected officials) should learn how to select and arrange buildings, roads, parks, election districts, and other things in ways that are fair, safe, efficient, and even beautiful.

Rationales for learning geography:

It helps us organize our place

And to understand other places

And here is an international rationale:

> Geography is about conditions in other places and connections with those places. Students (present and future citizens of the world) should learn about the land, climate, economy, and culture of other places. That knowledge will help them deal with an increasingly interconnected and often highly competitive world.

Are these rationales linked? Definitely – fair, safe, efficient, and beautiful places tend also to be residentially desirable, culturally tolerant, and globally competitive.

A Road Map for the Next Chapters and the Transparencies

Welcome to the world of geography, the art/science that deals with where things are located, why they are located there, and what difference their location makes. In the next three chapters, we will describe a spatial perspective, examine two different kinds of maps, and see how geography deals with three strands of meaning at the same time. That will form the background for a look at four key ideas (cornerstones) of geographic analysis (Chapter 4).

Each chapter uses maps or diagrams to illustrate some points. To make these illustrations as worthwhile as we could, we made them as copier-ready transparency masters on the accompanying CD, so that a teacher could use them immediately in an elementary- or secondary-school classroom. The text accompanying each graphic is an outline of a classroom activity related to the subject of the transparency.

These activities, however, are just examples in a book on teaching geography – they do *not* add up to a course in geography! We borrowed and adapted ideas from many sources, including the National Geography Standards, curricula such as ARGWorld and Mission Geography, and lesson plans written by teachers in many workshops and summer institutes. We made no effort to be "consistent" in approach or grade level, on the assumption that teachers will organize and modify the materials according to grade level and how lessons are structured (i.e., whether they are demonstrations, discussions, cooperative groups, individual projects, etc.).

In most cases, the Transparencies and activities in this book were designed to stand alone (there should not be a need for students to have prerequisites or other materials). In general, we do not believe that a geography class should consist of a bunch of standalone activities (much more about that later). For that reason, we sometimes suggest ways in which the Transparencies can be combined to reinforce each other, perhaps even over many months.

Each of the activities has a fairly simple and straightforward "engine" (the specific thing that students *do* to make the activity run). For example, one activity asks students to find their position on a map. In another, students discuss the advantages and drawbacks of various sites for a new settlement. A third involves comparing maps in order to identify a possible cause for a disease.

The engine of a classroom activity should be interesting, easy to explain, and meaningful.

An *interesting* engine makes students want to learn. That is the basis for an old educational proverb: "if you want to solve a discipline problem in your class, spend 4/5 of your time thinking how to make the content of the class interesting; you may not even need to spend all of the other twenty percent dealing with problems."

An *easy-to-explain* engine lets students get to work quickly. Life is too short to spend so much time explaining the rules that there is not enough time left to play the game.

A *meaningful* engine helps students see that the learning is worthwhile. This is the springboard for the authentic-assessment movement (more about that in Chapter 7).

You might notice that this book does not have any crossword puzzles or find-this-word-in-that-mass-of-letters games. Those activity engines are certainly easy to explain, and some students find them interesting, but do they have a meaningful connection with solving real-world issues or satisfying human desires? The most charitable thing I can say is that students may like them better than a dry lecture.

It seems to me that the menu of pedagogical possibilities, however, ought to have more than these three choices: dry lectures, pointless games, or nothing!

1

One Perspective: A Way of Looking at the World

The Central Valley of California has the kind of land that provokes envy among farmers in less favored regions. It is awesomely flat, with fertile soil, hot summers, and rivers full of water from snow-capped mountains nearby. Along a highway in the Valley, a bill-board proclaims: "One in four Californians has a farm-related job – we grow food for the rest of the nation."

That statement is nonsense, as are many statements about the California economy. But to understand why, it helps to see what evidence people might have considered in coming to that conclusion.

Transparency 1A, for example, is a map of farm production, and it clearly seems to support that claim. It is a **choropleth map** – one that colors each state according to some measure of rank or importance. Specifically, this map uses different colors to show the total production of grains, vegetables, fruits, milk, meat, and other farm products in each state. It clearly shows that production of these products is greater in California than in any other state, by a wide margin.

Choropleth map

One that colors entire counties, states, or other political units according to their value

A map of total food production, however, tells only part of the story.

Moreover, the designer of this particular map is graphically illiterate (would you call that condition *iggraphicacy*?) By this we mean that this map maker used conventional map symbols, but not in an appropriate way – choropleth shading is not an accepted way to show total quantities of something. We'll explain why in another chapter. For now, it's enough to know that the design of this map is like saying "I hungered is" in English. A perceptive reader can still figure out the correct message, but the language is used incorrectly and the result can be confusing.

How do we cut through the confusion? One way is to learn the "words" in the language of maps well. If we can recognize when someone is misusing them (whether out of ignorance or deliberate intent to mislead), we can take the necessary precautions to get the message correctly.

Another way is to use other information to help put the data on a map into perspective.

The Census of Population, for example, says that California has the largest population of any state, by a wide margin. This means that California has many more human mouths to feed than any other state. To put California's food production into perspective, therefore, we should compare food production with population. One way to do this is to calculate food production *per person*. By that measure, California does not rank near the top – in fact, its food production per person is actually less than the average for the other 49 states (Transparency 1B).

Here is a simple way to illustrate this put-the-data-in-perspective principle in a classroom. Suppose Ray had 8 right answers on a math test, whereas Kay had 15 correct on her test.

Raw data

Counts or amounts: number of people, tons of aluminum, miles of railroad dollars, and so on

Does this kind of **raw data** give you enough information to decide which student did better?

Kay's performance is clearly better, *IF* we know that both students took the same test. Let's say we know they didn't take the same test. What else do we need to know in order to find out who actually did better on the tests they did take?

For one thing, the tests may have been different in length. If Ray's test had 10 questions and Kay's had 30, then Ray's performance seems to have been better: he got 80 percent of his questions correct, whereas Kay had only 50 percent correct.

Processed data

Percentages, yield per acre, income per capita, miles per hour, and other ratios.

Calculating PERcentages of correct answers (a logical equivalent of farm production PER capita or yield PER acre) helps to put the test scores into PERspective. The calculated ratios (should we call them **processed data**, as opposed to raw data?) can help tell a more complete story. No single number, however, can ever be the whole truth. For example, Kay's test questions may have been more difficult; that makes her lower percentage more understandable.

Likewise, many of the farms in California produce vegetables, wine, nuts, and other specialty crops, which are harder to grow in other parts of the country. That makes California's crops more valuable, in money terms. At the same time, it makes another gap even wider; in fact, on a calories-per-capita basis, California is a net food *importer*, rather than a supplier, to the rest of the country.

Each additional attempt to put raw numbers into perspective helps people understand things better. That is a key goal of liberal education: to enhance the ability to put things in perspective. A map is an exceptionally effective way to put some kinds of data into national or international perspective.

This "misleading-data-detector," however, is not the only kind of perspective we are trying to teach in a geography class. A liberal education should also foster awareness of (and appreciation for) other perspectives. That use of the term "perspective" can apply to different cultures (on a global scale) or different population groups (on a local scale). It can even apply to different scholarly disciplines. In fact, that is an exceptionally good way to look at what we are trying to do in a geography class.

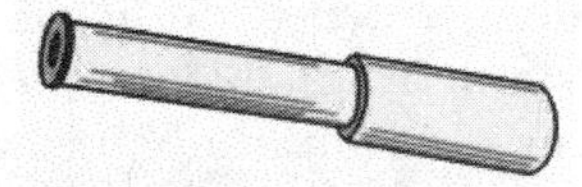

Disciplinary Perspectives

Geography, history, the humanities, and science frequently deal with the same topics, but they look at the world from different perspectives:

Academic perspective

A specific way of asking questions and finding answers

- *Scientists are concerned with process.* The focus is on causes and effects that occur regardless of time or place. The key questions often begin with "how."
- *Historians are concerned with time.* The focus is on the time of events and what happens before and after them. The key questions often begin with "when."
- *Humanists deal with ethics and aesthetics.* The focus is on how to judge things like morality and beauty. The key questions often begin with "should" or "how important."
- *Geographers are concerned with space.* The focus is on locations of things, conditions in a particular place, and connections among places. The key questions usually begin with "where."

A geographer may borrow knowledge from other disciplines, but the focus is always on the locations of things and the connections among locations. We do not claim that this is the only path to truth, or even the best path. It is just one part of the truth. In some cases, however, it is an essential part – one cannot get a valid picture without a geographic perspective.

The value of a geographic perspective rests on three simple ideas:

1. There are reasons for things being where they are.
2. There are advantages for things being in appropriate places.
3. There are penalties for things being in the wrong places.

Transparencies 1C to 1H provide some ways to illustrate these points in class.

After looking at these examples and many others like them, here is my favorite potluck-dinner-or-other-social-occasion rationale for geography (in 25 words or less):

> Geography is the discipline that helps us understand
> why something we know here, in this location,
> may be wrong in some other location.

That is a simple-sounding sentence, but it can be applied to a wide variety of subjects, ranging from table manners and courtship rituals to the design of freeways and the administration of public health policies. A skillful geographer (just like a good historian or scientist) has to draw on a wide range of knowledge and skills. Some of that necessary background is factual: to understand why a given action may fail in one place even though it works in another, we must know something about the places. Factual knowledge, however, is not always enough; we also have to understand the ways in which people organize facts and ideas.

Deductive versus Inductive Approaches

Inductive and deductive modes of teaching

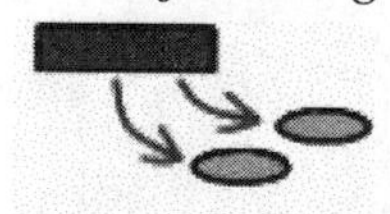

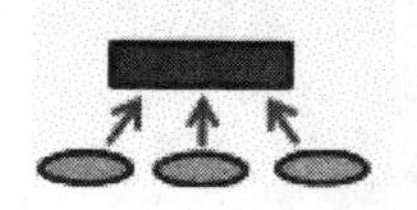

We could keep on talking about the geographic perspective in this verbal, abstract, **deductive** mode. I submit, however, that much of geography is easier to teach **inductively**, by making observations and then drawing conclusions about them. This "back-door" is an especially useful perspective for teachers, because it helps bridge an age-and-experience gap between teacher and student.

That gap is part of the reason why deductive teaching does not always work well. A nice, logical, deductive outline might sound fine to a teacher, who has a bunch of personal experiences and mental images that the outline ties together into a neat package. The fit with those experiences is precisely what makes the ideas in the logical outline seem so compelling, so "right."

Young students have fewer experiences, however, and the ones they have are often individual rather than shared. For that reason, they sometimes "just don't get" an idea that a teacher might think is almost self-evident. Geography teachers, therefore, have to make a conscious effort to build a "vocabulary" of shared images as well as an explanatory worldview.

Having made that assertion, I suspect that it is probably wise just to end this first chapter and proceed to the rest of the book. We will have more to say about images, explanations, and perspectives later. Those comments, however, will come after we assemble some shared experiences with the transparency masters and other concrete examples that illustrate this book. The delay is for a sound pedagogical reason: in the long run an inductive approach is better for most geography, including books about teaching geography!

▪ ▪ ▪

On the next few pages are (1) thoughts about the importance of the little word "where," and (2) a kind of editorial on American perspectives. You could safely postpone reading either of these "postscripts." They are here because they are related to the idea of perspective, and because a number of teachers in our focus groups said that they would appreciate a few comments about these topics. To help you decide whether to read them now, here is what they are about:

Postscript 1-1: This gives teachers a few more examples of why public understanding of geography is important in a fair and efficient society.

Postscript 1-2: This postscript is for reading if (or whenever) you wonder why schools in the United States seem to lag so far behind their counterparts in other countries in trying to foster the kind of public understanding of geography discussed in Postscript 1-1.

POSTSCRIPT 1-1

Thoughts about the Little Word "Where"

Where. Such a simple-sounding word, but it can mean so much.

We make decisions about **where** to put or do things every day. These decisions usually have consequences, which depend again on **where** you are. The consequences can often be predicted by using the tools of geographical analysis.

Think about **where** the bathroom is located in your apartment or house. If it is in the middle of the hall, people from the kitchen and the bedroom can get to it equally well. If, however, it is at the end of the hall, beyond the bedroom, it will take four seconds longer to get there from the living room and kitchen (**where** most people start more trips to the bathroom).

Four seconds, each way, for perhaps nine times a day, 365 days a year. That adds up to seven hours, which could be used for something else.

Big deal, you say? Seven hours' savings of time for each family in a city of a million people, if it could be translated into income, would be enough to build nearly 150 houses. It could repair more than a thousand miles of road. It could provide a year of day care for two thousand children. That is how much the people of a city would save if all of the bathrooms were located four seconds closer to **where** the people are.

On the other hand, there is a lavatory near my fifth floor office, but I often walk up to the one on the tenth or twelfth floor – people who write for a living should probably take a few exercise breaks during the day! The importance of **where**, therefore, depends on your perspective.

It also depends on what activities are being analyzed. Consider, for example, the task of trying to drive through a city **where** property lines had been arranged to fit life before the automobile. I was curious, so I paged to the "L's" in my *American Almanac* to compare Lowell, Massachusetts (an old city, by U.S. standards), with Lincoln, Nebraska (a younger city, of about the same size). Partly because of Lowell's having a less efficient road network, the U.S. Census reports that a typical worker in Lowell spends about 12 minutes more each day commuting to work. Over a year, the hours "wasted" in extra commuting in this mid-sized city are roughly equal to the annual total number of worker hours for all employees in a company the size of Microsoft, Rubbermaid, or Southwest Airlines (employment data from *Hoover's Handbook of American Business*).

All that extra cost, just because of **where** roads are located.

The costs or benefits of "where-knowledge" are not just financial.

- For example, imagine what it is like to get bitten by a strange dog in a city **where** you cannot speak the language. **Where** do you get help?
- Now imagine that you are choosing a trail and putting things in your backpack. Would it be useful to know **where** the blueberries are ripe this week?
- Finally, picture walking through a neighborhood with boarded-up windows, trash in the street, and graffiti on the walls, and not knowing **where** to find an open store or telephone if you needed one.

We could easily give many more examples of the value of where-knowledge when people are making decisions at the scale of an individual. As with the examples of the bathrooms and freeways, the value of such knowledge is proportionally greater (although often harder to measure) at the scale of a community or society.

Here is a poignant example. On a crisp November day, a child in northeast Detroit was jostling with other children and fell into a street, **where** she was seriously injured by a passing car. This accident happened **where** it did partly because 100,000 commuters use half a dozen city streets to get from **where** they live (in the northeastern suburbs along the river) to **where** they work (in the central city). Over a year, dozens of schoolchildren are likely to be injured on these particular streets. The basic geographic principle is easy to state: the riskiness of a given school crossing is related to **where** it is with respect to homes, jobs, and traffic patterns.

The consequences, however, can be magnified if suburban voters and politicians are guided by the benighted notion that people who live beyond a city-suburb boundary should not be required to pay taxes to help solve "the problems of the city." In other words, people's perceptions of **where** problems occur can sometimes keep them from taking moral and financial responsibility for problems that they may be physically responsible for causing.

At a global scale, a military decision about **where** to deploy an aircraft carrier battle group can involve extra costs of hundreds of millions of dollars, if the carriers are not in a strategic location when hostilities break out. On the other hand, the decision about **where** to send a peacekeeping envoy can mean the difference between a timely settlement of a dispute and full-scale war.

The fact is, every action on the planet has consequences that depend (at least in part) on **where** the action takes place. The proper response for an individual or a society depends on **where** the consequences occur and who gets hurt or helped by them.

Here's a simple example: suppose a Wisconsin shoe company opens a dozen stores in Florida and sends a truckload of insulated boots to each store to sell for Christmas. This action implies some considerable ignorance of the fact that the weather is different in Miami and Milwaukee. If the company goes bankrupt because its insulated shoes do not sell in a hot climate, the penalty for their ignorance falls directly on them, the ignorant people.

Here's a more complicated example: suppose a beach house is seriously damaged by a hurricane. Should the government declare it a natural disaster and provide a low-interest loan to help the owner rebuild, or should it refuse to issue a building permit for any further construction on such a risky site? The question has an important scale consideration: the benefits of a beachfront location for a house usually go to those few individuals who are lucky enough to live there, but the costs may be borne by everyone who pays insurance premiums or taxes (unless the tax and insurance policies are designed with precisely the kind of geographic understanding we are advocating).

In short, it is easy to find egregious examples of things that make sense in one place but are laughably inappropriate in another. The benefits of geographic awareness, however, are usually more subtle: efficiency, safety, fairness, and beauty often depend on arranging things in a slightly better way – bathrooms a bit closer, shopping malls in better locations, houses on safer sites, tax money transferred between municipalities to help pay for school crossing guards and other services, and so on. As with any subject, geographic understanding is a matter of degree.

POSTSCRIPT 1-2

A (We Hope) Soon-to-Be-Unnecessary Note about Standards and International Comparisons

Much of this book is based on discussions that took place in more than 60 teacher workshops in 23 states, as well as in Japan, Korea, and Russia. Many teachers at those workshops commented that the material being presented was different from what they were used to. Sometimes, though they thought the material was interesting, they doubted it would work for their classes. Comments like these usually came early in the workshop, and in nearly every case, after working with the material, the person later changed his or her mind.

This leaves me with a huge dilemma: an author needs credibility, and yet my experience as a workshop instructor tells me that many teachers' first impressions are skeptical. To get over this hurdle, it might be useful to try to figure out why American teachers often express doubts about teaching the kind of geography that is described in this book.

Even here, a geographic perspective is useful. When we use some straightforward measurements to compare the United States with other countries, we are struck by three facts that seem to be "causes" and one fact that seems to be a "consequence":

Cause-fact 1 – isolation. Compared with other countries, the United States has been relatively isolated. It occupies a large fraction of its continent, has borders with only two countries, and its economy is about eight times as large as the economies of Canada and Mexico put together. Until airplanes and telephones helped "shrink" the globe, Americans simply did not have to worry much about powerful neighbors with different languages and ideals.

Cause-fact 2 – wealth. Compared with other countries, the United States has been relatively rich. It occupies the most favorable part of its continent, and it has much more good land per person than other large industrial economies (Transparencies 1I and 1J).

Cause-fact 3 – power. Compared with other countries, the United States has been relatively powerful. As recently as the 1960s, this country exported more than it imported, loaned more than it borrowed, controlled more property in other countries than those countries controlled here, and held most of the patents on its industrial processes.

Consequence-fact: Typical U.S. high school students receive less than one-fifth as much instruction in geography as their counterparts in other countries.

So what should we reasonably expect? American students perform below the international average in geography. And many American teachers do not know what is attainable or desirable in a geography class in elementary or secondary school. That is not their fault – they have little experience with world-class geography. I repeat: it is not teachers' faults, but if we want to break the cycle of substandard geography education, we may have to ignore first impressions about what is possible or desirable.

We do, however, have a compelling reason to break that cycle, because the world is changing. The blunt fact is that all three of the cause-facts in the list above are less true now than they were a few decades ago. For example, ideas about land ownership that were acceptable when the country had 50 million people may not be appropriate as the U.S. population approaches 300 million. As parts of the country become so much more crowded, Americans need to think more about how they organize and use their land.

At the same time, connections among nations are becoming both more complex and more important. A century ago, there were few links between Tennessee and Japan. Now, a large Nissan factory near Nashville practically guarantees that events in Tokyo will affect people in Tennessee nearly every day, in many ways.

As a result, students in the United States need to learn how to analyze things spatially, not just try to memorize facts about places. They should be observing, hypothesizing, and evaluating, rather than just coloring maps and hunting for placenames in a maze of letters. In short, they need to do world-class geography. (This is a citizenship issue – the resource-rich United States has achieved a high standard of living; but inefficiencies and inequities in the way we use resources now seem to be hindering our efforts to maintain that momentum.)

So, try to ignore doubts, at least for a while. Read the next chapters, which are about two analytical blades, three strands of meaning, and four cornerstone ideas of geography, and think about how you might adapt these ideas for your students. It *will* work – hundreds of teachers have said so. This kind of geography is not the same as reading "all the facts about Zambia" in an almanac or atlas, but students actually know more about the world when they finish. Moreover, students usually think that this kind of geography is at least as much fun as watching yet another video about weird-looking animals, colorful costumes, or strange foods.

(We warned you that this was an editorial! It is offered in the belief that readers have a right to see where authors stand on issues that might influence the content of a book.)

2

Two Blades of a Scissors: Logical Cooperation

A cutting board is useful, because it has a big advantage over an unaided knife: a firm surface under the material you are cutting can usually make the cutting more efficient.

Unfortunately, a cutting board also has a huge drawback: cutting against a board can dull a knife.

> (This kind of dilemma is not unusual when thinking about geographic resources. Indeed, it is so common that it has been given a name: **geographical irony**. Here is a capsule definition of that concept: the very conditions that make a thing or place useful can also pose problems. For example, a hotel may attract more people if it is located right on the beach rather than a few hundred feet inland, but a beachfront location puts more people at risk when a hurricane comes. This book will have more examples of this concept later.)

Geographical irony

Conditions that make a place good can also cause problems
(see Chapter 4)

Viewed in the context of knives and cutting boards, a pair of scissors is an ingenious invention. For some kinds of cutting, scissors have the advantage of a cutting board without the drawback. By aligning two blades so that they cut toward but not directly into each other, it is possible to provide support for the material and still keep the blades sharp.

A scholarly discipline is basically a way of cutting into (analyzing) the world in order to see how it works. The efficiency of that cutting depends on the sharpness of the analytical blades and how precisely they can be aimed.

The Geographical Scissors

Geographers have a scissors-like method of analysis that works very well when used properly. Each blade of the scissors has a unique focus and a distinctive type of map, which work especially well when used together (see Transparencies 2A and 2B):

Regional geography

Study of how a variety of conditions and forces interact at a local scale (various authors link this approach with words like synthesis, idiographic, ecological, biography)

Regional geography is the study of many things in one area. This blade of the geographical scissors looks at all the conditions at a place – its climate, soil, population, economy, politics, religious beliefs, historical events, and so on. The specific focus is on how various features interact in a place to make that place what it is.

Someone using the regional approach might look at Afghanistan, for example, to see how mountains, road networks, vegetation cover, ethnicity, and building types interact with each other to allow for a guerrilla-war strategy that is hard to counter with conventional military force.

Reference map

Shows the locations of a variety of things within an area

The primary tool for the regional blade of the geographical scissors is a **reference map**. This kind of map uses different colors and symbols to show the locations of many things in an area. Armed with a good reference map, geographers can ask what is located in a place, what forces converge to shape the place, what resources are available there, and how people define, mark, and use the land there. (Transparencies 2C and 2E are reference maps.)

Topical geography

Study of patterns at a broad scale (various authors link this approach with words like analysis, nomothetic, systematic, culture)

Topical geography is the study of spatial patterns at a broad scale. The focus is on the arrangement of a single feature in a fairly large area – where it is abundant, clustered, scattered, scarce, or absent.

Someone using the topical approach might look at a map of Islam, for example, and notice that this religion is important across northern Africa and southwestern Asia, with extensions into central Asia, Europe, Indonesia, and several other parts of the world. (The extension into central Asia is of obvious significance for someone looking at Afghanistan; this linking of global pattern with local interaction is the scissors effect.)

Thematic map

Shows the pattern of a specific thing in a large area

The main tool of topical geography is a **thematic map**, which shows where a specific thing occurs within a large area. Armed with that kind of map, geographers ask why that feature occurs where it does, what else those places have in common, and how things may have spread to where they are now. (Transparencies 2D and 2F are thematic maps.)

Geography achieves its unique analytical power by using both the topical approach and the regional approach at the same time.

For example, you might notice a feature in a specific place and wonder why it is there (that's acting like a regional geographer). Look around for plausible causes, and identify a likely one. Then shift to the topical mode and try to find out where else that possible cause also occurs. You can then shift back to the regional mode and look at the interplay between the possible cause and other features in those places (as on Transparencies 4M and 4N, or 4T, 4U, and 4V).

The process can also start with the topical blade. Say you come across an intriguing thematic map (e.g., the map of Arab and Israeli settlements on Transparency 9C). Try to find some other things that have similar map patterns. Then shift to the regional mode to see how those factors interact with each other in a local area (as on Transparency 9D).

In short, regional and topical geography are like the two blades of a scissors. If they are separated, they can still cut, but the blades are far more effective when used together. (Keep this analogy in mind as you read the other chapters of this book; they have many examples of the scissors principle in action.)

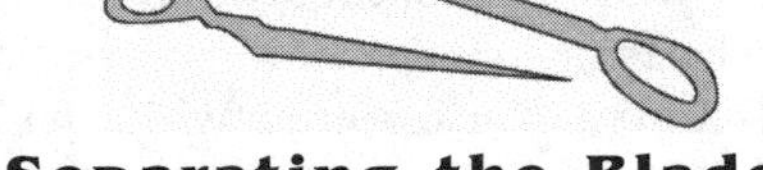

Separating the Blades

The blades of the geographical scissors work best when used together. Unfortunately, many geography curricula and textbooks are organized in such a way as to blunt this kind of cooperation. In effect, many geography classes and textbooks are more like knives and cutting boards; they work for awhile and then get dull.

This happens when textbooks or courses make **artificial divisions** within geography and then deal with each category separately. For example, universities typically offer a group of topical courses, with titles such as physical geography, climatology, economic geography, or urban geography. The curriculum also has some courses on the geography of various regions such as Africa, Australia, or North America. Along with these topical and regional courses, a typical university curriculum also includes technical courses such as cartography (map making) and quantitative methods.

Artificial divisions within geography

(and why they are counterproductive; see Chapter 8)

This kind of specialization can work well at a research university, where students take several courses and then enroll in graduate seminars that tie the various parts together and show how they all can work together to solve specific problems.

Unfortunately, many university teachers use the same categories even at the introductory level. They get in the habit of working primarily within one side of the discipline when they do their research. The other blade of the analytical scissors is usually in the back of their minds, where it stays sharp and ready to use in their seminars and advanced courses. Their students, however, do not have that background, and therefore they often see only part of the scissors, and it doesn't seem to work well. This may be one reason why a large number of student evaluations of university teachers give high marks for "knowledge of subject matter" but much lower scores for "overall teaching ability."

Doubly unfortunately, the university-level topical-regional split tends to "trickle down" into textbooks written for elementary and secondary school. This is especially true for texts (and unfortunately there are a lot of them) that are written or edited by social studies teachers, English majors, and other people who have had only a few geography classes. These people may not have experience using the scissors approach, and as a result they actually think that it is a desirable goal to keep the blades separate.

Triply unfortunately, publishers sometimes try to present this weakness as if it were actually a strength. They advertise that their textbooks "establish a general background with chapters on climate, landforms, agriculture, industry, and urbanization, and then go on to analyze various regions of the country."

That works, but not very well: it is like taking the scissors apart before trying to cut. The general topical chapters at the beginning of such a book seem abstract, even irrelevant, to students. They therefore often fail to master the material. Then, when they get to the regional chapters later in the course, they lack the background that presumably was "covered" in earlier chapters.

In short, a topical-and-then-regional organization loses one of the most ingenious parts of the geographic perspective, namely its awareness that the world has forces that operate on a global scale but interact with other forces on a local scale. That local interaction of a unique mix of forces is what creates the specific features we see in particular places, and in turn those local conditions add up to the global patterns we see.

Global forces, local interactions

(It's a variant of the old adage: think globally, act locally; that is the only way to achieve global change!)

To summarize the problem caused by separating the scissors, note that geography loses part of its "soul" when it uses only one blade:

Without the regional-interaction half of the scissors, geography becomes like a science, looking at a single factor in isolation in order to see how it works. This is valuable information, but if a geography class does only this, it is vulnerable to the criticism that other disciplines such as geology or economics do it better.

Without the global-topical half of the scissors, geography becomes like a catalog of features in a local area. This information is also valuable, but taken to extremes this kind of study can become a rather trivial pursuit.

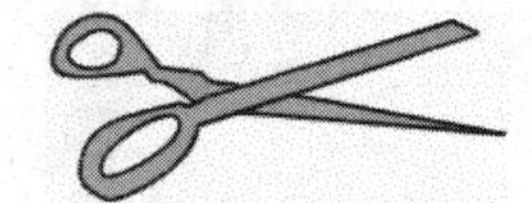

Keeping the Geographical Scissors Together

For geography teachers, the challenge is to figure out ways to keep the blades together, so that the scissors can cut efficiently. This is basically a technical problem in applied communication theory, and therefore it can be very useful to look at examples of well-designed advertisements, books, magazines, television shows, and other successful forms of communication.

Textbook authors, for example, sometimes solve the problem of a scissors approach by using **layout devices** such as boxes or tinted pages. A regional book, for example, might have chapters with titles such as "Australia," "New England," "The Equatorial Region," or "Yakut Homelands." The primary emphasis of each chapter is on the specific place noted in the title. To provide the other blade of the analytical scissors, the text may have "theory pages" or "analytical boxes" that treat topics such as Central Place Theory (see Transparencies 2G and 2H) or The Demographic Transition.

Layout devices

Mechanical "tricks," such as tinted pages, unique typefaces, or outline boxes, which book authors often use to help readers stay oriented

A textbook with a topical organization, by contrast, might have chapters with titles such as "Population Geography," "Economic Patterns," or "Environmental Problems." The primary emphasis of each chapter is on broad principles that apply anywhere. To provide the other half of the analytical scissors, the authors might make a set of extended figure captions, vignettes, or case studies. These offer local illustrations of the general principles that are discussed in the surrounding text. Outline boxes or colored pages can isolate these elements so that they do not interrupt the flow of logic in the chapter itself.

Matrix

A row-and-column table; this example has columns of topics and rows of regions

This kind of organization, of course, requires careful planning to make sure that important topics and regions are "covered." A synopsis of such planning often takes the form of a **matrix**, with topics on one axis and regions or local case studies on the other. Symbols in this matrix show which topics are emphasized in the context of which regions (or which local case studies help illustrate each topic – see Transparencies 2I and 2J).

Here comes a brief statement that some geography teachers might view as heresy: it really does not matter whether the primary organization of a book or a class is topical or regional. In fact, a teacher may choose to change which blade of the scissors is "up" from day to day. Some students may become disconcerted if the class does not "follow" the book every day. Anticipate that consternation and try to deal with it, but do not compromise on the basic principle: the benefits of scissors-like geography (if done well) far outweigh the occasional frustration.

Eight ways to organize a room to "be a scissors"

Using both blades of the analytical scissors requires only two things: the will to do it, and a few mechanical "tricks" to remind students of what kind of geography is happening at the moment. For example, a teacher can "zone" parts of a classroom to facilitate the creative tension between the regional and topical blades of the geographical scissors. Here are eight well-tested strategies:

Let readings be the other blade.

1. Use a topical outline for class activities and assign readings that have a regional perspective, or vice versa. For example, a teacher might design a unit that deals with recent changes in the locations of various kinds of retail services, such as outlet malls, gas stations, or drug stores. During class discussion, ask students to react by explaining how the general theory of retail trade fits with their readings on the construction of a new mall or the attempt to revitalize an old downtown area. (To convince hesitant parents or school board members about curriculum choices, the practical implications of these topics might also be noted. Most adults have lived through several revolutions in the geography of retail trade and have memories of both the excitement of a store-opening and the dislocations that a store-closing can cause for sellers and buyers alike.)

Have a specific place in the room for the other blade

2. Walk to a specific place in the room when you step from one role to the other. For example, suppose the topic for the day is the regional geography of Japan. A teacher could walk over to the "topical desk" in order to project a world map and to outline a theory about international movement of capital.

   (Yes, that kind of topic should be introduced in middle school. A teacher does not have to use a jargon term like "international capital flow" to help students become aware that Japanese investment is crucial to the economies of places as far apart (and as different) as Thailand, Toronto, and Middle Tennessee (where Nissan trucks are made).

3. Pick up a prop such as a book or flag (or put on a distinctive hat) when you step out of the topical role into the regional one, or vice versa. For example, if a day is organized as a topical discussion of natural hazards, it might include an occasional comment from "a resident of London," "a representative from the Tamil group," or "a voter in western Mississippi." This individual could provide insights into how the general idea for the day fits into the geography of a local area. (You might even try to use a regional accent and some colloquialisms, if you can do it tastefully; or bring a short audio recording or videotape to class.)

Use a prop to identify the other blade

4. Use different media to present different perspectives. In my university classroom, the slide projector's job is to provide images and other features about specific places (i.e., regional geography). The overhead projector's role is to show thematic maps and outline analytical theories (i.e., topical geography). With computer presentations, I use a different background and screen layout. Students soon realize that they should look for place-facts when material is presented with slides or screens that have frames around photos. They know that this kind of information will be tested with multiple-choice and matching questions (see Chapter 7). Material presented with the overhead, by contrast, is likely to be topical (theoretical) in nature. It applies to many places, and student mastery will be evaluated with short essay questions or in term projects. In short, the simple act of turning off the overhead and pointing at a slide can save me the trouble of verbally telling students what kind of information they are about to receive. Other media that lend themselves to being "typecast" in this way include the chalkboard, films, videotapes, guest speakers, or computer simulations. (Tables 3-1 and 3-2, at the end of Chapter 3, have tips on the design and use of slides and transparencies.)

Use different media for each blade

5. Assign roles to students. One way is to ask individual students or groups to do background research on some specific places or topics. After (or even while) making a topical presentation on world patterns of land use, the teacher can call on those students to offer relevant information on "their" specific country. Similarly, a teacher might talk about the regional geography of the Tokyo area, and interrupt the presentation to ask a student to summarize a reading about the kind of terrain and soil needed for rice production.

Assign roles to bring out the other blade

(It is only a short step to eliminate the teacher's presentation role and have students do both halves of the analysis. For example, one group can make a world map from United Nations data, while another group does research on a specific country or region. Bring the groups together, and give each the same instructions: prepare some clear illustrations of your point, present your findings to the class, and then put the ideas together by writing a paragraph that outlines how the material presented by the other group adds to your understanding of your own subject.)

Use field trips or homework for the other blade

6. Use field trips or homework to complement in-class activity. Present something in class (by any method, from lecture to individual or group activity), and then assign homework that uses the other perspective. For example, you might make a class presentation about how land value decreases as you go away from the center of a city. Then have students gather information as they travel to or from school or shopping areas. For example, you could ask them to focus on the size and spacing of buildings as an indication of the value of the land.

   (One intriguing student project is to note how much it costs to park a car in different parking lots around a city. This idea came from a fifth-grade student who did a geography-fair project on the prices people charged to park cars in their yards during the State Fair. As you might guess, the fees declined from about ten dollars right across the street to a mere two dollars seven blocks away from the entrance gate. Beyond that, it apparently was not worthwhile for people to stay close to their houses all day in the hope that someone would want to park in their yards.

   The student who did this project now has a powerful firsthand appreciation for what she might later learn is the Von Thunen model of location rent, a famous and very useful geographic theory. For some more tips on homework in geography, see Table 2-1, at the end of this chapter.)

Announce that a test will be used as the other blade

7. Use a test as the other blade. Present something in class from a topical perspective, then hand out a map and tell students that the test will ask how this theory applies in one of the places marked on this map. For example, a class might deal with birth and death rates, while the test asks about population change in specific countries. Give students a few days to think about an answer, discuss in their groups, do some Internet or library research, and/or ask questions in class. Praise students whose questions reflect an understanding of the scissors in action.

   Alternatively, describe several regions in class, perhaps with short dramatic skits or poetry readings as well as films or illustrated lectures. Warn students that the test will ask them to draw a thematic map to summarize the in-class presentations for the week. For example, one might read from diaries about the weather along a pioneer trail or military road and then ask students to draw a map to show the "geography of temperature" in that area. If carefully done, this activity can underscore a number of principles about the change in temperature with latitude, altitude, and season.

Ask students to reflect on what's happening in class

8. Ask students what's happening at the moment. Get them to think about whether the class has been looking at the world topically or regionally in the last few minutes.

You may have noticed that this book is doing something like option 4 (using different media for each blade of the scissors). The text itself is mostly general and theoretical, whereas the sample transparency masters often illustrate the principles in specific places. We reverse the roles in a few cases, deliberately, in order to make a point better.

The interplay between topical and regional perspectives is what stimulates thought. For that reason, we emphasize that you do not need to use the examples and transparency masters provided in this book. These examples have done their job if they inspire you to create similar ones for your classroom. For example, feel free to substitute another city roadmap for Transparency 3H, another mall for Transparency 4C, or another hill on Transparency 7F. You could also use temperature or growing season on another continent instead of rainfall in Africa on Transparency 4U. (In fact, if you think of a better example for any point in this book, sketch it and send it to the authors. We can always use new ideas for the next edition!)

A Contribution from Cognitive Psychology

One final note: cognitive psychologists now have a number of plausible theories about why a scissors-like interplay of local case studies and general principles seems to work better than staying within either the regional or topical mode. Current research suggests that learning is more efficient if multiple brain pathways are engaged.

Brain region

Specific area within the brain that "lights up" when a person does a specific thing

Here is a popular simplification of a complex process: The "analytical" left hemisphere of the brain is the part that deals with abstract theories and geometrical relationships. Meanwhile, the "intuitive" right hemisphere of the brain gets into the act when the topics are colors, face and shape recognition, and some spatial interactions among phenomena. Despite its simplistic tone, this analytical/intuitive division of the brain has more than a germ of truth in it. Spatial reasoning (the kind of thinking that geographers do – see Chapter 6) does seem to involve "cooperation" between radically different **"regions"** on opposite sides the brain. To be efficient, therefore, geography teaching should seek to promote involvement of several parts of the brain.

"Efficient," of course, is a loaded word in pedagogical circles. It is true that a teacher can probably get students to regurgitate more facts about places through rote memorization and drill. Assertion of that "truth," however, should bring up two questions, one general and one specific:

- The general question is whether that kind of learning has durability. In short, does it last?
- The specific question is whether rote memorization equips people with the kind of geographical skills that could prove useful when they download "information" from the Internet, watch the news on television, read newspapers and magazines, and make political choices and personal decisions about where to live, work, shop, and travel. In short, does fact memorization give them skills that can be applied in new situations for the rest of their lives?

The answer to both questions is "probably not." That is precisely why I advocate a scissors approach to teaching geography. Even though it may take more time and effort to set up properly, the gains in durability and applicability are worth the effort.

TABLE 2-1: Some Design Principles for Take-Home Worksheets and Projects

1. **Clear purpose**. Homework should have a point, an obvious relevance to goals that are already accepted or are clearly explained in a handout or web page. Some students, of course, dislike all homework. Many more students, however, resent homework that is ambiguously explained or (worse) is clearly busywork. As one teacher reviewer of an earlier version of this text said, "Students quickly figure out when a project is pointless, and it affects the way they view the rest of the class."
2. **Nonthreatening start.** The first few steps of homework should be relatively easy – perhaps a review question, a straightforward mathematical calculation, paraphrasing of a short reading, or making an observation that does not require fine distinctions. Many students get frustrated if the first step requires complex action or abstract thought, especially if it carries the possibility of being completely "off base." Some will look at the first question, decide that they do not know the answer, and refuse to continue with the homework. By contrast, a concrete first step that results in a product that can be tested immediately for its correctness will usually instill a sense of progress and a willingness to tackle more complex tasks. An acceptable alternative, some of the time, is a first question that calls for speculation, with immediate reassurance that there is no right or wrong answer.
3. **Cooperative logic**. The core of a geography project should feature an interplay of regional and topical information. That "scissors" requires careful thought to ensure that the various projects in a course fit together to "cover" the facts, theories, and skills that are to be emphasized in the course (perhaps by using a matrix, such as Transparency 2J).
4. **Time requirements**. The project should include clear statements of the amount of time that is reasonable to devote to each step. Few things are more frustrating for a student than misreading the scope of a project and doing much more or less than was expected. Describing the expectations in terms of the task to be done, the likely time requirements, and the product can help reduce anxiety by providing several mutually reinforcing statements about the scope of a project or other take-home assignment. Providing an example of a successful project (and maybe one that is less so) is even more effective.
5. **Easy-to-evaluate product**. The product should usually be uniform and easy to evaluate. For example, a standard checklist, graph form, or answer sheet can make the task of evaluation much easier. Even in cases where the product to be evaluated is a free-form essay or diagram, a clear set of instructions about the form of that essay or diagram can make the product easier for the teacher to evaluate and less stressful for the student. Having a variety of forms to fill out for a geography class can also help instill good work habits and a decreased fear of the forms that are a part of modern society.
6. **Repair mechanism**. A mechanism for repair and/or improvement after evaluation can help transform the "body language" of a project. Students are far more likely to master the material if they view the teacher as cooperating with them to build something for a personal portfolio, rather than just judging them on a product that is prepared for a deadline and then discarded.
7. **Extension**. Some clear suggestions about ways to expand a project are also useful, especially in a portfolio-based evaluation system. Extraordinary effort that is encouraged (and then acknowledged) can sometimes be a starting point for an informed career decision. The enjoyment felt in doing something worthwhile can become a clue to be noted and used in making choices about courses, educational options, and eventually careers.

TABLE 2-2. Some Design Principles for Personal "Geo-Diaries"

1. Think about a typical day in your life, and make a list of major events that occur in different places, for example, home, bus, school, shopping center, entertainment area.
2. Make a sketch map that shows where you travel and when you arrive at each place. Make the scale accurate, so that someone can easily see about how far you travel between each place.
3. Decide what 10-15 photographs, sketches, recordings, or other impressions you want to use to illustrate your personal geography. If time permits, discuss these with your teacher or classmates, especially if you want to compare geo-diaries with other people, perhaps over the Internet.
4. Take the photos and other impressions during a typical day (or on several days if that is more convenient). Try to avoid having people who can be identified in your scene. Unless you are a journalist working for a newspaper or television station, it is an invasion of privacy to publish pictures of identifiable people without their permission – so, unless you want to pay people for posing and/or get them to sign legal model releases, try to avoid taking pictures with identifiable people in them.
5. Mark the locations of the photos, sketches, and other impressions on the sketch map. Draw a small arrow on the map to show the direction you were aiming for each photo. Write a number or letter by the arrow. Use that number to link the photo with its caption, and note the time of the photo.
6. Check your photos to make sure they show what you want. You can retake photos that are ambiguous or don't show exactly what you want.
7. Write brief descriptions of the photos, interviews, or recordings on a separate piece of paper. "Brief" means 10-20 words, suitable for putting on screen in a class. If you want to add more explanation for a particular picture, write it in parentheses after your main caption. Here are several examples of good captions:

   "A house I pass at 7:35 am on the way to school. This house is a good example of the kind of two-story frame house that people built at the end of the 19th century."

   "Another house that I go by about five minutes later. The truck in front is for a tree-trimming business. (The tree trimmer started that business when he lost his job with a logging company.)"

   "The sound of traffic on ________ Avenue, which I have to cross about 7:45, very close to the peak of the rush hour."

   "A list of flowers and trees that I can see next to the road at point NN; in early July, I can even eat some mulberries right from the tree."

   "The mall where I go for music lessons. It's not very full at 2:30 on Thursday, but wait till Saturday!"

   "Gym where I'm on the volleyball team; practice starts at 3:30. We were second in our conference last year."
8. Discuss with your classmates how to identify your geo-diary, especially if you want to post some of them on the class web site. You might even want to have some e-mail discussions with students in other schools, so that you can share your geo-diaries to compare conditions in different parts of the country or even the world.

3

Three Strands of Meaning: Cognitive Psychology and Geography

Teaching geography is like teaching a foreign language.

Question:

How do people learn a language?

Effective teaching of any subject must start with an understanding of how people learn. For example, teachers in an English-language class for foreign students do not ask students to spend two weeks memorizing words that begin with the letters A and B. Students would rightfully call that numbingly boring. Moreover, both research and experience have reached the same conclusion: rote memorization of vocabulary does not work well.

Answer:

By learning mechanical details and abstract ideas at the same time

Likewise, people seldom try to teach a foreign language by concentrating on nouns for a month and then moving on to verbs, adjectives, and pronouns in the next three months. That approach would also be boring. Furthermore, it ignores the fact that grammar is about relationships among words, not the categories of words themselves.

So, language teachers rightly reject these two approaches. Instead, they try to get students involved with meaningful situations and interesting stories right away. In this way, learners see that mechanical details such as case endings and word placement are important in order to make sense of a story. In effect, the story provides the justification for learning the details of vocabulary and grammar, and the details provide the building blocks needed to convey the nuances of the narrative.

In the hands of a good teacher, the course continues with a carefully calibrated blend of abstract ideas, concrete details, grammatical rules, and semantic idioms. Each of these components is like a separate strand that the teacher helps the students weave together into a strong rope.

Indeed, today's foreign-language teachers may not even have students learn the entire alphabet before they start reading some simple sentences or learning some useful spoken phrases. These teachers realize that setting things up so that students do conscious (or even subconscious) linking of **cognitive strands** is one of the most powerful ways to promote learning. This is true for a very simple reason: the human brain is structured to acquire knowledge in precisely that way – through several pathways simultaneously. If those pathways are linked in a meaningful way, the learning is more likely to be durable.

Cognitive strands in language teaching

Words (vocabulary)

Sentences (grammar)

Narratives (semantics)

Why, then, do so many geography curriculum planners and textbook authors think that teachers should try to provide a "global frame of reference" and "necessary map skills" *before* "going on to study the traits of various regions?"

That just does not work well!

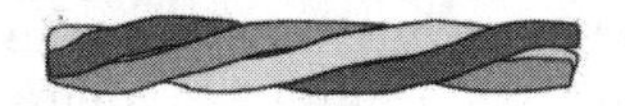

A Three-Strand Model of Geographic Education

Teachers in many fields, from archaeology and Bible studies to poetry and zoology, endorse a conceptually similar multistranded model of pedagogy. One strand has concrete "facts" (i.e., the artifact, the text, the couplet, the animal). Another strand has explanatory theories. A third strand consists of opinions and value judgments.

Hermeneutics

The science of extracting meaning from a text

(Many of these **hermeneutic** models of inquiry use the terms "level," "mode," or "plane" to express the core idea of this chapter. We chose "strand" for reasons we'll explain in the appendix, for those who are interested.)

Conceptualizing a three-strand model of cognition is hardly a novel approach, but it is a good way for teachers (at all grade levels!) to view teaching. Looking at geography in that way reminds us that we must try to foster "learning" through three different categories, more or less simultaneously:

Images (facts) – Visual images, other sensory impressions, measurements, and other facts associated with particular places. These are like the "letters" and "words" of the "language" of geography. You have to know them in order to make sentences.

Analyses (theories) – The concepts that help geographers interpret or explain the features they see in different places. These are the "sentences" and "paragraphs" in geography-language. You have to know them in order to make sound judgments about issues.

Evaluations (issues) – The value judgments and other opinions that people form about places. These are like the "metaphors" and "semantic overtones" that help communicate some of the most important ideas.

This chapter will explore some implications of this three-strand model of geographic knowledge. Because of the linear nature of a printed book, we devote a separate section to each strand in turn, but there will be plenty of intentional overlap. This is because we refuse to allow the details of book organization to deny the main point of the chapter: geography is a multistranded subject, and learning is most efficient when it takes place in several cognitive domains at the same time.

Images and Words — One Thread of the First Strand

The image/word strand

Learning words to represent the features in a place

One major strand of geographic education deals with building a "vocabulary" of words associated with places. As part of their attempt to understand a place, people need to observe and name the features that occur there. This is an essential part of trying to explain *why* the features are there (that's the second strand) and why their presence in that location may be good or bad, desirable or undesirable (that's the third strand).

This process of feature naming, however, is emphatically not a one-way street. A learner has to have some kind of theoretical concept of a place, some prior notion of what's going on there, in order to make valid observations about the features there. In effect, you have to know a little bit about something before you can efficiently learn more.

The Grand Canyon provides a good illustration of this seemingly circular process – a professional geologist tends to "see" more in the Canyon than a novice. The geologist's mind already contains a large stock of words about rocks and landforming processes. Whether standing on the Rim or hiking along a trail, the geologist can make use of words such as erosion, resistance, oxidation, perhaps even Redwall limestone and Vishnu schist, and so forth.

This process of naming and relating subfeatures of a scene takes time. That is why the geologist probably would still be there long after the novice had "seen it all" and left (Transparency 3A).

(Caveat: One should underscore the word "probably" in the previous sentence, because an interested and creative person can invent concepts while eyes and mind explore the scene. That apparent exception to the rule only reinforces the point: an interplay of concrete images, theoretical ideas, and value judgments is the best path to learning.)

An important part of geographic education, therefore, is the process of learning words for the unique features that occur in other places. From lush Amazon rainforests to the prefabricated concrete slums of suburban Moscow, various places on earth have distinctive landscape features that clearly say (with apologies to the *The Wizard of Oz*), "you're not in Kansas any more." It follows that a Kansas geography teacher must find some way of bringing those images and associated feelings into the classroom, so that Kansas students can learn about other places.

Image building

Technologies for bringing visual images about places into a classroom

Over the years, people have tried many different ways to accomplish this process of **image building**. Verbal descriptions, textbook photographs, enlarged prints, color slides, films, video clips, and now CD and DVD all work as image-presenters. Student activities that accomplish the same end include making photo essays, montages, videotapes, or place profiles, writing poems and journals, exploring data banks, and exchanging photos and other information with other schools (Transparency 3B).

Of all these image-building technologies, small digital cameras are at present the most flexible and cost-effective for geography teaching, *IF* suitable viewing or projection equipment is available. Otherwise, color slides or photos are the medium of choice, and they can always be scanned for later incorporation into electronic presentations.

These still images are also the basis for images in textbooks, CDs, DVDs, and photo essays. Videotape is beginning to make inroads as an alternative, but cost and flexibility still seem to favor still images.

Acquiring skill at taking photos (or videos) for classroom use is therefore a desirable goal, if only to enable geography teachers to recognize good images wherever they may be found. One might start by appreciating the fundamental difference between art and design:

The distinction between art and design

> An **artistic** photograph demands (and deserves) extended scrutiny to discern its many layers of messages. A **well-designed** photograph, by contrast, tries to bring an image of the world into the classroom efficiently. That does not mean that a classroom photo cannot be beautiful, even striking, but first and foremost it must be transparently intelligible. I suspect that many [educational] photographers resort to arty tricks to cover their design weaknesses. *Journal of Geography 84* (1985): 16.

Failure to distinguish between art and design is partly responsible for the plethora of "multimedia" materials out there that offer a smattering of exciting images, MTV style, but do not work together to create a durable whole out of their collections of parts. You want evidence? Look at all the television specials and picture books out there (you know, the ones with names like "Journey Down the Nile" or "Paris, a Photographic Tribute"). These have flooded the market at precisely the same time that U.S. students' understanding of geography has demonstrably declined.

In short, mere collections of gorgeous photos do not accomplish the goal of geographic image building. Students may enjoy looking, but videos or photos all by themselves don't seem to foster geographic understanding effectively, and as a result the visual memories they provide seldom endure.

So what kind of photographs (and, by implication, films and video clips) should a teacher look for? Ones that bring important images of the world into the classroom efficiently and unambiguously. In that way, students can learn the words while the teacher is helping in the process of weaving them together with the ideas and value judgments that are also part of the lesson. Table 3-1 at the end of this chapter has some basic design principles for taking still images and videos for geography classes.

Words and Images — Another Thread of the First Strand

Familiar words can (often do) have different meanings in different places.

The image-word strand has another thread, because a word can have different meanings in different places. Here is a simple example: think about the houses of Native American people at the time of Columbus. Depending on where you were on the continent, "house" might have meant a snow cave, log hut, bison-skin wigwam, stone cliff dwelling, long house, or bark lodge.

In a class or take-home activity, students can try to match photos or drawings of the Native American houses with descriptions of various environmental conditions. At nearly any grade level from primary to college, this simple matching activity can start a good discussion of explanatory theories and value judgments (Transparencies 3C and 3D):

> "What specific features can you see (in this picture) that make this house appropriate in this environment?"
>
> "How might someone who lives in this kind of house try to deal with an invader from Europe?"

The answers to these questions regarding Native American houses are different for houses in different parts of North America. That fact is an important aspect of the historical geography of the continent. If students engage in this discussion of house types early in a term, then just a brief mention later in the year can help encourage students to apply the different-images-for-the-same-word principle to other words, such as "factory," "ghetto," "store," or "wilderness."

In this way, students begin to appreciate that the very process of seeing and naming features is strongly influenced by the students' own personal history: the families, schools, and other aspects of the culture in which they were raised. (Why else would Saddam Hussein have put so many patriotic slogans and pictures of himself on billboards all over Baghdad? He wanted the citizens of that country to associate the word "leader" with his face!)

People who grow up in a given culture are not always aware of how strongly their culture is influenced by the environment around it. Awareness of the details of some mutual interactions of culture and environment is an important learner outcome in geography, partly because it helps keep students from accepting simplistic ideas of cause and effect.

This basically restates the main point of this chapter, which says that geographic learning may fail to take place if a book or class activity stays firmly planted within one cognitive domain. That, in a nutshell, is the fatal weakness of so many commercial films, videotapes, slides, CD-ROMs, or data sets. They may claim to have "all the information you ever wanted to know about a place," but they provide knowledge only within the factual cognitive strand. If students are given the task of memorizing a lengthy list of facts and numbers, one should not be surprised if they don't learn very much. Facts, all by themselves, begin to look like other facts and eventually get confused with them, unless they are woven together with concepts and issues, the other cognitive strands.

Before we go on to talk about the analysis and evaluation strands in the geographic rope, however, it may be useful to provide one more extended illustration of the basic point: the same word can (and often does) mean different things in different places.

Consider, for example, the word "tree" (and the mental image it is likely to bring up):

What does "tree" mean?

It depends greatly on where you are!

In central Kansas, "tree" means a cottonwood with spreading limbs and trembling leaves, often growing next to a usually dry creek.

In Ohio or central Europe, "tree" means a fan-shaped elm or ash, soaring upward among other tall trees.

In southern Alabama, "tree" means a pine with a straight trunk, pruned by fire and shade, with its branches far above ground.

In southern California or eastern Australia, "tree" means a gracefully twisted eucalyptus, part of a planted grove around some buildings or a row along the road.

In central California or Portugal, "tree" means a low oak, spreading its branches wide as it stands alone or in small groups on grassy slopes.

In the cold forests of Canada or Siberia, "tree" means one of a mass of skinny spruces or firs, growing close together with a thick carpet of needles underfoot.

And on poor soils from Maine to Texas, "tree" is a scrubby juniper (often called "cedar"). Juniper trees can tolerate soils too thin and alkaline for crops. They are the opposite of walnut trees that "told" settlers in Michigan what soils were fertile.

In short, the word "tree" means different things in different places. Moreover, people who know about trees can look at the trees in a place and infer other things about that place. For example, early European settlers in Michigan chose land that grew nut trees, such as walnut or hickory, because they had learned from personal experience (or had been taught by others) that such land was also capable of producing good crops. It follows that the more clearly people "see" features like the trees in a place, the better their inferences about that place are likely to be.

The foundation for this interplay between observation and inference is laid in primary grades. In a science class, teachers could show pictures of various kinds of trees, such as oaks and pines, and ask students to memorize their names. If they take the next step, however, and relate the traits of various trees to the environments they occupy, students will learn better. And if the environment of the tree is linked to local history, it works even better. Structured thus, a lesson makes use of several cognitive strands in what is rightly called a **neural network** in a child's brain.

Neural network

A model of the brain as a complex web of interconnected pathways

This conceptual model is the basis for a class of computer programs

This link between fact and context works in both directions. For example, to understand controversies about old-growth forests in Oregon or the "deforestation" of the Amazon Basin, a student must have a fairly accurate mental image of coastal or equatorial rainforests. To get that image, one must either travel to Oregon or Brazil or be exposed to high-quality photographic, film, video, or verbal depictions of the forests that grow in those places.

Principles that apply to something as simple as a tree are even more important when dealing with value-laden words such as "housing," "ghetto," "government," "church," or even "terrorist." These words also tend to have different meanings in different places, and those differences are important in trying to understand places and the people who live in them.

Beyond specific examples like these, a teacher should try to foster a general awareness that even familiar words can have different meanings in different places. Unwillingness (or inability!) to recognize this simple but vital principle is a big failing of many magazine writers and television and radio talk-show hosts. This is most unfortunate in a democracy: good intentions can combine with too-simple worldviews to form wildly erroneous analyses and prescriptions for social problems. Examples can come from many topics, from curing disease or stopping crime in crowded inner cities to drilling for oil in the remote wilderness of northern Alaska.

In short, appreciation of the fact that familiar words can have different meanings in different places can legitimately claim a place near the top of the list of citizenship values engendered by a study of geography.

Geographic Theory — The Second Strand

The theory strand

Learning theories to help explain the locations of things

The second strand in the "rope" of geographic learning deals with explanatory theories – the "sentences and paragraphs," if you will, that help people figure out why things are the way they are in a particular place. All people use geographic theories, whether they know it or not. Here are some examples: "Let's go to store A – it's closer to home." "Store B costs too much – too many rich people go there." "Store C is on the way to Aunt Tilly's house." "It's dangerous to park around Store D." "Store E is locally owned, so shopping there is better for our economy." And so forth.

Gravity model

A theory to explain the amount of traffic between two places

The first statement in this list ("Let's go to store A . . .) is an expression of the **gravity model**, a good example of a geographic concept that people "know" even before they are told it has a name. This useful theory tries to explain the amount of traffic, telephone calls, or other interaction between places. In basic form, the theory says that people are more likely to travel to nearby places than to places farther away. (That is subject, of course, to that well-known caveat: all other things being equal.)

The greater attractiveness of nearby places seems so transparently obvious that one wonders why some politicians and planners were surprised when travelers rejected the spacious modernity of the Dulles airport in suburban Washington, DC. Busy travelers seem to prefer the close-in convenience of Reagan National Airport.

Of course, variables such as noise, crowding, land value, personal tastes, or alternative destinations also have an impact on the amount of traffic we expect between places. The gravity model has many more elaborate forms that take those variables into account (Transparencies 3E and 3F).

Thus, like important theories in any discipline, the gravity model cannot be "covered" at a particular point in the geography curriculum and then assumed to be "known." On the contrary, it should be introduced in primary grades, expanded in middle school, and extended in high school. Indeed, the gravity model has so many implications that it will probably form the basis for doctoral dissertations for many years to come. One thing is certain: it has become an essential part of the toolbox of urban and regional planners, store managers, and traffic analysts.

Increasingly elaborate versions of the gravity model usually require more and better input data. This means that the skills of observation and measurement must continually improve in order to keep up with advances in theory. The strands of fact and theory are therefore tightly twisted together, in both research and teaching. The result, over time, is an improved theory that can predict traffic with greater accuracy, thus allowing people to build things like roads, phone lines, and fiber-optic cables efficiently.

Geographical diffusion *The spread of diseases, innovations, rumors, and other things from place to place.*

This twisting of the cognitive strands is also apparent in another important group of geographic theories, those that deal with **diffusion** (the spread of things from place to place). Something as familiar as the arrival of a new strain of flu still comes as a bit of a surprise each year, because the exact pattern of spread depends on a complex interaction between the invading organism and a host of local conditions, such as environment, political borders, cultural rules, personal mobility, and so forth. Ironically, topics like environment, politics, culture, and migration often are used as separate chapter titles in geography books, thereby making it more difficult (although still necessary) to bring the ideas together when we are dealing with something like the spread of a disease.

Here, let me just underscore the fact that geographical theories are qualitatively different from geographical images. The two strands of meaning appear to be processed in different parts of the brain, stored in different forms, and retrieved in different ways. Not surprisingly, they also seem to be most efficiently taught with different techniques (and assessed with different instruments).

Thus, while a slide projector or photograph is well suited to presenting images, an overhead projector is more useful for presenting geographic theories. Like an old-fashioned blackboard or a state-of-the-art-next-year electronic easel, an overhead projector lets a teacher create or add to the visual message "in real time" while explaining a theory orally. Unlike a blackboard, however, an overhead projector allows some of the message to be prepared beforehand and saved from class to class, and even from year to year. The overhead also allows better "body language" for dealing with complex ideas, because the teacher faces the class and can watch for signs of confusion. Table 3-2 at the end of this chapter has some design tips for making overhead transparencies for geography classes.

Personal Theories — Another Thread in the Theory Strand

The scientific test of a theory

Does it handle the observed data better than any other theory?

Geographers should not just blithely borrow the word "theory" from scientists without bringing along another basic principle of science: *theories are not, cannot be, right or wrong.* At most, a theory can be better or worse than other theories. You cannot "prove" a theory right; you can only test it by seeing how well it handles the observations you make and the data you gather. Those tests, in turn, can provide additional support for a theory or cast doubt on it, but they cannot "prove" it.

This book is not the proper place for a long exposition of some currently fashionable philosophical debates about positivism, paradigms, and postmodernist epistemologies. Moreover, your author is not terribly fond of debates that too often resemble a bunch of medieval theologians arguing about how many angels can dance on the head of a pin. Here, let us just concede that geographic theories are not like Newtonian physics, all simple actions and reactions, causes and effects.

It is true that many geographic hypotheses do sound like simple cause-and-effect assertions: "farmers prefer fertile soils with the potential for high yields" or "railroad builders follow stream valleys in order to avoid hills." One does not need to look far into either agricultural policy or railroad history, however, to see that locational decisions are not always made on the basis of simple measures of soil fertility or average slope.

For example, as I write this paragraph the United States Department of Agriculture is trying to solve the "problem" of overproduction by paying Iowa farmers to stop growing corn on some of their fields. Meanwhile, a different government department is subsidizing farmers in other states by helping them reclaim salt-laden desert soils or drain coastal swamps, thereby enabling them to plant more crops. This kind of policy schizophrenia seems to indicate one of two things: either a significant amount of geographic ignorance, or deference to a "higher" authority, such as "national security" or "political clout." The result is a geographic pattern of land use that must be interpreted in the light of the personal theories of government officials as well as of the "scientific" theories of geographers and economists.

A glance at a railroad map can add some more support to the idea that simple cause-and-effect theories rarely explain all of what we see in the landscape. Let's start with an obvious fact: the purpose of a railroad is to move goods from one place to another. Slope and distance are therefore of great concern for railroad engineers, who usually try to minimize both.

Other factors, however, can also have an influence on where railroads are built. For example, governments in many countries encouraged the construction of railroads through specific regions in order to tie the regions together politically. The resulting pattern of routes may not be the most efficient way to move goods from producers to consumers (Transparency 3G).

Vernacular theory

Cause-and-effect ideas that people in an area believe are true

Many of these seeming exceptions to the rule of commonsense geography are rooted in an equally commonsense fact: *the people in a place have their own theories about how things happen*. These **vernacular theories** (also called "folk theories") can also be topics for geographic study (that is one inevitable implication of geography's being a "social science").

For example, we can make a map by shading all of the counties where more than 50% of the people think that tax cuts are a sensible strategy for stimulating economic growth. That map, in turn, can be compared with maps of other topics, such as household income, stock ownership, average age, educational level, or family size. All these personal traits might influence people to hold different opinions about tax policy.

Good examples of these behavioral differences are abundant near many national or state borders, where people on opposite sides of the border have similar land but do different things with it. For example, Montana farmers grow wheat on the same kind of land that Canadian ranchers use for raising cattle. The difference is due mainly to the way two governments define land and provide price supports and subsidies to farmers. (The National Council for Geographic Education publishes a satellite image that clearly shows this contrast; contact NCGE.org; see also the ARGWorld CD unit on Borders and History).

Other parts of the world provide many other examples of vernacular theories in action. Together, these make up a huge outdoor "laboratory" that allows research geographers (and students) to study the interplay of human action and environment.

One result of this study is an ever-growing body of interrelated theories about how people use land. The goal of these theories is to help us explain both the causes and the effects of particular kinds of land use. Specifically, they try to answer questions like:

- Why do people use land in the way they do? How does their environment influence their land use?
- What consequences are likely to occur if people use land in specific ways? Are those consequences different in different places?
- How might the use of land change if other things change?

Many of the new national and state geography standards deal with ideas about the relationships between human activity in a place and the environmental conditions there. Let us, therefore, spend a few pages looking at several key aspects of modern land-use theory:

- Environmental conditions

  A simple version of land-use theory says that climate, slope, soil, and other natural features in a place have an influence on land-use options and therefore on land value. For example, look at the relationship between soil and livelihood in Indonesia. The soils on the largest island, Sumatra, are mostly nutrient-poor red oxisols (that's a redundant phrase, since "oxisol" means nutrient-poor red soil in the USDA's Soil Taxonomy!). The neighboring island of Java has dark volcanic andosols, which are easy to work and very productive. The inherent richness of the Javanese soil helps farmers become more productive in Java, which in turn helps explain the higher population density of Java. The rich soil also makes it worthwhile to invest in terraces, fences, and other structures that will further enhance the productivity of the land and soil. This logically leads to the next theoretical idea:

Soil Taxonomy

Set of terms to put soils into groups according to key features

- Cultural values and know-how

  Knowledge of how to use a resource has an influence on land-use options and therefore on land value. A fertile soil or rich mineral deposit does not just hand money to people. Someone has to recognize the potential value of a resource and learn how to use it. If the knowledge of how to use a resource is missing, the resource is of little value. History has many examples of inventions – such as deepwell pumps or copper smelters – that suddenly allowed people to use a new resource and thus stimulated migrations into formerly uninhabited areas. This observation leads to a useful generalization: any idea, rule, or invention that helps people use a new resource (or, conversely, a law or cultural idea that prohibits its use) can dramatically change the relative value of places (see the ARGWorld CD unit on Changing Technology).

- Infrastructure

Infrastructure

Roads, railroads, power lines, sewers, and other features that help move people, goods, energy, and ideas

The existing **infrastructure** in a place has an influence on land-use options and therefore on land value. Even knowledgeable farmers with exceptionally fertile soil and an ideal climate are still not likely to grow grain to sell in the world market unless there is a solid foundation of property rights, machinery parts, seed producers, storage bins, transportation links, and financial services. Once in place, any major infrastructural feature such as a lumber mill or Interstate highway has a profound influence on the value of land around it (Transparency 4O is an example of a student activity based on this idea). The infrastructure in a place is usually the legacy of far-sighted people who put together the political and financial means to survey land and build roads, powerlines, schools, welfare systems, and other support structures.

- Prior use

Awareness of the importance of infrastructure leads logically to a fourth aspect of land-use theory: all previous land uses can have an influence on land-use options and therefore on land value. For example, the systems of land division used by Spanish priests, British lords, French farmers, and U.S. Public Land surveyors have each been different. Each of their methods was designed for its own place and time, yet the resulting property descriptions and boundaries can persist for a long time. In fact, some of these **legacies** of prior use still govern the arrangements of streets, locations of traffic jams, and the appearance of neighborhoods in modern cities such as San Diego, St. Louis, Detroit, Mexico City, Hong Kong, and Saigon (Ho Chi Minh City).

Legacy

Persistent feature of previous use

- Political borders

Acknowledging the importance of artificial lines can lead directly to the next theoretical idea: political borders have an influence on land-use options and therefore on land value. This is especially obvious in places where governmental rules or cultural values are different on opposite sides of a border. For example, look at the countries of the former Soviet Union and their European and Asian neighbors. In those areas, long-suppressed differences about the role of religion in politics are now emerging from Bosnia and Chechnya to Kazakhstan and western China.

- Number of people

Finally (in this short list of important aspects of land-use theory in geography): the number of people in an area has a powerful influence on land-use options and therefore on land value. Ask a second grader in a small town what he or she would like to see built there. The answer is likely to include a wish-list of interesting things: Disney Worlds, shopping centers, sandy beaches, and so forth. Reconciling a person's wish-list with the geographic realities of a given place is one purpose of a geographic education. That education should include discussion of a conceptual framework called **central-place theory**, which deals with the sizes and spacing of centers that can provide particular services to people in an area. The blunt fact is that most communities have neither the resources nor enough people to support an infinite variety of big, interesting things. People have to make choices.

Central-place theory

a theory to explain the size and spacing of service centers such as stores, malls, and towns (see Transparency 2G)

It's easy to illustrate these principles of land use with absurd examples. For example, a society that tries to build a wine vineyard in every valley or a Disneyland in every small town is destined to bankrupt itself. It becomes more difficult, but more important, to raise the same questions about features that we take for granted in our own surroundings. For example, how much of the current investment in new minimalls, outlet malls, and strip malls really makes sense? If teachers are not going to try to give their students at least an appreciation for the need to ask such questions, will American consumers have anyone else to blame if they have too many malls and not enough factories, power plants, or good universities to compete in the real world?

Location-allocation analysis

Deciding how many of something (e.g., convenience stores) to put in an area, and where to put them

End of sermon: the practical benefits of geography include a healthy appreciation for the conditions that influence land use, especially the effect of scale and balance. In sum, a community cannot expect to be efficient if it errs in any of these ways:

- By using land in ways that do not fit environmental conditions
- By failing to invest in appropriate infrastructure
- By providing too few services for its people
- By trying to build features in places that do not have enough people to support them adequately

The goal is not just to see what is impractical in a given place. It is equally important to explore what *might* be attainable there, if people want it enough.

The foregoing pages of theoretical statements about gravity models, diffusion, and land use had two purposes. First, they provide a summary of some key ideas that should be introduced in primary grades and reinforced throughout elementary and secondary school. (Choosing a proper vocabulary for these ideas is an important question in Chapter 7.)

Second, the list helps illuminate an important truth about geography as a discipline. What sets geography apart as both interesting and important for a smoothly functioning society is *not* the blinding conceptual richness of its theories. Rather, it is the elementary idea that even "obviously correct" theories are profoundly place-dependent. A plausible explanation of landscape features in one place can be ludicrously wrongheaded in another.

In short, explanatory theories may be different in different places. As was said in Chapter 1, a study of geography can help us see why something we know, here, might be wrong in another place (or, conversely, why something that works in another place might not work where we are). The next chapters have more illustrations of this idea. But first, we should look at one of the most powerful analytical tools of geography:

Map-Pattern Analysis and Map Comparison

Map-pattern analysis

Looking for clues to help understand the density, spacing, alignment, imbalance, and other aspects of the arrangement of features on a map

Geographers have many ways to try to identify what features may be related to each other. One of the most powerful of these is **map-pattern analysis** and its close companion, map comparison.

Map-pattern analysis begins by making a map of a feature of interest. For example, a company might make a map of existing coffee shops or ice-cream stores. Careful study of the pattern can help the company find "gaps" where a new store might be profitable. This is a widely used kind of map-pattern analysis, but it is hardly the only one. In fact, many of the most useful geographic theories began with someone just looking at a puzzling map and thinking about reasons for the pattern on it.

Spatial imbalance

Tendency for features to be located mainly in one part of an area

The first step is to look for **spatial imbalance** in the pattern. If a majority of features are on one side of an area, perhaps it should be investigated whether there is something influencing that discrepancy in placement. Perhaps there is something on the other side of the area preventing the spread of the observed features. In short, map-pattern analysis almost always leads to map comparison, either with a printed map or with an analyst's mental one. Transparencies 9C and 9D show how this technique can help people interpret the patterns of Arab and Israeli settlements on the West Bank. Other map pairs that could be used for pattern comparison include Transparencies 3I and 3M, 4U and 4V, 4W and 4X, and 4Y and 4Z.

To build a set of useful mental maps:

Try to remember one key line, rather than a whole map

This kind of geographic analysis can take place by just blindly comparing maps in an atlas. It is much more efficient and effective, however, if the analyst has a set of well-chosen mental maps that can be used in trying to interpret a new map. Some useful "comparison maps" for a geographer who is looking at a puzzling map of the United States include:

- The zero-moisture-balance line. On one side of this line, surplus water is available to form rivers, grow crops, leach nutrients out of the soil, wear hills down, and perform many other "tasks" in the environment (Transparency 3I; compare with 3M).
- The four-month frost-free season, which is basically the limit for farming. Most grains cannot mature with a growing season of less than about 80 or 90 days, and the effective growing season is several weeks shorter than the "average" frost-free season (Transparency 3J).
- The seven-month frost-free season, which is the effective limit for cotton and rice. These important plantation crops could not grow in places with a shorter-than-seven-month growing season (Transparency 3K; compare with Transparency 4X).
- The maximum extent of the **Pleistocene** glaciers, which covered much of North America and altered the terrain and drainage in ways that continue to have many effects for farming, road building, and construction (Transparency 3L).

Pleistocene

The "Ice Age" in recent geologic time, when glaciers covered much of northern Europe and North America

- The position of the settlement frontier at various times in the past. An understanding of the historic movement of European people across North America can help us understand many present-day features, such as town history, population density, ownership patterns, ethnic make-up of the population, and architectural styles (Transparency 3M).
- The states of the Confederacy. This historic pattern continues to have an influence on things such as education, income, religion, and voting, even though more than a century has elapsed since the Civil War (Transparency 3N)
- The states where the Federal government still controls most of the land. This pattern has an apparent relationship to maps of population density, railroads, parks, suicides, and beliefs about timber and grazing rights (Transparency 3O).

Someone looking at another region of the world could also start with a set of simple climate maps. The length of the frost-free season is a key variable in most places. The 12-month line is especially important in tropical countries, because knowing where it never freezes can help explain many things about the patterns of crops, weeds, and diseases. The amount of precipitation is also a crucial factor in most places, especially those that do not have the technology for drilling deep wells or moving water long distances. (Transparencies 4T-4V and the CD unit on Malaria deal with relationships among precipitation, population, and disease in Africa.)

In making short lists of simple maps for students to learn for comparison purposes, keep in mind two important differences between the United States and other countries. First, U.S. towns and cities are relatively young, compared with those in many other countries. Second, the United States has had fairly strong political control over its own territory throughout its history. A teacher should underscore the fact that most of the world's people live in places that have been under outside political control at different times in the past. For this reason, students should learn the general extent of the Roman, Mongol, British, Dutch, French, Turkish, Japanese, Soviet, and other previous empires. Many parts of the world still have land-division systems, road networks, and buildings that are legacies of their former controllers (Transparency 3P, and the CD unit on Borders and History).

Map-pattern comparison is such an important teaching and research technique that it is the primary reason why we made all of the overhead transparencies in this book half-size – that allows the simultaneous projections of two images (see Table 3-2).

It is also why many geographers prefer simple maps, often with only one or two colors, to allow for easy comparison.

Cartographic convention

Accepted way to use map symbols to communicate a particular idea

A reader of simple maps, however, must be able to judge their cartographic integrity – one does not want to spend too much time studying patterns on a map that has questionable accuracy. This is the reason for the long list of National Geography Standards that deal with map symbols and other **cartographic conventions**. Mapping is a language, which can be used well or inappropriately, and students should learn how to recognize inappropriate use.

It is not appropriate, for example, to use color shadings to show the total number of employees in each state (recall Transparencies 1A and 1B?). This use of the choropleth technique is the cartographic equivalent of saying "I is thirst" or "hunger her am." Those strange "sentences" may be intelligible, after some thought, but they are not examples of the proper use of English words or grammar to communicate ideas.

To avoid misunderstanding, a person who is studying maps must be especially careful to check whether the maps handle and display data correctly. The risk of confusion goes up whenever more than two maps are combined or compared. For example, suppose someone is trying to find a good location for a solid-waste disposal facility. That person might try studying a map that combines information about slope, soil, and underlying geology. The composite map may use a particular color to show the acceptable areas. This may make the pattern of acceptable areas easy to see, but it provides no information about the reasons why some sites were unacceptable – did they "fail" because of bad slope, soils, geology, some combination of the above, or just because one of the input maps was copied wrong or made too simple? And are we sure that slope, soil, and geology are the only factors that should be considered? What about land ownership, historic sites, endangered species, or Native American claims?

To sum up, geographers often rely on mental maps to compare with new maps they encounter. The human brain, however, has limits. A teacher should therefore choose the list of maps-to-be-memorized with great care and make sure they are appropriately designed. (Chapter 7 will show some ways to evaluate how well students have learned key map patterns.)

Before we leave this discussion of geographic theory, we should repeat one point for emphasis: map comparison cannot *prove* a connection (see Transparencies 4Y and 4Z for an illustration of this idea in a policy context). At best, map comparison can help support or refute an idea. More often, it can only suggest ideas for further investigation.

That caveat is true of all science. Experiments may support some ideas and cast doubt on others, but even the most well-designed experiment cannot prove that a theory is absolutely true. Indeed, real-world scientific experiments often raise as many questions as they answer.

In short, students should learn that definitive answers are much rarer than tentative conclusions. Appreciation of that fundamental idea is a worthwhile goal in any class. At some point, theory-forming tends to grade almost imperceptibly into a third cognitive strand, the realm of personal opinions and value judgments.

Value Judgments—The Third Strand

The issues strand

Learning how to evaluate efficiency, fairness, beauty

The third strand of geography deals with people's personal opinions about topics such as landscape beauty, fairness, and tolerance. For example, a young child might evaluate some places as scary and others as safe. An older child might classify places according to the likelihood of having fun there. Citizens might evaluate the locational decisions of large corporations in terms of labor equity, environmental impact, or architectural integrity (to name just a few criteria that might be applied).

The relative importance of these criteria may well vary from one place to another (or from one group of people to another in the same place). In effect, geographic opinions can also have a geographic pattern! That means that people in different places can have different ideas about where to put things such as stores, factories, offices, festivals, and bike trails.

Some geographic opinions have the force of law. For example, the Federal government once had a policy of refusing to guarantee home mortgages in racially mixed areas. This led to a process called **redlining**. Banks and insurance companies drew (red?) lines around neighborhoods where they would make fewer loans or charge higher insurance premiums or interest rates.

Redlining

Process of drawing lines around areas where banks can refuse to make loans to some people

It does not take a genius to predict that home deterioration is a likely consequence if people are unable to borrow money for home purchase or repair. In time, this became evident on maps on which dilapidated housing, unsafe or abandoned buildings, and arson were mapped. These often became severe problems in areas that had been redlined.

Some people blamed the individuals living in those neighborhoods for the deterioration – "Those people just don't keep up their properties." That blame, in turn, could be used as a rationalization for the policy – "Why should we lend money to people who live like that?" Other people who studied the issue more thoroughly realized that a no-loan policy had logical impacts that could be analyzed and even predicted. When confronted with the obviously bad effects of the practice, legislatures and courts eventually outlawed redlining (although it seems to persist in some subtle forms in some cities).

The topic of redlining is a good example of the interplay between theory and opinion. It is not hard to formulate a plausible theory to explain why banks might do redlining. For one thing, old buildings can develop structural problems. Crime and vandalism may be higher in neighborhoods with poor people. Property value may not rise as fast in old city neighborhoods as in other areas, because newer homes usually attract more potential buyers. In short, loans in some locations are inherently risky. That is a geographic theory that has substantial basis in fact.

People, however, can form evaluative opinions based on other geographic theories and facts. For example, they may realize that every person who lives in a risky neighborhood is not necessarily a risky investment. Or, they may conclude that redlining is a violation of civil-rights laws, to the extent that housing in a city is segregated by race. Finally, they may simply be aware that a redlined neighborhood has almost no chance of recovery, whereas a law that prohibits redlining may afford at least a chance of rebuilding.

Here is a capsule view of the situation, stated in two short paragraphs:

> If there is no law against redlining, a bank that refuses to redline is likely to be at a disadvantage. More of its loans will be risky, and therefore it may be less profitable than other banks.
>
> If there is a law against redlining, however, all banks are on an equal footing. Every bank will have some risky loans, which will raise the costs for all borrowers.

The important point is that citizens in a society have to choose whether the benefits of an antiredlining law outweigh the costs. In other words, they have to make informed decisions (whether individually or through representatives). Those decisions will have an effect on the geographic patterns that we are likely to observe in the future.

For an international example of the same principle, consider the practice of "China bashing" (blaming Chinese competition for the loss of jobs when factories in the United States close).

On one hand, a factory that closes may be just a number in the annual report of a global corporation. At the same time, it is also a provider of jobs and income at a local scale. Moreover, it usually is connected to and thereby supports jobs in other parts of the U.S. economy – in factories in other places, as well as the jobs of truck drivers, shipping clerks, warehouse owners, employees in financial institutions, and so on.

On the other hand, consumers usually prefer cheaper products. The money that people save by buying cheaper products can help solve other problems, thus creating more jobs. In short, paying more to get a locally made product does not necessarily result in a better standard of living for the people in an area.

The issue becomes even more cloudy when we consider the effects of **capital flows**. Some of the profit from a Toyota made in Kentucky goes overseas to Japan. Much of the profit from a Ford made in Michigan stays in the United States. Many of the parts for the Ford, however, come from other countries. Meanwhile, many of the parts for the Toyota come from companies within the United States (Transparency 3Q).

Capital flow

Movement of money among corporations and their suppliers, workers, or investors

In short, a good understanding of the flows of goods, services, and money is essential if citizens are to form valid opinions about international trade, environmental quality, and residential justice (or even that perennial political buzzword, taxes). All of these topics have a strong geographic component, because designers, workers, investors, and customers are linked to each other by flows of goods, money, and influence. People should understand these facts before automatically mouthing slogans such as "buy American" or "tax companies that send jobs overseas."

People's opinions about these topics can be formed at a surprisingly young age. This is especially true if students watch dramatic television shows about evil businessmen during the same formative years in which their geography classes feature nothing more thought-provoking than crossword puzzles and finding the names of states in a maze of letters.

A little sermon about values in a democracy

(Of course, in a free society, people are entitled to whatever opinions they choose. Evaluative ideas about the geographic arrangement of things can be formed by doing meaningful geography in school classrooms, or by participating in town meetings, listening to talk-show programs, or watching television dramas. People in a democracy get to choose, and one can derive some comfort from noting that the great virtue of a democracy is that people have no one else to blame when they get their way. Here endeth the minisermon: one pervasive theme of this chapter is that people deserve what they get if they are willing to accept sound-bite opinion-forming, in which opinions stand in isolation, not woven together with supporting facts and theories.)

Here is the key question: how can a geography teacher at any grade level encourage thoughtful evaluation and other higher-order thinking skills? (That phrase, abbreviated "HOTS," is in vogue now, but one who accepts the idea of a three-strand rope might question whether evaluation is really "higher" than careful observation. That is the topic of Postscript 3-1 at the end of this chapter.)

Dogma

"This statement is the only truth"

Relativism

"One statement is as good as any other"

Teaching about value judgments can be like walking a tightrope between **dogma** and **relativism**. On one hand, a teacher should try to engender respect for other points of view. The phrase "reasonable people might disagree" is useful to underscore the possibility of honest differences of opinion.

On the other hand, respect for diversity does not necessarily imply acceptance of any idea. To be acceptable, an opinion should be founded on sound analysis of honestly gathered data. In other words, the other two strands of meaning, theory and fact, should also be strong. Encouraging students to form, defend, and evaluate opinions about fairness and beauty is not only part of geography, it is one of the main philosophical foundations for all study.

(Table 3-3 has some techniques for promoting healthy evaluation of ideas in a classroom.)

Before leaving the topic of value judgments, however, we might tease out one more useful thread. One of the most valuable roles of geography is to provide effective illustration of the fact that there may be more than one valid way of viewing a particular landscape feature or idea. As we said earlier, one must seek a medium ground between dogma and relativism.

In practice, this means trying to foster an acceptance (or at least tolerance) of other reasonable ideas, and at the same time trying to instill an ability to see logical flaws and to reject unreasonable ideas. Careful search for and study of reasons why people might hold different opinions in different places is one of the best ways to develop this balance.

(Role-playing is an effective learning exercise in this context; Table 3-4 shows how role-playing can help students wrestle with value differences.)

Conclusion

Teaching geography is like teaching a foreign language. You have to teach words (facts and images about places), grammar (geographic theories and concepts), and narratives (opinions and value judgments about geographic issues) more or less at the same time. When that happens, students engage several parts of the human brain. Those parts are preadapted to facilitate learning, and the resulting learning will be more durable and useful.

POSTSCRIPT 3-1

About Those Taxonomies of Learner Outcomes

At the time the first draft of this chapter was being written, the phrase "higher-order thinking skills" (HOTS) was in vogue among curriculum designers and school administrators. The life expectancy of this particular phrase may not be very long, but the concept of a hierarchy of qualitatively different thinking skills is likely to persist.

This chapter used the somewhat awkward terms "strand," "domain," and "realm" instead of the more familiar "level" or "order" of cognition for a very important reason – to counteract an unfortunate inference that one could make by reading some popular books on "the scientific method." In describing what they think scientists do, some of these authors imply that scientists go "up" a logical hierarchy from lower-order observation through theory to higher-order evaluation. According to that model, scientists spend some time (or that of their assistants) gathering information through measurements and field observations, and then the scientists formulate brilliant theories to "explain" the data.

In the real world, however, it's more like the chicken and the egg: sometimes it works as described above, and sometimes a lab assistant has a flash of insight about how things might be. That theory in turn suggests a new way of making observations. And sometimes, someone invents a new measuring device (like radiocarbon dating) that provides a different stream of data for consideration in making theories. In short, good theories enable people to observe better, and good observations are the raw material for better theories.

Once we abandon the notion that observations are a "foundation" of "higher thinking," we can begin to appreciate that each strand of cognition depends on the others. It is true that evaluations are based on theories, and theories depend on facts for support. It is also true that images and facts do not just flow into the human mind. On the contrary, they are filtered through a person's prior ideas and opinions. This is why people with strong opinions may have trouble "seeing" contrary evidence.

A good first step in trying to open a closed mind is to cast some doubt on the "simplicity" of the process of observation or image-making. Activities that require students to gather facts about a place in a systematic manner can increase their willingness to accept evidence even if it contradicts their previous opinions. For this reason, it seems like a good idea, at least occasionally, to set the "merely mechanical" process of observation up as the "higher" authority over abstract processes of theory formulation or value judgment (Transparency 3R).

In sum, making a geographic observation (i.e., naming and/or describing something in a place) is no "lower" a form of cognition than explaining a theory. Image-building, theory writing, and evaluation can all be done well or poorly; one is not necessarily "higher" than another. Any of them can legitimately be described as a "high-order" thinking skill.

Or, better, try to choose language that describes various strands of cognition as different but not necessarily higher or lower than each other. A refusal to accept the logical (or moral) superiority of theory, observation, or value judgment is the first step toward a sound model of the process of inquiry. To paraphrase John Gardner: "A society that elevates philosophy as a noble occupation while demeaning plumbing as a low-level job is heading for trouble, because, in time, neither its theories nor its pipes will hold water."

TABLE 3-1. Some Design Principles for Slides or Electronic Images

1. **Attention**. Focus on just the object of interest. Keep foreground and background clutter to a minimum, and maybe even use the depth-of-field limits of a large lens opening to keep foregrounds and backgrounds deliberately out of focus (unless, of course, clutter *is* the point you are trying to make!)
2. **Redundancy.** Use several ordinary pictures of the same feature rather than one "best" one. Faced with multiple images of something, the human mind extracts the common elements, which are inevitably fewer in number and therefore easier to remember than all of the details of a "perfect" photo.
3. **Timelessness**. Avoid elements – cars, movie marquees, billboards – that date the picture too narrowly (unless establishing the time is part of your purpose). Few things damage credibility more than a photograph of a timeless object (say a farm crop, landform, or architectural style) that is partly ignored amid chuckles about a finned DeSoto, Beatles concert, or Ronald Reagan billboard on the screen.
4. **Duplication**. Carry at least two cameras. The cost of getting somewhere is many, many times greater than the cost of film or memory chips. Take every important picture with several cameras, perhaps with different framing or exposure setting. Then, you have at least some pictures if a battery or memory chip fails, the developer makes a mistake, you lose a roll, the camera falls into a puddle or opens by accident, the exposure was wrong, the colors faded on outdated film, or a thief stole the camera from a motel (to list just some of the things that have happened to my cameras in 20 years).
5. **Identification**. Mark each image in a unique way. With slides, I use a permanent marker to paint a patch of color in the upper right-hand corner of each one as it is put in the tray for projection. I then write an ID code on the patch with black marker. The colors tell me instantly whether the slides are all in the tray right-side-up. The particular coding system does not matter much – almost inevitably, some slides won't fit your categories, and some will seem to fit several of them. The important thing is to have a unique marking system and therefore a single place to put each slide, so that you can find it again.

   With electronic images, I have a master file and then make copies for other files. I use an ordinary word-processor to list the codes, locations, dates, and several keywords for each image; the search command in the program is usually able to find any photo out of the many thousands I have. A database would perhaps be more efficient, and I might use one if I were just starting out, but it does not seem worth the effort to convert at this time.
6. **Role**. Most important, take pictures to illustrate a narrative, not to *be* a narrative. A truly boring way to show pictures is to introduce each one by saying, "and this is a photo of x." Students soon learn that they can doze, wait for the magic phrase, snap up to scan the picture, and nod off again. It is much better to tell an interesting story and use pictures to illustrate the story in a varied way, with some fast sequences and some lingering over especially good examples. This implies that the proper time to think about the story is when you are taking pictures, not while you are sorting them the night before class. (A presentation outline that is written before a trip is also a great thing to have at tax time; properly documented, a teacher's camera, film, batteries, memory chips, and many travel expenses are deductible – I have the attempted audits and refund checks to prove it!)

For more hints along these lines, look at "One Commandment and Ten Suggestions: Teaching the Video Generation with Slides," *Journal of Geography 84* (1985): 15-19.

TABLE 3-2. Some Design Principles for Overhead Transparencies or the Screens in Powerpoint-Style Presentations

1. **Size**. Design transparencies to be *half* of an 8-1/2 by 11 page. This has four advantages. The first is financial: transparency film is expensive, and this effectively doubles your budget. The second is pragmatic: a half-size transparency will work even with a small screen or a room that prevents some people from seeing the whole screen. The third is substantive: it is easy to compare maps when you can project two at a time. The last is organizational: you can put an outline on one transparency and leave it on screen while exchanging other transparencies; other forms of organization will suggest themselves once you get used to that much flexibility.
2. **Highlighting**. Use color to highlight specific features. Filling an area with a yellow color, for example, can transform a hard-to-follow wiggly black line into an instantly recognizable continent. A water-color marker can be used to highlight features "on the fly" as you discuss a transparency or respond to questions; a few seconds with a moistened tissue can then clean the slide for the next use.
3. **Cartographic conventions**. Follow the standard rules for cartography. Use larger symbols or darker shades for large amounts of something that is counted or measured. Use graduated symbols for "raw" data (absolute measurements or amounts, such as iron-ore production, population, or earthquake intensity). Use choropleth or isoline shading for "processed" data (per-capita averages, percentages, or other PER ratios, such as yields per acre or children per family – remember the California food-production example on page 6?). Use symbols of different shape but equal size for nominal data (things that are different in kind but not in importance or amount, such as copper and iron mines or Catholic and Baptist churches). WARNING: many newspaper and magazine editors, alas, are cartographically illiterate; using their maps "as is," without encouraging students to evaluate them and suggest improvements, will send a mixed message and help perpetuate this particular form of ignorance.
4. **Type size**. Use large type. Putting a textbook page through a transparency machine is technically feasible, but it is pedagogically indefensible. Here is a simple rule of thumb: write letters so big that they appear grotesque, and then double their size. Remember one fact: you are not designing transparencies to be aesthetically pleasing to a teacher looking at them from the front of the room. You are designing them to be easily legible by a tired, vision-impaired student in the back of the room. Moreover, once you have taken the time to make a transparency, you will probably use it several times, perhaps in different rooms, and therefore you should really plan for a vision-impaired student somewhere in the worst room in which you may be asked to make a presentation. In practice, that usually means a minimum type size of about 16 points, or roughly double what is typically used in textbooks or as the default size in word processors.
5. **Completeness**. Design transparencies so that they communicate a full message without much explanation. People are not attentive all the time, and a presenter should make it possible for someone to "catch up" by examining a transparency. In practice, this means including a clear title, legend, scales or north arrows (if necessary), and the source of information. The last item is the only one that can violate the bigger-type-is-better rule – 10-point type (about letter standard) is OK for citing a source, so that you can read it to the class if necessary and/or find the data again to check or update the transparency.

For more hints along these lines, look at "Ten Suggestions and One Commandment: Using Overhead Projectors in Geography Classrooms," *Journal of Geography 84* (1985): 75-79.

TABLE 3-3. Some Design Principles for Values Discussions

1. **Prepare**. Have students do background reading on an issue; then have them flip a coin to decide what side of the issue they will present, either orally or in writing. This works best if the teacher makes it very clear that students will be evaluated on their research and logic rather than the "correctness" of the opinions expressed.
2. **Play devil's advocate**. Occasionally explain an idea in a persuasive way, and then admit that you personally do not believe it. This is a seductive and dangerous ploy, because it can promote cynicism, but that is no reason to avoid using such a powerful teaching device when appropriate (i.e., rarely, and carefully!). Introducing and then criticizing a persuasive idea is better citizenship training than setting up obviously wrong straw people and then demolishing them. The key is to admit that reasonable people might hold the idea you have just explained, but that for good reasons you do not. Then explain your reasons, while admitting that others might reasonably disagree.
3. Take a **geographic opinion poll** and analyze the results. For example, ask students to write the names of ten states (or five parts of the city, etc.) they would most like to visit. Then have them (in groups or as a class) count the votes for each state. This becomes a good topic for analysis when students take the next step, which is to map the results. For example, in a class of 20 students, you might hand out base maps and have students shade a state with a dark color if it received ten or more votes. Shade states that got 3 to 9 votes with an intermediate color, and use a light color for states with fewer than 3 votes (adjust numbers to fit different class sizes – the goal is to have about a third of the states in each color category). The final step is to analyze that map of travel preference. What might explain the pattern on it? This activity also helps to reinforce knowledge of state or other locations, and it can take a wide variety of forms. For example, ask advanced students to list five African countries that they think are *least* likely to have civil wars (like the one that was raging in Sudan at the time I wrote this), and give them a few days for research before compiling the results for class discussion. Analyzing maps such as these can also help students learn to draw a distinction between facts and "metafacts" (e.g., the kind of horse-race opinion-poll reporting that substitutes for analysis of issues in some political campaigns).
4. Use a **semantic differential** to elicit opinions about a place. This is a kind of test in which respondents are asked to judge whether they think the place is closer to one or the other of an opposed pair of words. For example, you might project a picture and ask students to put a mark on one of the five lines in each row of a list such as the following:

hot	___	___	___	___	___	cold
ugly	___	___	___	___	___	beautiful
fair	___	___	___	___	___	unfair
dangerous	___	___	___	___	___	safe
strong	___	___	___	___	___	weak
poor	___	___	___	___	___	rich

Tabulating results can provide materials for discussion and further research. For example, ask, "What in this picture made so many of us mark it as hot rather than cold?" "What evidence in the picture might get someone to express [name an opinion held by only a few respondents]?" By hiding individual responses in a summary of group opinion, a teacher can get students to address controversial issues with less emotional involvement, defensive posturing, or tacit assumption of consensus (e.g., "everybody thinks such-and-so"; they obviously don't, and it can be revealing to find out why).

TABLE 3-4. Some Design Principles for Role-Play Simulations

1. **Purpose**. Decide in advance whether the primary goal of the simulation is to enhance research skills, presentation skills, or awareness of other points of view. It is all but impossible to design a simulation that accomplishes all three goals equally well.
2. A **research simulation** uses open-ended directions to get students started. Unequal success in finding good information, however, tends to give one side a decided edge in a debate. Evaluation criteria should therefore focus on research and data analysis, rather than unthinking tolerance of other points of view.
3. A **presentation simulation** requires a set of fairly evenly matched cases. Give each team or individual a background role-card with some suggestions about graphs or maps that could be prepared to help the case. For example, teams may prepare presentations to the International Whaling Commission. Each team is given some key facts:
   - American ships are old and slow, and therefore that team might argue for a "bag limit" (a limit on the number of whales that each country is allowed to catch).
   - Japanese ships are big and fast, and therefore that team might argue for a season-length limit (they'll catch more than other countries can in a short season).
   - Russians have developed a harpoon that can catch whales at a distance, and therefore that team might argue for reducing the number of boats for each country.
   - Norway has a long history of whaling, and many villages depend on it for jobs, and therefore that team might "pull on the heart strings" for preferential treatment.

   With cases roughly equal in strength but qualitatively different, evaluation should depend on which side argues its case most persuasively.
4. A **values-comparison simulation** is designed to get students to think beyond winning a debate and to consider the validity of other positions. Here is a famous example:
   - One group has 12 red cards and is told that people in their culture refuse to negotiate with anyone who does not look them in the eye and shake hands when they meet.
   - One group has 20 blue cards and is told that it is a mark of respect in their culture to avoid eye contact and to speak softly.
   - One group has 30 green cards and is told that in their culture a person should avoid bodily contact with strangers and should always start a negotiation with a gift of a card.
   - One group has 40 yellow cards and is told that they should try to make even trades and never accept a gift without making one in return.

   The heart of this simulation is a four-minute trading session – tell students to trade cards until the cards of each color are divided equally among the groups, so that each group has the same number of a given color (I know that rule is ambiguous, but that's part of the simulation). Each person in each group acts individually to try to make trades. After each trading session is a four-minute planning session, in which people meet with their own color group to compare notes and plan strategy. Repeat as often as seems worthwhile, and add, subtract, or change rules if it seems appropriate. Students soon learn that "sensible-sounding" cultural rules can still lead to misunderstanding.

Role-play simulations work best if the "engine" is easy to explain and the rules can be modified to fit specific objectives. Moreover, a teacher must be willing to spend almost as much time debriefing as running the simulation. The debriefing does not have to happen immediately. For example, one might recall the simulation a month later to help make a point: "remember back when you were trading those cards? How you had to learn about the negotiation rules of other cultures in order to reach your common goal?"

4

Four Cornerstones: Foundation Ideas of Geography

Knowledge often begins with a creed, a statement of faith in something that is essentially unprovable.

Science, for example, starts by assuming that everything in the universe is the effect of one or more causes that can be identified and understood. Philosophers, liberal-arts graduate students, and college sophomores (of all ages!) rather gleefully point out that it is impossible to "prove" the idea of causality. Scientists just assume that it occurs and proceed from there. Despite the unprovability of its basic assumption, science has developed a sizeable body of useful ideas.

For example, automobiles move "by themselves" (that's what "automobile" means) because car builders accept a theory about combustion. Our highways are full of cars that work, even though the assumption on which the theory rests cannot be "proved."

Geography also begins with a statement of faith.

Geographic creeds

A "pure" geographer (one who uses geographical methods to learn about the world) starts with a creed-like statement about location: "I believe there are reasons why things are located where they are."

An "applied" geographer (one who uses geographical methods to do things in the world) puts a slightly different twist on the same creed: "I believe that people should have good reasons for locating things where they do."

Both of these creed-like statements underscore the importance of location, because a focus on location is what makes someone a geographer, as opposed to a scientist or philosopher. In other words, geography is not all of knowledge; it is just the kind that deals with location (remember the discussion of disciplinary perspectives in Chapter 1?).

> Warning: Although geographers believe there are reasons for things being where they are, they also have to realize that those reasons may not always seem reasonable.

Reasons, real reasons, and unreasonable reasons

Here is an example. A teacher starts by pointing out a plausible reason for a town being located where it is: its site was defendable in a militarily sense, perhaps, or is near a valuable mineral deposit, a good river crossing, fertile soil, at a strategic location in a mountain pass, has proximity to a freeway exit, and so on.

Occasionally, however, a town is located where it is for purely quirky reasons. Imagine, for example, that some individuals decide to start a new town, even though they have no discernable survival skills, and their neighbors even question their sanity. Those individuals travel to a distant land and decide to start a town in a particular place. When asked why, they respond, "because we felt a warm purple glow in our big toes when we walked across the site, so we figured it would be a good place for a town."

This "reason" makes sense to those people, though probably not to anyone else. Its "unreasonableness" is unsettling to anyone who hopes for a tidy logical explanation for any landscape feature. But that's the way it is: some features just do not have reasonable reasons for being where they are.

This requirement for reasonableness can result in a difficult balancing act for a teacher. On one hand, we want students to appreciate that the world is organized in ways that are understandable. At the same time, we have to acknowledge that some parts of it do not seem to make sense. Some of those parts may become understandable in the future, after we learn more. But the locations of some things in the world will never make sense, because in fact they made no sense when they were first put there.

Chaos theory

Big effects often follow tiny (and therefore unpredictable) causes

This whole discussion could be called a geographical variation of **chaos theory**, which says that big observable events are often consequences of processes that are set in motion by tiny, hard-to-observe events. For example, an Oklahoma storm may ultimately be triggered by a pink butterfly landing on a flower in Indonesia – that butterfly disturbed the wind just enough to guide a bigger flow of air in a slightly different direction, which in turn gathered a little more moisture from the Pacific Ocean and crossed the Oregon coast a little further south than it otherwise would, which in turn . . . You get the idea.

Taking this idea to its extreme, some people suggest that the fundamental principle of the universe is the random radioactive decay of atoms, the ultimate in unpredictability. This kind of philosophical speculation can be addictive, because it lets someone sound "scientific" in a seminar or at a party while making sweeping statements about the essential unpredictability of the universe.

In fact, however, even radioactive decay is quite predictable at some scales of analysis. We may not know which specific atoms will disintegrate in a given second, but we are able to predict approximately how many will, and we can design nuclear reactors to harness the resulting power. So, even though questions about the ultimate causes of individual events cannot always be answered, we can still draw useful conclusions about what is typical behavior and predict large numbers of events.

The rest of this chapter (and the entire discipline of geography) is devoted to the parts of the world that do (or at least might) make sense at geographical scales (i.e., from communities to continents). This philosophical introduction is just a caution for two groups: those who might hope that we could find a geographical theory that explains everything they see, and those who think that chaos theory means that nothing is explainable.

Four cornerstones of geography:

Location

Place

Link

Region

The analytical study of geographical phenomena rests on four tightly interrelated ideas: location, place, link, and region.

Each of these terms has a specialized meaning for geographers. It is also true that the use of each is fuzzy enough that some geographers have trouble with them – they prefer to use other words for the same ideas (e.g., "site" instead of "place"). As a result, there is an ongoing dispute about basic terminology in some geographical journals.

My advice to teachers is simple: don't worry too much about it. Terms at this basic level are almost always words like "cause" and "effect," which seem clear enough until you realize that nearly every effect is in turn a cause for some other effect farther along the logical chain. At that point, logic seems to lose its clarity.

Such lack of clarity is not easy to explain to a child who is looking at you with the peculiar expression children reserve for adults whose density seems to transcend all understanding. For that reason, and others we will discuss in Chapters 5, 6, and 9, I would prefer just to say that the set of cornerstone ideas in this chapter may be of some value in writing curricula or talking with administrators, but I would not insist on using exactly these words in an elementary-school classroom.

With the reader's permission, then, let us just go on to consider these cornerstone ideas – location, place, link, and region – and how to present them in the classroom, without getting hung up on disputes about the choice of the words themselves. In any case, students should be tested on their ability to think of examples, not their ability to define terms.

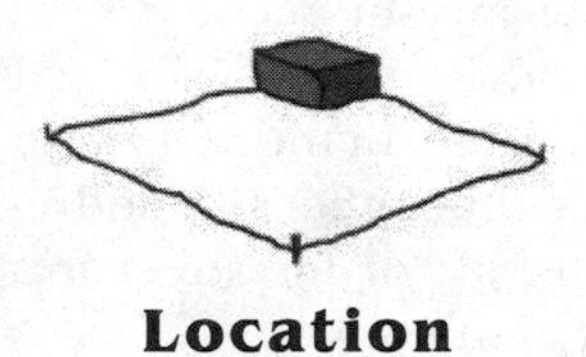

Location

Location

Position in space

The first cornerstone of geography is the concept of **location**: you have to know where something is before you can study it geographically.

Spatial relationships

What geographers study

Once you know the location of something, then you can study what is **spatially related** to it: in other words, what is next to it, upwind from it, between it and the ocean, just beyond it, and so forth. Those relationships, in turn, have effects that we can measure, analyze, understand, and perhaps control.

So, to repeat, you have to know where something is before you can study it geographically.

Unfortunately, the seemingly simple task of knowing where something is is circular, because you can know the location only if you have at least one other piece of basic spatial information – distance, direction, adjacency (next to), or enclosure (within). Go ahead – try to describe where you are sitting right now without using any of those other spatial ideas!

Some textbook authors (and, alas, the authors of geography standards in a number of states) make things even more confusing by dividing the concept of location into two categories:

Relative and absolute location

Relative location – Position expressed in terms of one or more known locations: "the library is between Smith and Jones Hall" or "the pool is half a mile north of the high school."

Absolute location – Position expressed in terms of a coordinate system or mathematical grid: "Memphis is at 35 degrees north latitude and 90 degrees west longitude" or "my house is at 1920 First Avenue."

Having explained the distinction, I have to admit that I do not believe it is either true or very useful. For one thing, even something as precise-sounding as longitude is not absolute. The dictionary defines **longitude** as angular distance east or west of an astronomical observatory in Greenwich, England. Think about it: that is an arbitrary measure of distance away from an arbitrarily chosen line. In other words, it is a form of relative, not absolute, location (Transparency 4A).

Longitude

Angular distance east or west of Greenwich, England

Likewise, house numbers in a city are usually based on distance from some central point, but the center is almost always arbitrary. Moreover, the scale of measurement is rarely uniform. For example, 1925 First Avenue is probably located somewhere between 1920 and 1930 First Avenue. In the real world, however, it seldom is exactly halfway between the other two addresses, and in fact it usually is on the opposite side of the street from them (since it is an odd number and they are even).

In other words, there really is no such thing as absolute location. There are just a large number of different ways of expressing relative locations. Some of those ways are more precise than others, but none of them is absolute in any meaningful sense (is Einstein nodding his head in his grave?).

For this reason, the textbook distinction between relative and absolute location seems at best artificial, and in most cases it is pedagogically unproductive. It is better for students to master a number of different ways of describing location. One could say, truthfully, that the idea of location is so important that people have invented many different ways of expressing it (Transparency 4B).

In any case, children are usually much better than adults at just "doing location," without concern for its logical ambiguities (as long as teachers and other adults do not impose artificial distinctions such as the one described above). Here's why: human children depend on others for food and shelter, and for that reason getting lost may be detrimental to their survival into adulthood! Stated in more "scientific" terms, locational skills have been an essential survival tool for the human species, and the human brain is therefore quite well adapted to acquiring and processing locational information.

The fact that locational terms and concepts are basic or innate does not eliminate the need to teach them. Here, the sheer size of the world looms up to complicate a teacher's life. This is because cultures that occupy different parts of the world have also developed different ways of communicating locational information. People who live in forested regions, for example, might use distinctive trees as landmarks – examples include lob pines in the Canadian lake country, council trees in the savannas of East Africa, and witness trees in the U.S. Public Land Survey. By contrast, people who live in deserts (or on islands) often use stars as tools to help them find locations. And people who live in urban areas have street names and building numbers that provide even more ways of describing locations. As a result of all this variety, a typical primary curriculum in geography (in any culture!) devotes a lot of time to mastering various ways of describing and manipulating locational ideas.

Instruction about location will usually be more effective if it is presented in the form of meaningful questions and refers to familiar features. Transparency 4C is one time-tested way of showing students how to use different terms to describe the same location. It is a simplified map of a shopping mall in upstate New York. (It's OK to use this transparency for class demonstration, but for student practice it is definitely preferable to use a map of a local mall if one is available, or to look for a map of a local theater, stadium, golf course, etc.)

Similar strategies can be used to teach the conventional "vocabularies" for other **spatial primitives**, such as distance (Transparency 4D), direction (Transparency 4E), and enclosure (Transparency 4F).

Spatial primitives, basic spatial concepts, such as

Position

Distance

Direction

Enclosure

The activities that go along with Transparencies 4C to 4F all have two features in common.

First, students learn to use several conventional "languages" to express ideas about location, distance, direction, or enclosure. Translating between languages is the highest level of mastery, because it demands a thorough understanding of each of the languages. Moreover, an activity that requires translation can be a useful diagnostic tool, because it helps teachers identify students who have trouble with specific languages and could use a review.

Student activities should involve real places

Second, these activities involve real places (even though they may be introduced with simplified demonstrations). This provides an intuitive rationale for the activity – students can hardly say that activities are meaningless busywork when they are obviously related to real-world issues. Moreover, society expects geography teachers to instill in students a common mental map of important locations in the local area, the nation, and the world.

Some magazine and newspaper articles seem to imply that the major purpose of geographic education is building this mental map of the locations of places. You've seen the headlines: "High school seniors fail geography: 57% can't even find China on a world map."

Most professional geographers, by contrast, heartily disagree with the idea that geography is primarily about locating places. The fact that this book didn't stop on the previous page suggests that I agree, but I also think that it is counterproductive to go to the other extreme and to teach fundamental spatial concepts primarily with maps of imaginary places (and unfortunately, often with cutesy names such as Big City, Middletown, and Farmburg).

Using maps of imaginary places does not make a teacher's job any easier. In fact, it may be more difficult, because students may subconsciously (or even consciously) question whether the maps deal with ideas that are worth knowing. So, absent a clear and compelling rationale for creating imaginary places, let me assert a general principle: in general, it is preferable to use real places in geographic instruction, and to choose those places with care.

> (Third edition note: no other statement in this book has generated as much reaction as this one. Many teachers have commented or written that their students "really enjoy" making or reading maps of imaginary landscapes.
>
> I don't doubt it.
>
> But I submit that students enjoy working with *interesting* maps. The valid test, therefore, is not whether they like making imaginary maps more than a lecture, video, or other typical class activity. The valid test is whether they would enjoy working with interesting maps of real places just as much, or even more. In other words, it is imperative that we make a fair comparison before deciding what approach is better.
>
> Unfortunately, few people have studied this rigorously, but the tiny amount of research that is available suggests that "fun" depends on whether the maps are interesting, not on whether they are about real or imaginary places. If that is true, why not use real places? Students would use more of their brains, learn on several strands [see Chapter 3], and still have fun!)

Should teachers use maps of imaginary places?

Sometimes, perhaps, but not as often as some do

We could also look at this question from the other side, by asking whether there are any drawbacks to using imaginary continents, islands, and so forth. When we do that, it becomes obvious that a book or student activity that uses imaginary places to illustrate basic principles is sending two unintended but clear messages:

"We're too lazy to do the homework needed to show how this skill or principle applies in a real location."

"Knowing real locations isn't very important after all."

Those are powerful messages –

do we really want to send them?

With at least two good reasons to use real places, and at least two good reasons not to use imaginary ones, let me restate the principle from the previous page: in general, it is preferable to use real places in geographic instruction, and to choose the places with some care.

The latter point ("choose the places with some care") is also important. The list of locations you choose to mention is a clear indication of what you think is worth studying. One guide for selection of real places for classroom activities might be their importance in performing other everyday tasks, such as navigating around town or reading a newspaper.

I would even go so far as to say that this should be a criterion for the selection of textbooks and other instructional aids. As you review various books, ask: "Do the places in them seem to have been selected with some care, so that at the end of the class students have a reasonable mental map of the world? Also, do the placenames in the book betray a significant bias?"

Placename bias can take a variety of forms:

Eurocentric or Anglocentric bias. Exclusive focus on landscapes made by English-speaking Europeans is not appropriate in today's world. After all, a large fraction of the people of the United States did not originate in Europe, and many of the Europeans who came here did not speak English.

Exotic bias. It is also inappropriate to focus exclusively on places inhabited by headhunters, polygamists, pearl divers, ice-builders, or others whose lifestyles are as far from the American suburban norm as possible. Students can emerge from such a class with vivid impressions of Lapland and click languages but no idea of the location of London, Singapore, or Tehran, or how those cities fit into the world.

Categorilla bias. Mention of exceptional features (the highest mountain, the most dangerous rapids, the largest factory, the biggest ball of twine) can grab attention, but too much focus on **categorillas** does not help students put real-world events into perspective.

"Categorilla"

Term coined for the longest, deepest, highest, or otherwise most exceptional feature in a category

Upscale bias. Stunning vistas, beautiful places, and lifestyles of the rich and famous are hardly the whole picture.

Problem bias. At the other extreme, an endless litany of catastrophes and other problem places may promote blame-seeking, cynicism, or a kind of numb fatalism.

Headline bias. Waving an appropriate headline from today's newspaper in front of a class can help make the day's topic seem more relevant and worth knowing. But scanning headlines to get ideas for today's class can easily make a course seem disjointed, because, in fact, it will be. Events don't happen in such a way that topics would be "covered" in appropriate sequence and depth!

Personal favorites bias. This one is the hardest to deal with. It is true that teachers are likely to be more enthusiastic (and therefore perceived as more interesting) when they focus on topics they find interesting – hiking, food, opera, whatever. At the same time, some people find hiking, food, opera, or whatever boring and irrelevant.

So, I admit that I devote almost an entire lecture in my college class to the role of barbed wire and the need for simultaneous invention of other tools such as fencing pliers, posthole diggers, cattle breeds, and fence laws. I do this because I enjoy building fences; I have a nice collection of tools to use in demonstrations; and that particular topic energizes me. But I also make considerable effort to tie fence technology to other issues such as westward migration, culture shock, and the geography of inventions. Moreover, I put some serious fences around my own interests – this topic is never allowed to consume more than 26 minutes of a semester-long class, and I deliberately mention places in 11 different states.

In short, the placenames that are emphasized (or even just mentioned) in a geography course happen to be a good measure of the quality of the course, even though the mastery of placenames is not the sole (or even the primary) purpose of the course. The list of placenames is a prescription that tells students what kinds of places are important to know. The list is a goal that tells teachers what ideas are important to emphasize. Finally, it is a summary of course content, for scrutiny by administrators and legislators.

The point mentioned earlier, in favor of using real-world places, does not mean that one cannot simplify a real location in order to illustrate a point. Geographers do that every time they make a thematic map (recall Chapter 2?). Simplification, however, should not violate the basic principle: at the end of a geography course, a student should have a fairly good mental map of the locations of countries, large cities, major rivers, and other important features within the area being studied.

(You might even tabulate the places mentioned in this book and critique that list – we practice what we preach!)

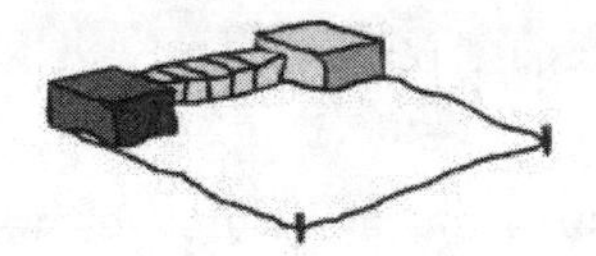

Place

The second cornerstone idea of geography is encapsulated in the deceptively simple words "place" or "site." These terms refer to a location's characteristics, the features, conditions, processes, and human actions that make the location meaningful.

Place (site)

The mix of "natural" and artificial features that give meaning to a location

Here's the difference: a location is a position in space; a place is the sum total of all the conditions and features at a given location. That may seem like a fussy distinction, but it allows a geography teacher to highlight the difference between the skills needed to define a position in space and the quite different skills needed to describe the conditions there.

Each location is unique; no two places have exactly the same set of natural processes and human actions. That realization is an important learner outcome even for children of a very young age.

The ability to discriminate between places is a prerequisite for the process of **classification**, which is one of the methods that people use to organize the world around them.

Classification

Putting things into categories

> Warning: Teaching about classification has been criticized by some educators as politically incorrect: they call it Eurocentric, male-dominated, hegemonic.
>
> That is true, *IF* people are taught to accept a particular classification without question. But if people learn that a classification is nothing more (or less) than a human attempt to reduce the bewildering complexity of the world in order to see it more clearly, then the chance of blindly accepting another person's categories will be reduced. Teaching the logical basis of classification can therefore help liberate us from those who would try to impose their categories as if they were the only truth.

The purpose of classification is to simplify the world so that we can see relationships better. The process of putting places into categories begins with the realization that some places are more alike than others.

Someone trying to classify places has to strike a balance between two opposing ideas:

Each location on earth has a distinctive mix of dissimilar forces that interact to produce unique features. For that reason, no two places are alike.

Each location on earth has some dominant forces that in turn have a number of predictable consequences. For that reason, places in similar locations tend to have similar conditions (these places are called **geographic analogs**).

Geographic analogs

Places with similar locations (see Chapter 6)

Subsidence zone

Zone of sinking air ~25 to 30 degrees north or south of the equator

Consider a location in North Africa as an example. At a latitude of about 27 degrees north of the equator, this place is in one of two broad areas of sinking air known as the **subsidence zones**. Atmospheric subsidence is a predictable consequence of a few simple but very powerful forces:

1. The noon sun is almost directly overhead at the equator (technically, the noon sun at the equator is never more toward the poles than 23 degrees north or south).
2. A sun that shines directly down can add a lot of heat to the ground (that is why the South Pole is cold, because the sun there never gets very far above the horizon).
3. The warm ground heats the air immediately above it.
4. Heated air rises, causing clouds and rain, often at very predictable times – a classic example is that an area in New Guinea has a storm between 2 and 4 p.m. about 300 days a year.
5. The air that has risen high above the equator is then forced to move north and south by continued updrafts, day after day, as long as the sun keeps shining.
6. Air moving away from the equator at high elevations (say 50,000 feet up) is cooled, squeezed together (because the earth is round), and turned to the east (because the earth is rotating – Transparency 4G).
7. Cooled, compressed, and deflected air sinks toward the ground, and this sinking is most pronounced about 2,000 miles north and south of where it was pushed up.

This is where we should probably stop the explanation, even for high-school students. The reasons why the subsidence occurs about 2,000 miles away from the equatorial updrafts have to do with some complex principles of fluid dynamics. That being the case, the distance would be different if the earth were bigger (or smaller), the sun hotter (or cooler), the air denser (or thinner), the surface rougher (or smoother), or the rotation faster (or slower).

At some point, however, the details threaten to get in the way of the main point, which is that large amounts of air will be sinking downward at approximately 25 to 30 degrees north (and south) latitude, as long as the sun keeps shining and the earth keeps spinning (Transparencies 4H and 4I).

Sinking air, in turn, has a whole bunch of predictable consequences for a place:

Consequences of sinking air

1. Sinking air gets warmer. It is no accident that the high-temperature record on each continent is located in the subsidence zone.
2. Relative humidity decreases as air is warmed, and therefore the place tends to be dry. It is hardly surprising that the list of areas affected by subsidence reads like a Who's-Who of the world's great deserts: the Sahara of North Africa, of course, and the Arabian, Atacama, Kalahari, Mojave, Thar, and Great Australian deserts (Transparency 4J).
3. Dry air puts stress on plants, and therefore plants in the subsidence zone tend to be sparse, short, and often equipped with moisture-gathering or moisture-saving structures (e.g., deep roots, water-storage tissue, photosynthetic stems, protective thorns or spines, leathery leaves, or chemical methods of reducing competition from other plants).
4. Dry soils tend to accumulate salt, which is left behind as water evaporates. Soils in subsidence areas, therefore, usually have an excess of sodium, potassium, and/or calcium salts.
5. Dry winds create distinctive landforms by moving sand and dust around. A place in a subsidence zone probably has rocky slopes, sand dunes, sculpted cliffs, **fans**, and playa "lakes" that are often dry.
6. Dry weather, rocky slopes, and salty soils are not good for farming or for grazing livestock. For that reason, human populations in subsidence zones tend to be sparse, except near rivers or in places with enough wealth to bring water from elsewhere. (Irrigated fields in the valley of the Nile River in Egypt and air-conditioned casinos in Las Vegas seem to have nothing in common, until you see them as different ways to solve the same problem: how to live in a hot, dry place.)

Fan

"Apron" of loose rock around a mountain in a desert climate

In short, a basic cause – sinking air – has a whole set of predictable consequences: cloud-free sky, low precipitation, high temperature, salty soil, sand dunes, sculpted cliffs, fans, sparse population, and problems for human use.

One can start to study these interactions by noting the similarity of map patterns of various phenomena. The assumption is that features that occur together in similar places may also be related in other ways. (We'll say more about this later in this chapter, in the section on regional comparison, and in Chapter 6.)

The point

Here's the point of this long discussion of subsidence and its consequences: the features of a specific place should be taught as an interrelated set of causes and effects, with a minimal number of important causes and a greater number of plausible effects. Unfortunately for students seeking easy explanations, the important causes are not the same in every place. For example, sinking air is an important cause in subsidence areas, but it is not important near the equator or on windward coasts such as Norway or New Zealand. In other places, one should seek other causes to help organize a list of consequences.

Selecting traits that seem to have important causal influences is part of both the science and the art of geographical analysis. If we describe place analysis as partly a science, we then must agree that a hypothesis about relationships in a place should be tested in the same way as it would in any other science:

Scientifically acceptable tests of a hypothesis

Does the theory explain the observed data?

If the theory explains the data, does it also fit with other generally accepted theories?

If several theories explain the data, which one has the fewest undesirable side-effects?

If a theoretical statement cannot pass these three tests, we should either reject it outright or modify it before we act on it.

The task of seeking plausible theories to help organize our teaching about places is not easy. The alternative, however, is a truly gruesome prospect: a geography class that consists of rote memorization of the traits of a seemingly endless number of places. In other words, the ideas of map-pattern similarity and geographic causality are useful tools to help keep the total mass of information manageable.

> Caution: This does not mean that the conditions of every place on earth should be treated as simple consequences of individual powerful causes. In the real world, many places are better described as resulting from the interplay of a fairly sizeable number of mostly independent forces. This observation has led some people to propose that the best way to teach geography is through the use of **checklists**.

Checklist

Data form with some labeled blanks for students to fill in for each country

According to this proposal, students should be taught to organize information about places by filling the blanks on a form like this:

climate ____________________	rock __________________
high temperature ___________	soil __________________
low temperature ____________	minerals ______________
precipitation ______________	______________

Other blanks ask for information about crops, language, largest cities, major exports, and so on.

This checklist approach, unfortunately, tends to work well only for the first few days of a class, though many students do enjoy using a checklist to impose order on a mass of information they can find about a place.

The problem with the checklist method is the same as with any standardized kind of instruction: it is basically like a cutting block, and the analytical knife soon gets dull (recall Chapter 2?). After doing checklists for 5 or 10 states or countries, students reach a stage of numb acquiescence, if not outright revolt over the tedium of it all.

The solution to this type of problem is to maintain the rigor of the checklist while varying the form. The variety can help students acquire the information they need to compare places without the boredom involved in actually filling out a checklist for every place. For example, students can:

1. Match photographs of places with verbal descriptions of the places' vegetation or landforms, or with statistical profiles of populations or economies (this can be made into a game with an appropriate handout or bulletin-board display);
2. Use Census data to construct population pyramids for several countries and match the pyramids with descriptions of the economies of those countries (instructive examples might include Afghanistan, Japan, Kuwait, and Rwanda);
3. Use a spreadsheet to calculate per-capita values by entering and dividing absolute numbers for various places to be compared (e.g., calculating the available space per person of a country by dividing its total area by its population; do this for a range of dissimilar countries such as Brazil, Indonesia, Japan, Mongolia, and Singapore; see Transparency 1J);
4. Use a web browser to search for information to compare several places (e.g., counting the hotel rooms, mobile-home parks, and real estate agencies in a high-income resort-and-retirement area such as Palm Springs, California, and in a middle-class area such as Bella Vista, Arkansas, or Deming, New Mexico; see Transparency 9B);
5. Gather information from thematic maps and atlases and record what one might observe along a **transect** across an area (e.g., the kind of natural vegetation one would expect to see along a road from Cairo to Capetown, or the size and shape of houses one would expect to see along a specific bus route in a city; see Transparencies 3R and 7G);

Transect

Line along which one observes and records data

6. Compile information and prepare a one-page profile of a place (like Transparency 3B; see also 7D). This works better if it seems like a plausible real-world situation, such as a company president who is thinking of building a new restaurant in a particular area. The desired student product is a community profile that resembles what many towns and counties now make as part of their economic-development effort. Local Chambers of Commerce will often give free copies of their publications to traveling teachers. A box of reports and brochures from a range of communities can serve as good examples of applied geography and as a data resource for students (and it can also help defend the tax deductibility of some of your vacation expenses!);
7. Prepare posters, mobiles, computer presentations, or other descriptions and exchange them with other students, either in their own class, in other classes, or even in other states or countries (through mail, e-mail, or a school web site);
8. Make a "country-in-a-bag" by filling a paper bag with items that come from a specific country or that are reminders of key features of a country – for example, a coffee bean, a handful of red soil, a woolen scarf, a Spanish-language dictionary, a Quechua phrasebook, a photo of a Catholic shrine, and a CD of pan-pipe music as aspects of Bolivia);
9. Gather information and write the first few pages of a mystery novel set in a specific place; those pages should help set the scene by noting the features of the place that are likely to influence the action or otherwise affect the main characters;
10. Prepare **reference maps** that show how various features of a place are arranged and connected with each other. The form of the report can range from a detailed map to an impressionistic cartoon. The goal is to record the features that are most important in making the place what it is.

Reference map

Shows the locations of a variety of things in a specific area (see Chapter 2)

In short, there are many ways of imposing discipline on the process of gathering, organizing, and presenting information about specific places. This part of a class has obvious practical value, since the description of places is one of the jobs professional geographers are asked to do. After all, someone has to prepare all the different kinds of community profiles that are used out in the real world:

- Local-issue briefs for political candidates,
- Market analyses for store managers,
- Background papers for trade or labor negotiators,
- Context outlines for journalists or television reporters,
- Neighborhood reports for real-estate appraisers,
- Area profiles for church or school administrators, and
- Relocation packets for families or factory owners.

The list of 10 options on the previous pages did not include what is perhaps the single most important method of geographic place-analysis: the process of making **thematic maps** as a way of gathering and organizing information about places. As described in Chapter 2, this is the other blade of the place-analysis "scissors" of modern geography. That realization could serve as a transition to the next cornerstone, the exploration of *links* between places.

Thematic map

Shows the patterns of a few features in a fairly large area

We should not leave the section on place, however, without noting three other very important aspects of place analysis: the cultural definition of resources, the legacies of past occupation, and the pervasive influence of land division.

Cultural definition of resources

People decide what aspects of their surroundings are resources

The cultural definition of resources is the principle that people, not nature, define what is a resource. For example, a rich deposit of uranium ore is of no value for a culture that does not use uranium. People may live in a place for generations, growing food, building houses, raising families, and leaving other imprints on the local landscape. The tabulation of important landscape features in that place, however, will not include uranium mines until the invention of nuclear reactors (or the arrival of a culture that uses uranium, or some kind of trade with people who want uranium). Until there is a demand for it, an ore deposit is not a resource.

This principle is an extremely important part of geography. It helps explain why different cultures can end up creating very dissimilar landscapes in apparently similar environments. Understanding this, in turn, helps guard against adopting a simple **determinism** that puts nature in charge of "determining" what human beings are able to do in a particular area. Awareness that other groups of people might do something different in an environment can help broaden the range of options people might consider as they decide what to do with their own landscape. This broadening of perspective is one of the most important citizenship outcomes of geographic education. For that reason, it should be introduced early and reinforced often throughout elementary and secondary school.

Environmental determinism

Simplistic idea that environment determines what people can do in a place

The second principle is that previous activity in an area often leaves features (**legacies** or relics) that can influence what future generations can do there. These features may have positive or negative effects. For example, a sturdy old factory building can become a good place for a new mall or office. On the other hand, a toxic chemical spill can contaminate a site and make it harder to use for any other purpose. A decision to build an airport in a specific place will have an effect on what kinds of things are likely to be built nearby – small factories or office buildings, perhaps a hotel, but probably not a classical music recording studio!

Legacies (relics)

Landscape features left by previous inhabitants of a place

A geographic description of a place, therefore, should include an attempt to identify prior uses and their legacies. (You might hear this idea described by the term "sequent occupance," a well-known geographic theory that describes human occupation of an area as a sequence of phases, with relics of prior forms of use usually visible in the present landscape.)

Land division

Process of marking areas that "belong" to or are controlled by particular groups or individuals

The third basic principle of place analysis that should not be overlooked is the role of **land division** in place characteristics. Different groups of people divide land in different ways. The first effective division of land in an area, however, usually leaves long-lasting imprints and exert a powerful influence on what future generations can do there. These effects range from obvious to quite subtle.

Some of the imprints are big physical features, such as roads, fences, or irrigation canals. These big features are easy to map, and it is often fairly easy to see their constraining effect on future land-use decisions. For example, look at the potential for traffic jams in the Detroit street patterns in Transparency 3H, or the CD unit on Measuring Distance, which has a section on the costs of delivering mail in places with different road patterns.

Other imprints of past use have present-day legal or technological implications. For example, the Homestead Act of 1862 allowed people to claim 160 acres of land if they built a house on it. The result is a pattern of scattered houses and relatively small fields (by the standards of modern mechanized farming). The goal of the Homestead Act was to promote orderly settlement, but its legacies include some rather costly impacts on farms and small towns. Many of these impacts are related to the burden of delivering mail, running electric lines, and providing other services to a population that is dispersed throughout a large area (Transparency 4K; see also the discussion of spatial patterns in Chapter 6).

Range

Distance people are willing to travel to get particular goods or services

Other costs become apparent when we consider the effects of locating small towns at an optimal distance from scattered farms with horse and buggy as transportation. Small towns were located 6 to 8 miles apart in many Midwestern states. That was because 3 or 4 miles was about the average **range** of daily travel – the distance farmers were able go and return in one day.

The invention of the automobile allowed farmers to travel much farther in the same amount of time. This deprived many small towns of their former market function (as illustrated in Transparency 4P).

Later, the construction of bypasses and Interstate highways had similar effects, because they made places near their exits more valuable and took away traffic from places that were located on the previous generation of roads and railroads.

Here is one way this can become a problem: it is a fact that abandoning houses, neighbors, and communities has a very high psychological cost. As a result, people in thousands of small towns and bypassed communities have spent millions of hours trying to convince stores, doctors, and factories to locate (or stay) in them. The goal is to provide jobs for people who otherwise would move away.

Though extended discussion of rural development strategies would obviously be out of place in this book, the ferocity of those debates (and of the parallel city/suburb debates about mass transit or housing policy) only serves to underscore the basic principle in this section: *the first effective land division and the resulting arrangement of roads, city limits, and other landscape features can have profound impacts on what people do in the future.*

It follows that we are unlikely to solve certain problems, in both rural and urban areas, unless we understand both the present forces at work and the legacies of the past occupation of the area.

Here is a summary of this entire section on place: To understand the characteristics of a place, we have to know something about the resources there, the cultural definition of those resources, the previous uses of the land, and especially the method of land division that was used there. This is a tall order, and it is part of the reason why teachers must always seek to simplify their tasks by organizing place information into logical packages.

Better yet, give students the tools they need to perform such organization themselves. Fortunately, some of the same tools can be used for the third cornerstone idea, links between places.

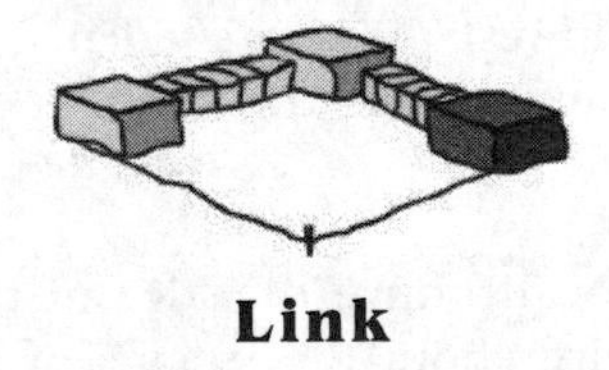

Link

As if describing and explaining features within a place were not challenging enough, each location has another whole category of traits that depend on its connections (**links**) with other locations. This cornerstone of geography has its focus on the analysis of ways in which locations are connected to each other.

Link

Connection between places

Some connections between places are "natural," which means that they occur without human thought or intervention. For example, air moves from high to low pressure, water flows downhill, and tectonic plates shift position. As a result, winds tend to blow in predictable directions, rivers flow to the oceans, and mountains rise in some places while erosion wears the land down in other places.

Natural links between places

Those natural movements, in turn, help create specific features in specific places: trees on windward slopes, deserts in rainshadow areas, earthquakes near plate boundaries, and so forth. In that way, understanding of **natural links** between places can help us understand the traits of the places.

Artificial links between places

In addition to these natural links, many connections between places are created by humans as they move from one location to another, trade with people in other places, send messages to other people, and alter the natural flows of energy, water, or other materials. These **artificial links** include interpersonal ties among families, economic connections among corporations, and political links to governments.

As said before (maybe too often, but it is really important!), there is a scissors-like back-and-forth relationship between the conditions in a specific place and the connections between places. For this reason, the cornerstone idea of links can be explored in the same two ways as the features of a place:

1. By cataloguing landscape features that are caused by or otherwise related to the links – river valleys, canals, railroads, repair shops, grain elevators, roads, gas stations, telephone wires, radio towers, and so forth, and
2. By analyzing logical consequences of the most important links – for example, the acid precipitation that occurs in New England as a result of winds moving polluted air eastward from factories in the Midwest, or the sudden growth of a formerly obscure town that is now at the junction of two new Interstate highways.

Where did this come from?

One common way to introduce the idea of geographic links in a classroom is to inquire about the origins of familiar objects. For example, where do your backpack, shoes, or even something as small as a pencil come from? The answer is often a surprisingly large number of places all around the world – a typical pencil might have rubber from Malaysia, tin from Bolivia, wood from Texas, graphite from Ohio, paint from Delaware, and so on. When the focus is on a more complex object, such as a modern automobile or a jet airplane, the list of sources can be truly staggering (Transparency 4L).

This complexity of origins lies at the root of an intriguing puzzle that has played a role in several recent political campaigns. The puzzle consists of trying to determine whether a Ford car assembled in Mexico with parts from six countries is more "American" than a Toyota assembled in Kentucky with parts from nine countries and many states (see Transparency 3Q).

The answer to that puzzle has implications for several aspects of U.S. foreign policy. Therefore, geography teachers should help their students gain some understanding of the basic patterns of world trade and corporate control. This understanding should include awareness that the name of a company does not always reveal its product, its home base, or its links to other places.

Students in primary grades can begin by tabulating where their parents work and what is produced there. Middle-school students can profile local companies or make world thematic maps showing the origins of their backpacks and the items in them on a given day. High-school students can select an industry and describe its local and worldwide links. In each case, the goal is to see that the everyday lives of people in a place are often linked with what happens in other places, sometimes very far away.

Awareness of connections between locations around the world is a useful learner outcome, but it is not the only goal of this part of geographic education. Citizens, directly or indirectly, have to make decisions about the amount and kinds of roads, airports, telephone systems, and other **infrastructure** they want built in a place in order to link it with other places. This infrastructure is usually expensive, and therefore the decisions can have a great deal of impact on the success of a community.

Infrastructure

Roads, railroads, storage buildings, water lines, sewers, television and radio towers, and other features built to facilitate links between places.

Here is a classic example, something that happened almost two centuries ago and still has an impact today. In the 1820s, the citizens of Philadelphia were facing a crisis. Their former position as the number-one city in the United States was under serious threat from a relative upstart, New York City.

To understand the nature of the threat, however, we have to look at the connections between the two cities and two key destinations: the "old countries" of Europe and the "new frontier" in the middle of North America.

On the European front, both Philadelphia and New York had good ports with sheltered harbors and deep water, but New York had a shorter route to the ocean. Both cities had some nearby forests and waterfalls for power, but those in Philadelphia were closer to the center of the city. In short, when it came to local conditions and connections with Europe, the two cities were just about even.

On the frontier side, however, things were dramatically different. New York had the good fortune of being located on the Hudson River. This river, and its main tributary, the Mohawk, flowed mostly through a broad valley of fairly weak rock. Building a canal in this valley was quite an engineering feat for the technology of the day, but it was *possible.*

Once completed, the Erie Canal soon proved its worth. The existence of an inexpensive route to the ocean made it possible for farmers in upstate New York to produce grain for export. It is not at all surprising that a map of farm production in 1840 shows a major grain-producing region along the Erie Canal (Transparency 4M).

Moreover, it's not surprising that Buffalo, New York, should emerge as a major grain-shipping and flour-milling center in the middle 1800s. After all, it was located in a place where it was easy to transport grain on lake ships or canal boats from upstate New York, northern Ohio, and southern Ontario.

The flow of grain and other products down the Erie Canal and out through the port of New York had important side-effects. Traders moved to New York, and financial institutions arose to serve the traders (why else would the New York Stock Exchange be there?). Population and wealth began to grow rapidly.

At this point, the story gets messy. Pennsylvanians noticed that wealth and power were shifting to New York. Reciting what seems to be the bumper sticker of the geographically ignorant, they said, in effect: "New York built a canal, and they're growing faster than we are, so we ought to build a canal, too."

The bumper sticker of geographically ignorant people:

It works for them, where they are, so it should work for us here

One glance at a map of mountains and rivers should be enough to show why this plan was questionable at best. Where New York has a wide valley through the middle of the state, Pennsylvania has a bunch of mountain ridges that run generally from southwest to northeast (in other words, at right angles to the intended line of travel). The Pennsylvania canal followed the twisting Juniata River through dozens of rapids-filled gaps in the long ridges of the Appalachian Mountains. Building locks to get around those rapids added millions to the cost and days to the travel time. Then, when the canal reached the imposing cliff at the edge of the Allegheny Plateau, it could go no farther.

The Pennsylvanians could have seen the light and changed their plan, but instead they repeated another famous adage of the ignorant: “We’ve already invested so much in this, we ought to finish it.” They proceeded to build an expensive system of cables and rails to drag the canal boats up the mountainside, as shown in this picture from the National Park Service visitor’s center:

The whole enterprise had no real chance of competing successfully with the nearly level canal through New York. Philadelphia then had to face the next half century with two handicaps: a location that was less well linked to the expanding West, and a huge debt incurred by the attempt to build a canal link through unfavorable terrain (Transparency 4N).

This kind of misinvestment in geographically inappropriate links continues to happen. Along the Atlantic and Gulf coasts, for example, cities from Houston to Baltimore have recently invested in expensive high-volume **container port** facilities, even though only a handful of such ports would be needed to handle all of the cargo for the entire eastern half of the country. As a result, most of the ports are underutilized, which makes them inefficient and therefore more expensive per ton of cargo.

Container port

Port that can move containers of cargo from ships to trucks or railroads efficiently

Another example comes from the snowy north, where Minneapolis is building a rail-based mass-transit system with open-air stations. Proponents cite the success of such systems in Portland and San Diego, without paying due attention to the differences in climate, terrain, population density, and employment patterns.

The same kind of logic led planners in Moscow to build a system of multiple airports spaced all around the city, even though the population of the city was neither large enough nor rich enough to use even one international airport fully. Finally, countries from Albania to Zimbabwe are now making decisions about the kind of computer and telecom networks they should install to link their people together and with the rest of the world. It seems most likely that there will be some mistakes among these decisions.

Why be concerned about infrastructure?

All of this concern about infrastructure stems from a basic fact: The existing networks of roads, airports, fiber-optic cables, and so forth have a tremendous impact on the usability of specific sites. The effects can be either positive or negative. For example, a busy highway intersection is a good site for a gas station. On the other hand, a new airport runway can be a disaster for a nearby outdoor motion-picture studio or opera-under-the-stars.

In sum, study of the ways in which places are connected to each other is one of the cornerstone ideas of geography. The study of links should therefore be introduced early and reinforced often throughout a geography curriculum:

Ways of evaluating infrastructure; student activities that deal with the effects of roads or other forms of infrastructure

In primary grades, students can make travel diaries, trace historic journeys, examine license plates in a mall parking lot, or tabulate and map features that occur near Interstate highway exits (Transparency 4O).

In middle school, students can identify the origins of products in stores, trace the movements of their ancestors, map commodity flows, or use Census data to examine what happens to small towns that used to have good railroad or canal connections but were bypassed by the Interstate highways (Transparency 4P).

In high school, students can examine the trade area of a mall or use the gravity model to predict traffic between specific places and thus evaluate places along the route as likely sites for convenience stores or outlet malls (remember Transparencies 3E and 3F?).

At all grade levels, students can look at historic events in terms of how they were influenced by geographic links among places. On an intercontinental scale, for example, places such as the Straits of Gibraltar or Hormuz have strategic importance because they are **chokepoints**, locations where traffic along a major route has to go through a narrow "funnel."

Chokepoint

Ocean strait, mountain pass, or other place where all traffic has to go through a narrow channel

In the case of Gibraltar, the nature of the chokepoint has shifted several times in the past. At some times, the Strait was viewed as important because it was a relatively easy way for land armies to get from Africa to Europe; at other times, it was valued more because it was the only water route between the Mediterranean Sea and the Atlantic Ocean (Transparency 4Q).

Other important historic chokepoints included the Straits of Dover, the Isthmus of Panama, Harpers Ferry, the Strait of Malacca, the Khyber Pass, and cities such as Istanbul, Singapore, and Vienna (see the CD unit on Locating a Fort – any phenomenon that can be expressed as a list like this is a natural topic for role-play simulations and/or individual poster or computer presentations).

Chokepoints can be economic or political as well as physical. For example, a city that controls an important aspect of the flow of money (like New York does through its stock exchange) can make a lot of money by charging a fee for every transaction. Likewise, political capitals control the flow of policy influence and tax dollars. It is no accident that the capital city of nearly every state is also one of the places with the highest average income per capita in the state.

We should not end this section on links without repeating that connections between places are not all human-made. Many places get their character from the natural links they have with other places. For example, note the tendency for smelly industries and low-income neighborhoods to be located on the east sides of American towns, because the prevailing winds blow from west to east in most parts of the United States (recall Transparency 4G?). At an international scale, one of the persistent sources of conflict is the simple fact that rivers flow downhill, carrying pollutants from one country to another (see the CD unit on Water in Mesopotamia).

Finally, to tie this section on connections (links) back to the previous one on conditions (places), we should repeat that the distinctive features we see in a place are often consequences of links with other places. For example, the traits of Subsidence areas are really consequences of an atmospheric link between the equator and places about 2,000 miles away. It is the persistent upward, poleward, and then downward movement of air that makes the Sahara so dry.

Indeed, if we want to trace the causal chain backward even farther, we could say that the dryness of the Sahara is really a consequence of the steady output of energy from the sun, 93 million miles away. The flow of energy toward a spherical earth initiates a sequence of air and water movements that create the climatic zones of the earth. Noting those global flows and the resulting local traits is a good way to summarize the ideas of link and place; it is also a good transition to the fourth cornerstone idea, region.

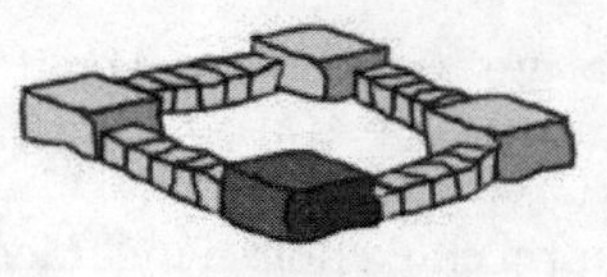

Region

To geographers, the word "region" serves the same function as the words "era" or "epoch" for historians. These words are handy labels for the results of an imaginary division that helps us make sense of a complex world (in other words, regionalization is basically a geographic form of classification). Geographers draw a distinction between two major types of regions, which are directly related to the previously discussed ideas of place and link:

Formal region

Group of places with similar traits

1. A **formal region** is a group of places that have similar conditions. For example, the Corn Belt in the United States is a region of similar farms. Likewise, the Sahel is a region of dry winters, short grass, and cattle grazing in North Africa; Yasenevo is a region of tall apartment buildings in Moscow. As it says in the ARGUS text: "When a fairly large number of people respond to fairly uniform conditions in fairly similar ways, the result is fairly homogeneous territory that we call a region."

   There is a reason for using the word "fairly" four times in that sentence. No region is entirely homogeneous, because no two places are exactly alike. Nevertheless, some places are more similar than others, and making a map of generally similar places can make it easier to remember them, study them, compare them with other places, or identify the factors that seem to mark the edges of regions (Transparency 4R).

Functional region

Group of places linked to each other

2. A **functional region** is a group of places that are linked together by a flow of something. For example, the Amazon Basin is connected by the flow of water toward the Atlantic Ocean. The Tokyo Transit Region is linked by a flow of commuters on underground trains. The Peruvian settlement region consists of several high valleys linked to lowland cities by traffic on major roads (Transparency 4S).

Drawing lines around regions is mostly for the purpose of comparison and classification. The key question is: What other places resemble this place or are connected to it? The result of drawing lines around regions is to simplify a map and make it easier to compare with other maps. And as I have already said several times, comparison of map patterns is one of the most powerful analytical tools of geography (more about this in Chapters 6 and 9).

Map comparison can help us communicate well-known relationships. For example, asking children to compare maps of rainfall and population in Africa can help underscore a basic geographic fact: people tend to have more options for livelihood in areas that have at least a moderate amount of rainfall. Most of the major cities, farming areas, and industrial centers are in the well-watered regions in West Africa, in the highlands of East Africa, and in the rainy-winter climates along the northern and southern coasts (Transparencies 4T and 4U; one might also refer to Transparencies 4G-4J in order to trace some logical connections between the effects of sinking air and the patterns we see on regional maps of population and rainfall).

This kind of visual map comparison cannot *prove* anything. It can, however, give us reason to reject erroneous ideas. One could not look at these maps of Africa, for example, and argue that desert climates are invigorating and help to promote population growth. If that were true, we would expect to find more people in the subsidence regions.

Exceptions to patterns are often as interesting as the generalizations that can be made by comparing maps. In North Africa, for example, the narrow ribbon of densely settled land along the Nile River is a striking exception to the rule of sparse population in the latitude of subsidence. Generations of schoolchildren have grown up with history books that describe ancient Egypt as "the Gift of the Nile." Comparing a precipitation map with a population map helps to show why that gift was so miraculous.

Epidemiology

The study of human diseases

Map comparison has enormous practical value in **epidemiology** (the study of human diseases). One well-known example involves a disease that is a consequence of the subsidence-based climatic patterns we have been using as a recurring example throughout this chapter. That disease is malaria, a serious illness caused by a tiny organism that is carried by mosquitoes. The relationship of this disease with climate is obvious, because mosquitoes are more numerous in hot and wet climates.

Another disease, sickle-cell anemia, is found in the same areas, but it does not seem to have a direct link to climate. The similarity of map patterns, however, helped persuade researchers to investigate further. They discovered that people who were genetic carriers of the sickle-cell trait also had a degree of immunity to malaria. Sickle-cell anemia thus continues to occur more in areas where the gene can help people survive the more serious disease, malaria (Transparency 4V and the ARGWorld CD unit on Malaria).

In a similar way, comparison of a map of landforms with a map of slave-holding plantations can help us refine our understanding of historical events in the United States. The regions containing large plantations coincide almost perfectly with a few fairly small regions of fertile soil and level land (Transparencies 4W and 4X and the ARGUS CD unit on Landforms and Plantations).

Subregion

Small distinctive area within a larger region

As a result, the large political region known as the Confederacy actually consisted of a number of smaller and quite dissimilar economic and social **subregions**. The images of huge plantations and cotton fields that have been so popular with novel writers and movie producers were typical in only a very small part of the South before the Civil War. That fact, in turn, helps us understand:

- The patterns of military campaigns during the Civil War
- The difficulty of forming effective state governments during the Reconstruction after the War
- The pattern of investment in education, infrastructure, and industry in the early 20th century
- Some patterns on modern maps of topics such as income, education, and voting behavior

Transparencies 4Y and 4Z offer a final example that can highlight both the advantages and the pitfalls of regional map comparison. The example involves an ongoing controversy about high taxes and unnecessary welfare programs, two issues that are on the list of favorite topics of some late night radio talk shows.

Different kinds of social scientists would take different approaches to studying this topic. An economist, for example, might tabulate the proportions of taxes used for welfare programs. A sociologist might interview welfare recipients to see how long they stayed in the program. A demographer might examine the age and family status of welfare recipients.

Map correlation

Measure of the degree to which high values on one map are in the same locations as high values on another map

A geographer could contribute to the discussion by using the idea of **map pattern correlation**, as applied to a few well-selected maps, such as the ones in the Transparencies. Such maps are not hard to make. Data come from the easy-to-get *Statistical Abstract of the United States (published annually), and a middle-school student could easily make the maps. Even a casual observer can see that the maps are visually similar (there is a high degree of map pattern correlation). Nine of the 12 states in the high category on the map of unemployment payments are also in the top 12 states on the map of inventions. And of the 17 states in the lowest group on the inventions map, 12 are also in the bottom category for unemployment dollars.*

This high degree of map pattern correlation implies that it may be worthwhile to search for a relationship between welfare programs and inventiveness. As with many interesting topics out in the real world, it is possible to examine these maps and come up with several plausible hypotheses to explain the patterns on them.

Theory 1: Inventive states tend to become rich enough that they can afford more generous welfare and unemployment programs. This, in turn, might stimulate a movement of poor people into those states. In time, the high taxes for welfare programs would drain the local economy and decrease its inventiveness and growth potential. To maintain their competitive edge (according to this theory), these states should try to reduce their unemployment payments.

Theory 2: A strong unemployment program helps to promote inventiveness. This idea is also plausible, because a wide variety of people – not just inventors, but also janitors, truck drivers, machinery repairers, and so forth – may want to make sure that the "safety net" is adequate, *before* they decide to take a job with a new company that might not succeed. Thus (according to this theory), high taxes in order to provide good unemployment insurance are a necessary part of the foundation for inventiveness and economic growth.

Clearly, these theories cannot both be correct. The comparison of regional maps tells us only that there may be a causal link between inventiveness and unemployment payments. It does not say what is cause and what is effect. One might continue the investigation by comparing maps of other factors, other places, and other times.

One fruitful way to identify good topics for future research is to study the exceptions to the apparent rule that we observe when we compare two maps. For example, look at the states that are low on the inventiveness map but rank higher on the unemployment payment map. Do Alaska, Hawaii, Kansas, Maine, Nevada, West Virginia, and Wyoming have anything in common? Well, for one thing, you would have trouble finding many other states that are farther out of the transportation and communication mainstream in the United States (try to drive across the middle of any of them, or use a cell phone along the way, to see what I mean).

Is remoteness or isolation important in trying to understand inventiveness? Maybe, but . . .

The purpose of this book is not to argue social policy. Even though I have some fairly strong opinions about inventiveness (which I am more than happy to share with my classes and local newspapers), I think it is appropriate right now to back away from this topic and underscore the more general idea that map comparisons like these are surely relevant in an age of tightening budgets and rising concern about economic competitiveness.

For that reason, the summary statement for this section is phrased in terms of the abstract concept of regionalization rather than the concrete topic of inventiveness. Here is that summary: The geographic theme of region is primarily of academic interest, in and of itself. It has practical significance (and therefore gets a boost in "teachability") when we realize that comparison of regional maps can suggest relationships in the real world that might be worth investigating in order to solve real-world problems.

Of course, to do that efficiently,
people must have easy access to
a good stock of regional maps,
or be able to make them!

That's a good reason to teach geography!

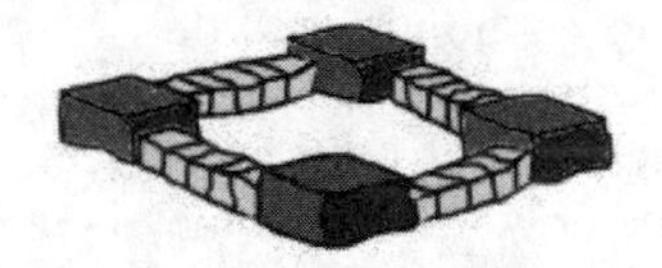

Summary

Salt Lake City can provide a good concluding example for this chapter about cornerstone ideas. The relative *location* of the high Wasatch Mountains just to the east of the city helps explain why a *place* in the Salt Lake Valley has the climatic *conditions* it does. The *movement* of river water from the mountaintop to the valley is a natural *connection* between these *locations*. The *site* of Salt Lake City was chosen by a group of people who saw the rivers as an important resource. The result is a small but important *region* of rivers, canals, irrigated fields, and cities, all *located* in (surrounded by) a larger *region* of desert shrubs, salty soils, and sparse population.

Geographical irony

How what makes a place successful for some purposes can pose a threat for others

Salt Lake City is also a superb illustration of **geographical irony**. Repeated earthquakes along the Wasatch Fault built the mountains, which capture the snow that fills the rivers with water and thus makes the valley livable. Those earthquakes, however, can also pose a threat to houses, gas lines, water mains, and roads (see the ARGUS CD unit on the Wasatch Fault).

In short, precisely what makes a place a boon for some people or purposes can pose a threat for others. That is a recurring theme in geography (as noted by Ptolemy, Herodotus, Magellan, Lewis and Clark, and me, about 40 pages ago!).

What are appropriate ways of presenting the cornerstone ideas that have been described in this chapter? How can teachers help students learn the concepts of location, place, site, cultural definition of resources, human-environment interactions, legacies of prior occupance, infrastructure, link, movement, region, determinism and free choice, and geographical irony?

These questions are extremely important, because these ideas should be introduced in the primary grades! The question is not, When should we present these ideas? Rather it is, What is the appropriate vocabulary for teaching these ideas at a given grade level?

The search for ways to present cornerstone ideas is what leads people to write scope-and-sequence outlines. Once we reach a consensus on what to teach (as embodied in the national and state standards), then teams of authors and teachers can start to figure out how to teach it (and what materials are needed to support the teachers and students). These efforts at development of materials often begin with an attempt to translate ideas into the type of language suitable for students at specific grade levels, which is the subject of Chapter 5, on pedagogical themes and standards, and of Chapter 6, on the skills of spatial thinking.

5

Five Themes: Meeting the Standards

In 1984, a group of geographers met to decide how to describe the core ideas of geography for classroom teachers and administrators. One tangible result of their deliberation was a 28-page brochure entitled *Guidelines for Geographic Education: Elementary and Secondary Schools*. This document asserted that teachers should organize geography classes around five fundamental themes: location, place, relationships within places (interaction), relationships between places (movement), and region.

Ten years later, this modest document had achieved the status of the Bible or the Koran among some educators. Conferences were being organized around the use of "the five themes" in geographic education. Teachers were being exhorted to design units on each of "the five themes." Textbooks were being written with "the five themes" as chapter titles or section headings. Courses were being approved or rejected on the basis of their use of "the five themes." When the National Geography Standards were published in 1994, many people criticized them because they lacked the clarity and simplicity of the five themes.

Eight years later, the No Child Left Behind Act of 2001 named geography as a core discipline and mandated the creation of state standards. Many people wanted these standards to use the language and focus of the five themes.

I hate to say, therefore, that much of the attention given to the five themes has been seriously misguided.

People have been talking about themes, using theme language, and recommending ways to teach according to themes without understanding what a theme really is. In doing so, they have been producing materials and suggesting strategies that actually subvert the idea of theme-oriented pedagogy.

That misunderstanding is clearly evident in discussions that have surrounded implementing the National Geography Standards. People in many states have compounded the error as they tried to write state standards that would be easier to implement than the National Standards (more about that later). Unfortunately, we are just at the beginning of a cycle of new textbooks that will address those new state standards. If people continue to behave as they have in the past, things are not likely to get much better.

Let's talk about it, OK?

What Is a Pedagogical Theme?

(Paragraphs adapted from the *Journal of Geography* 1992)

To learn about the use of themes, it might be worthwhile to listen to some music. Composers, after all, have been using themes to organize their work for centuries. For example, listen to the third movement of Tchaikovsky's Sixth Symphony. After a rather long introduction, a clarinet plays a four-note phrase. This phrase is part of the theme of the movement, but we do not know it yet. It is just there, amid the swirling strings, tootling flutes, and percussive snaredrummery.

Played by a lone clarinet, the phrase is understated, barely audible. Then, after more orchestral diversion, the phrase comes back, now seven notes long and louder, more insistent. A while later, another instrument replays the first four notes, and then a small group repeats the longer phrase. This pattern continues, and the movement is almost half over before we hear the entire theme.

At that point, all of the previous hints take on a deeper meaning. Tchaikovsky's strategy was to use fragments of the theme as portents, so that the theme itself would be all the more memorable when it finally arrives. The full statement of the theme, in turn, helps to tie the previous music together into a coherent unit.

What is the role of a theme?

To tie things together

That is precisely what a musical theme does: It holds a piece together and helps define what ideas are important. By analogy, a pedagogical theme is something that pervades a class and helps hold it together by defining what images and ideas are important. The theme provides guidance about what kind of questions should be asked and what kind of answers are valid. But the theme is not necessarily announced as the topic for the day. Nor is it necessarily a prominent part of the discussion.

Note: I said "not necessarily," which is not the same as saying "not." Beethoven bluntly stated the theme of his Fifth Symphony in the first four notes. Sibelius often waited half a movement before releasing his themes. Dvorak liked to bring themes from previous movements together in the final movements of his symphonies. Bach and Shostakovich used the initials of their names as recurring themes in a number of compositions.

The use of themes is even more pervasive in musical cultures that do not make such a point of extolling "originality" and attaching individual composers' names to music. For example, the three-note "sakura" phrase pervades a whole body of Japanese music. Particular rhythms denote specific dance steps in medieval and modern Spanish music. Many themes in Chinese music bear an uncanny resemblance to Celtic tunes from Ireland and Scotland, since they all use the same five-note scale. And in modern cities, guitar riffs are freely borrowed to acknowledge professional kinship among blues musicians.

One plausible conclusion from the musical analogy is that the selection of ways to present a theme is truly a choice, with many options.

The Disciplinary Use of Themes

This chapter is not about "the five themes" of geography. It is about how to use themes in geographic education. Chapter 4 was actually about the five themes, though it used different words. A key point of both chapters is that themes are an important organizing tool for a teacher, but geography teachers do not all have to use the same words (or even the same number of themes).

In fact, the process of thinking about themes might be even more important than the actual list of themes. The list of cornerstone ideas in Chapter 4 has only four items, yet the four cornerstones and "the five themes" are almost identical in scope and intent.

The basic equivalence of the lists of themes and cornerstones is comforting. This is how things should be – a discipline should strive for enough similarity of approach that outsiders can readily identify its members as part of a single team. This is best accomplished by lengthy discussion, the same messy process that helps any democratic group achieve consensus.

That is what happened when 350-odd people worked (in varying ways) for many months to draft the National Geography Standards (summarized in Postscript 5-1). The prolonged discussion brought a greater degree of consensus than there had been before. The document itself, however, is long, redundant, and sometimes contradictory. I don't particularly like reading it (except perhaps as an aid to sleep on stormy nights!). Nevertheless, I think it is very important, because it is the *process* of thinking about themes that matters, not the particular wording that a certain person or group favors.

To avoid misunderstanding, therefore, let me underscore that the process of continually seeking consensus is exceedingly important for a discipline. That is why the cornerstones in the previous chapter of this book are essentially identical to "the five themes:"

Equivalence of the five themes and the four cornerstones

1. Both lists begin with the term "location," because we have to be able to express where something is before we can analyze it geographically. Some people use the word "position" for the same concept.
2. "Place" and "site" are essentially similar concepts. They both focus on the complex mix of factors that occur together in a given location. Of the two short words, I mildly prefer "site" because it is a slightly less fuzzy term. To be truthful, however, I think the word "conditions" is better than either of the one-syllable words. That's the word I use in my own classroom, though I sometimes mention the others in order to emphasize the parallelism of the ideas.
3. "Movement" and "link" are obviously related. One denotes the process of changing position, while the other includes the process, the mode, and the consequences of movement. Recently, I have tended to use the term "connections" for this theme, partly because I think the phrase "conditions and connections" has a nice poetic ring (and in fact it denotes the two blades of the scissors described in Chapter 2).
4. Finally, the concept of "region" has the same meaning and position in both lists. It helps bring the concepts of place and link together.

Only the theme of "human-environment interaction" appears to have no equivalent in the chapter on cornerstones.

Why "the five themes" included the phrase "human-environment interaction"

That absence, however, is more apparent than real. As interpreted by many users, the five themes had four pithy one-word terms and one surpassingly awkward three-word phrase. The phrase is both grammatically and conceptually different from the other four terms. Moreover, the theme of "human-environment interaction" does not add anything that is not already included in the theme of place, since a place is defined as the sum of natural and human features as they interact in a specific location.

This admission raises the question of why the five themes include this unwieldy term at all. I think it is because geography exists in a world that has many academic disciplines, which often get embroiled in unseemly turf battles. Moreover, if we understand how the phrase came to be one of the five themes, it will be much easier to see why the National Geography Standards include some of the things they do.

The group who first proposed the five themes tried to draw a neat conceptual distinction between "relationships within places" and "relationships between places." They saw the words "location" and "place" as leading up to the scissors-like pair of relationship themes. (I agree with what they were trying to do, but I still think "conditions and connections" is a better pair!)

Meanwhile, out in the real world of curriculum development, a major threat to the "domain" of geography seemed to come from people arguing for a separate environmental studies course. At the same time, a formerly powerful coalition of historians, political scientists, and social studies teachers was under fire from a mix of parents, legislators, and radio and television talk-show hosts. Geography advocates did not think it was necessary to define much of a "boundary" with social studies, but they apparently did see a need to "mark territory" against the environmental studies contingent. The result was a rewording of the third theme as "human-environment interaction." This term stayed in its strategic position in the middle of the list, even though it is both grammatically and conceptually different from the other four.

One purpose for writing this book is to help teachers and administrators devise strategies for implementing the National Geography Standards. With that goal in mind, I'd rather not make a big issue about "the proper home" for environmental studies.

In sum, I feel fairly comfortable in quietly burying one of "the five themes" as they were known. We really don't need it.

(See? Lightning didn't strike!)

For the record, then: human-environment interaction in a given location usually involves movement of several kinds, and the nature and results of the interaction are part of the essential character of the place. Indeed, human-environment interaction is precisely what defines a place as unique and worth studying.

In short, human-environment interaction is like a "super-theme" that sprawls across the other four themes (Transparencies 5A and 5B).

> Caveat: If parents or administrators in your school want to see "human-environment interaction" in your lesson plans and outlines, by all means put it there. If they prefer the language of the National Standards, use that instead. If your state standards use different terms, feel free to use them. The issue is far too important to fight about terminology. Well-conceived themes are crucial in geography teaching, but the actual wording of them is much less important than how they are used.

Within limits, use whatever terms your supervisors prefer – it's better to focus on the use, not argue about the terminology

One exceptionally healthy step would be to eliminate the capital letters, quotation marks, and reverential tones that are often employed when people print or speak of the five themes or the national standards. After all, it is not the theme that makes Tchaikovsky's Sixth Symphony memorable, it is how he used it.

Using Themes in the Classroom

It is much more important to use themes and standards properly than to quibble about the exact wording.

On this score, geography has at best a mixed record. Far too many authors and publishers jumped onto the theme bandwagon without really thinking about the role of themes in a classroom. The result is a whole passel of books and lesson plans that use the language of the five themes in a shallow and self-defeating way. Most of those books were printed on high-quality paper, with good bindings, according to detailed specifications that were adopted by various states in order to ensure that expensive books would last a long time. For that reason, many of them are still out there in classrooms, and will likely be there for some time.

Moreover, if history repeats itself (as it often does), we can expect to see the same thing happen with the new National Geography Standards, unless we pause for a moment and think about the role of themes.

So, how should themes and standards be used?

In two words: inductively and cooperatively.

Inductive learning

Looking at cases and then drawing a conclusion

Why use themes inductively? To help promote the kind of "Aha" that leads to durable learning.

As stated above, the purpose of a theme is to hold a lesson together and help it fit with the rest of the course. The theme is subliminal and permeates a class, giving it coherence.

In the hands of a master teacher, however, the actual content of a given lesson may not even mention the theme. Avoidance of mentioning the theme is deliberate, not accidental, because the teacher is structuring the lesson in order to lead students to seize the main idea and make it their own. The most durable learning is that which occurs when an idea is so compelling that students grab hold of it and use it as if they thought of it themselves.

> To people who hold this view, teaching is nothing more (or less!) than a conscious effort to structure experiences so that the desired themes emerge out of guided manipulation of realistic data in compelling situations.

That kind of "Aha" is hard enough to set up in any circumstances. It becomes almost impossible to achieve when the theme for the day is announced at the beginning of the unit (or, worse, is used as a title for an activity or a section of the textbook).

In short, rote announcing of themes as lesson topics or chapter headings implies that learning consists of rote memorization of ideas. That is rarely effective pedagogy.

So why is theme-labeled pedagogy so often advocated?

A theory to explain why theme-labeled pedagogy is encouraged

Here is a theory that strikes me as fairly convincing —
speakers on the lecture circuit can get away with it!

They arrive with considerable fanfare. They dispense their ideas to an audience entranced by a new and fresh face, and they leave before they wear out their welcome.

I've done it myself, fairly often, and it feels good.

The problem emerges whenever a lecturer walks away convinced that what works as a one-shot deal would also work on a day-to-day basis. Classroom teachers have a different task. They have to come up with creative ways to maintain student attention for 180 days, more or less.

That schedule demands a more thoughtful approach than simply announcing a theme and then doing a lesson about it. In short, teachers have to deploy information about themes as one of the tactics they use in designing lessons.

Dealing with themes one at a time is like using only one blade of a scissors! (recall Chapter 2?)

Why use themes cooperatively? Because using them in isolation is like taking a pair of scissors apart. Geographical themes usually work better when they work together:

- Analyzing movement helps us understand place.
- Describing conditions at specific locations can clarify the reasons for movement between places.
- Drawing regional maps helps us compare places and predict movements; and so on.

Using themes as lesson or book-chapter titles tends to constrain this interplay between themes, and that, in turn, actually makes it harder to learn about places and how they are connected.

This point does not need to be belabored here, because it has been built into this book all along. Chapter 2 talked about cooperative use of thematic and reference perspectives. Chapter 3 dealt with a simultaneous focus on multiple strands of meaning. Chapter 4 showed how the key concepts of location, place, link, and region interact. Thus, when Chapter 5 says that the themes of geography work best when used together, it can point to three previous chapters that did precisely that.

(P.S.: And wait 'til you see Chapter 10!)

What about Teaching without Themes?

If teaching with themes is so complicated, why do it? Because someone who tries to teach a subject as complex as geography without a strong set of organizing themes can hardly avoid falling into one of two sizeable traps:

Pitfalls of themeless pedagogy

Overgeneralization – Proceeding "up" from one broad generalization to an even broader one as you try to build a logical view of the entire world, and

Trivia-listing – Compiling ever more complete lists of facts about places as you try not to omit any part of the truth.

These traps are the context for two fairly strong prescriptions:

Prescriptions for theme-driven pedagogy

Do not use themes as chapter or unit titles. That puts the cart before the horse and usually makes teaching less effective. It might work for a well-publicized one-shot speech or a short workshop, but not for a full year.

By all means use themes. Keep them in mind as you select topics, photos, data, basemaps, discussion questions, and evaluation instruments. In other words, let themes guide decisions about every aspect of the course.

Summary

Themes and standards are extremely important in instruction.

What is the role of a theme?

To tie things together

They are what hold a subject together and show how ideas are related. In an era when many people think that "research" is nothing more than an Internet search based on a few keywords, and where the result is an unsorted and unevaluated mixture of documents and images, themes are even more important to keep a subject from becoming disorganized trivia.

That is why the current bumper crop of shallow "theme-oriented" or "standards-driven" books, lesson plans, and course outlines (in geography, history, science, and many other disciplines) is so deeply discouraging, disgusting, disheartening, dismaying, and distasteful (to use just a few words that popped out of what is apparently a long and well-alphabetized list in my mind).

OK, I may be overreacting. Like most committed teachers, I truly despise cheap pedagogical rip-offs. Unfortunately, many hastily written but well-marketed books based on "the national standards" or "the key themes" in various disciplines will be just that – cheap pedagogical rip-offs. To keep overworked school boards and administrators from blindly adopting those apparently "on-target" books, teachers have to be able to articulate how they want to teach geographical themes.

To get there, teachers have to wrestle with ideas, think about how to deploy them in the classroom, and seek consensus. That, in turn, involves knowing about cognitive skills and the phrasing of educational objectives and standards, which are the topics of the next two chapters.

As one master teacher said in reviewing an earlier edition of this book:

> It's not OK for a teacher to know just as much about a subject as the students should learn. You have to know *more* than the students will learn, so that you can figure out how to put what they learn into a proper perspective.

That's one of the main reasons why an ongoing discussion about geographical themes is important.

POSTSCRIPT 5-1

National Geography Standards, Grades K-12

(*Geography for Life*, Washington, DC: National Geographic Society, 1994)

For each major grade level, the National Geography Standards suggest an organizational framework that has 18 goal-statements in 6 broad groups, with dozens of specific illustrations. Here is a short summary of the 18 major goal-statements:

According to the Standards, a geographically informed citizen knows:

The World in Spatial Terms

1. How to use maps and other geographic representations to interpret the world and to analyze world events.
2. How to use mental maps to organize information.
3. How to analyze spatial organization and spatial interaction, and how to use those ideas in making locational decisions.

Places and Regions

4. About the physical and human characteristics of places.
5. That people create regions of various types to simplify and thus help interpret the Earth's complexity.
6. How culture and experience influence people's perception of places and regions.

Physical Systems

7. Physical processes, patterns, and cycles that shape the surface of the Earth.
8. Characteristics and spatial distributions of ecosystems, their productivity and diversity.

Human Systems

9. Characteristics, distribution, and migration of population, impacts of migration on physical and human systems.
10. Characteristics of cultural mosaics, how cultures change, how cultures influence regional characteristics, how technology affects standard of living.
11. Patterns and networks of economic interdependence, how people earn a living, issues in local and global economy.
12. Patterns and functions of human settlement, locations and internal structure of cities, causes of change in settlements.
13. Forces of conflict or cooperation, how spaces are divided, how external forces can conflict with internal interests.

Environment and Society

14. How human actions modify the physical environment, how societies can devise solutions for environmental change.
15. How physical systems affect human systems, and how perceptions of natural hazards affect responses to them.
16. Changes in meaning and use of resources, policy related to resources, how resources can be recycled.

Uses of Geography

17. How to apply geography to interpret the past, how geographic processes affected history.
18. How to apply geography to interpret the present and plan for the future, to solve problems and make decisions.

6

Spatial Thinking: Geographical Skills

For most people, skills are not ends in themselves. They are just the means to those ends.

This book has already described the primary goals of a geography class:

1. Knowledge of other places, to guide us when we travel, and to help us understand why people there act the way they do,

and

2. Knowledge of how to arrange things (borders, roads, houses, malls, stadiums, election districts, etc.) in our own place so that the results are fair, safe, efficient, and beautiful.

This chapter is about the analytical skills of geography, which are the means that geographers use to reach those ends.

Geographical skills fall into two main categories – general skills that are shared with other disciplines, and specific skills that are the distinctive contribution of the discipline of geography. The National Geography Standards used language that tends to focus on the first group of skills. They organized their treatment of skills around five major topics:

1. Asking geographical questions
2. Gathering geographical information
3. Organizing geographical information
4. Displaying geographical information
5. Answering geographical questions

1. Expressing Location — Where Is It?

Location

Where is it?

As noted in the chapter on cornerstones, the concept of location is surprisingly difficult to communicate without employing at least one other spatial concept, such as distance, direction, adjacency, or enclosure.

Elementary-school students can learn several different ways to express these basic spatial ideas. They can also improve their ability to describe topological relationships among objects, using terms such as next to, left, right, above, below, between, beyond, near, far, and so on. Suggested activities include finding one's position on a map, finding objects shown on a map, placing symbols in appropriate places on a map to represent objects in the real world, and giving directions to others.

Middle-school students should use more sophisticated ways of identifying locations and communicating that information to others. These include various kinds of compass directions, map grids, and mathematical coordinates. The concept of scale becomes an important part of the discussion at this time.

High-school students should be able to understand the use of triangulation and other geometric procedures to determine locations. Tools that might be introduced at this time include sighting compasses, surveying instruments, and GPS receivers. Activities might include basic navigation, various kinds of orienteering, and finding earthquake epicenters by analyzing seismic records from different observation stations.

2. Describing Conditions at a Location — What Is There?

Condition

What is there?

Once a person has figured out how to determine the location of something and to communicate that information to others, the next step in the logical chain of spatial thinking is to describe the features and conditions at that location. At first, this usually means some kind of personal observation and data recording. One simple way is for a person to choose words to describe what is seen, heard, or smelled in a place. Other condition-recording tools include interviews, sound recorders, cameras, and measurement devices such as thermometers and decibel meters. This information can be recorded with words in a journal, numbers in a table, or graphic symbols on reference maps, scaled-symbol maps, or pie charts.

> Caution: Because facts like these are easy to test, far too many textbooks, teachers, curriculum designers, school

administrators, and journalists tend to overemphasize this form of spatial thinking. In view of that tendency, we'll say it again: Learning facts about places is just one form of spatial thinking. Indeed, we are tempted to say that if the human brain were capable of remembering everything a person saw or heard, there would be little need for other forms of spatial thinking. Unfortunately, our heads would also have to be hundreds of feet in diameter to hold that much information, and we probably couldn't move!

The fact that human brains are able to forget (and thus to preserve some precious memory storage space for things we consider important) is what makes other forms of spatial thinking necessary. In effect, most of the other items on this list are basically ways to organize and store information about places. It follows, therefore, that a geography curriculum should not focus too much on this "concrete-level" aspect of spatial thinking – it should not emphasize the memorization of spatial facts too much. That means, bluntly, that we dare not assume that students have learned about the geography of Albania just because we have shown a video that contains some testable facts about that country.

3. Tracing Connections with Other Locations — How Is It Linked?

Connection

How is it linked with other places?

Connections are logically similar to conditions – they both are basically just lists of spatial facts to be recorded or memorized. Where conditions apply to a single place, however, connections always involve two or more places. People would not go to all the expense of building a railroad just to see two long strips of metal exactly 4 feet, 8-1/2 inches apart. The whole point is to arrange those rails so that they connect your place with some other place, in order to move yourself or something else from one place to the other.

The human brain can store information about connections in two ways: by making lists of origins and destinations or by forming a mental image of the connections themselves.

Elementary-school students can focus on the reasons for movement and the structures people build to make movement easier. A surprisingly large amount of instruction and practice is often needed to master the skill of using maps in order to locate destinations, identify connections, select travel routes, and estimate the time needed to get to a specified place via a particular connection.

Middle-school students can broaden their scope by examining conditions in the areas between places, and especially how various kinds of barriers, chokepoints, or channels can make connections easier or more difficult. Explicit links with history are easy to make, because many places became historically important precisely because they were strategically located near channels or chokepoints of some kind – such as Gibraltar, Istanbul, Panama, Singapore, Harpers Ferry, El Paso.

High-school students can evaluate the volume of flow along a particular connection, by making flowline maps to show international trade or migration, or by applying geographic theories of movement, such as the gravity model of traffic forecasting or push-pull ideas about human migration.

4. Comparing Locations — How Are Places Similar or Different?

Comparison

How is it similar or different?

At several times in this book, I have said that conditions at places and connections between places are the main "stuff" of geography. That assertion has almost always been followed by an immediate caveat: simple memorization of long lists of conditions and connections is not only boring and difficult, but in fact it violates some basic principles of human brain structure and function. The skill of comparison is a good illustration of this: it is an easily demonstrable fact that people can remember much more about a new place if they explicitly compare it with a familiar place. For this reason, geographers have developed a large number of methods for making fair and meaningful comparisons of places. Many of these methods can involve curricular links with reading, writing, mathematics, or even music, art, or dance.

Elementary-school students, for example, can make verbal or pictorial comparisons of two or more places in order to decide which ones are colder, higher, older, more crowded, or more productive. It is also useful for students to begin to understand the concept of "typical" or "average," even though they may not perform the actual mathematical calculations yet.

Middle-school students can begin to learn to use analogies and to calculate averages and ratios such as population per square mile or food production per person. They can also learn to show the results of these comparisons with choropleth maps, bar graphs, and various kinds of scaled symbols. This is also a good age to introduce the kind of add-up-the-points index that many people use to rate places such as ski resorts, residential neighborhoods, malls, or restaurants. One goal of this study is to become appropriately skeptical of such rankings!

High-school students should wrestle with more complex comparisons, such as measures of trends in per-capita income, adjusted for inflation in different places. It is not hard to justify such lessons in applied mathematics, because comparison skills like these are needed in order to get an accurate message from many tables and graphs in newspapers, magazines, television news shows, or online data analyses.

5. Determining the Zone of Influence around a Location — How Far from a Feature Is Its Influence Significant?

Aura

What is its influence on other places?

Whenever you (or someone else, or a company, a government, or nature) put something in a place, it changes the character (the conditions and/or connections) of that place. It also has an influence on other places. The influence is usually stronger for nearby places than for those far away.

Elementary-school students should learn that things such as malls, factory smokestacks, and airports can have effects on surrounding areas, both positive and negative. They should also learn that the extent of the effects can be observed and often predicted. Such observations and predictions are an essential first step in treating neighbors fairly.

Middle-school students can extend that idea to consider ways to lessen the bad effects of new features in a given location. For example, they can learn how laws now require people planning a new airport runway to try to identify houses that would be subjected to excessive noise. Although the basic principle of spatial influence is easier to see with airports and other large-scale items that have obvious negative impacts, it is important to emphasize that it also applies to smaller features and to things that have positive effects.

High-school students can refine their concept of the "aura" of a feature (its zone of influence) by recognizing that its extent may be different in different directions, often for good reasons. For example, prevailing winds tend to blow smoke farther away from a factory in some directions than in others. Applying similar logic to a completely different topic, a subway or other mass-transit line can give people who live along it better access to a new stadium. These nonsymmetrical influences can make it quite challenging to assess the spatial impact of a new feature.

6. Delimiting a Region of Similar Places — What Nearby Places Are Similar to This One?

Region

What nearby places are similar to this one?

The world is much too big and complicated to learn all the conditions, connections, and influences for every place in it. The concept of region is basically a reflection of the human brain's natural tendency to put similar objects together into groups and then to remember the group rather than all of the individual objects.

Elementary-school students can divide a map of their home community into regions based on land use: house regions, store regions, and regions with other uses of land (e.g., for parks, factories, fields, etc.).

Middle-school students can extend this idea to divide a map of their state, nation, or other continents into regions of various kinds: areas with similar vegetation or geology, for example, or where people speak specific languages, cheer for particular sports teams, vote for particular candidates, and so forth. As they do this, students will become familiar with some of the "map vocabularies" (e.g. bounded-area maps, choropleth maps, etc.) that geographers have developed to describe the results of their "regionalization," that is, their division of the world into smaller areas that are more homogeneous and easier to remember.

High-school students should learn how to "deconstruct" a published map of regions and describe the complexity that underlies its simplified pattern of lines and colored areas. Someone who does not understand the abstraction that is inherent in mapping is likely to be doomed to a lifetime of misperception and stereotyping. On the other hand, inability to organize knowledge of the world into regions is a prescription for information overload and a related tendency to avoid grappling with important issues.

7. Describing the Area between Places — What Is the Nature of the Transition between Places?

Transition

How do things change between two places?

The borders around regions are seldom sharp and distinct. In the real world, dissimilar areas often have a zone of transition between them. For example, the area between a commercial district in a city and a residential neighborhood sometimes has a mixture of old houses and newer office buildings and stores. The ability to identify and describe transitions between places is a different kind of spatial thinking that should be taught at the same time as the skill of regionalizing.

One might ask **elementary-school** students, for example, to think about what they see on their way to school. Do they go through different kinds of places, or does everything along the route mostly look the same? If there are different areas, are they separated by a sharp dividing line or a gradual change? It may be useful to let students make their own sketches, or have them fill in a form like Transparencies 7F and 7G.

Middle-school students can apply their new math skills by doing activities that involve measuring distance, change in elevation, and slope, as well as simple interpolation between measurements. They can also learn that the spatial concept of transition can apply to other kinds of regions, such as those that demarcate property values, or population density, vegetation, language, temperature, and so on.

At **all levels** from elementary school through college, students should be encouraged to examine real-world landscapes and to see the world not in terms of discrete, separate boxes (rainforest, desert, city, countryside, etc.), nor as completely random mixtures of features. In most cases, something in between those extremes is appropriate, and the complementary concepts of region and transition are some of the ways the human brain tries to organize its impressions and communicate those impressions to others.

8. Finding an Analog for a Given Place — What Distant Places Are Similar to This One?

San Francisco and Reno, Nevada, are barely 200 miles apart, but they are located in radically different climate regions. San Francisco has a Mediterranean-style climate with mild temperatures and foggy weather, especially in winter. Reno, by contrast, is in a high desert, with hot, dry summers and cold, dry winters. The road between these two cities goes through other dissimilar regions: the flat farms of California's Central Valley, a dense conifer forest on the slopes of the Sierras, and some mountains that are high enough to have snow on them all year. As you continue east from Reno, conditions keep changing, but no other place in the United States quite matches San Francisco's unique pattern of mild summers and humid winters.

If you get on a boat or plane and travel across the Atlantic, however, you find that Lisbon, Portugal, has a situation that is remarkably similar to San Francisco's. Both cities are located on hilly peninsulas, both have sheltered harbors, and both are about 38 degrees latitude away from the equator. To a geographer, it is not surprising that their weather is almost identical: in

January, both places have an average temperature of about 48 degrees Fahrenheit, with about 4 inches of rain, followed by a mild summer with almost no precipitation. These cities are climatic *analogs*: places that occupy similar situations on different continents and therefore have similar weather conditions. Not surprisingly, people in both places have learned that some activities are more successful than others in their environments. For example, some people in both Lisbon and San Francisco looked at their superb natural harbors and decided to focus on ocean-going trade. Likewise, some people in both places choose to grow grapes and make wine, because those can be very profitable enterprises in a warm, dry-summer climate.

Analog

What places have similar conditions?

Elementary-school students can learn some analogs for places that may come up in reading, history, and current events: Shanghai and Houston, for example, or Seoul and New York, Iraq and Arizona, Tashkent and Salt Lake City, St. Petersburg and Anchorage, Alaska.

Middle-school and **high-school** students can explore the scientific reasons for these climatic analogs – the global regularities of sunlight, wind, and ocean currents that are responsible for the world pattern of climate. In time, students should also become aware of other kinds of geographical analogs. For example, different cities can have suburbs that are similar: similar distance from the urban center downtown, similar age and income of residents. These *urban analogs* tend to share other characteristics, such as street patterns, architectural style, range of store choices, and so on. Likewise, places in different countries that have similar histories of investment often have similar industries with similar labor issues.

Here is one more example: Many of the historic migrations to the United States showed that the idea of analogs was at work, as people chose to move to parts of the United States that were similar to their home countries. For example, Spanish explorers looked at steamy Florida and Louisiana, but most of them chose to settle in California, which is climatically similar to Spain. Scandinavian people moved from the lake-strewn forests of Sweden and Finland to the lake-strewn forests of Wisconsin and Minnesota. Urban Chinese people moved from crowded apartments in Hong Kong to crowded apartment areas in New York City. In short, the concept of geographic analogs is a powerful one: it provides a different way of organizing our mental maps of the world, and it can sometimes even be more useful than the concepts of region and transition.

9. Identifying a Spatial Pattern — Are There Biases, Clusters, Strings, Donuts, Waves, and Other Distinctive Patterns?

Pattern

What distinctive arrangements can you see on a map?

Few things in the real world are truly random. A random pattern is not organized in any way; in a random pattern, things occur where they do purely by chance. It is very important to remember that a fairly uniform arrangement of features in a place is not (NOT) random. Evenness is actually a distinctive kind of pattern, and geographers usually think there has to be some reason when features are evenly spaced.

For example, many rural areas in the United States have roads spaced one mile apart and four houses on each square mile of land. This pattern makes sense when you remember that the Homestead Act required the land to be surveyed into square-mile sections, and then people were allowed to claim farms that were 160 acres (one-fourth of a square mile) in size.

Other features are arranged in tight clusters or long strings. These kinds of patterns are also not random; examples of such patterns are scattered deposits of a valuable mineral in a particular place, or a river, road, or other transportation artery that forms a line.

Elementary-school students should look at dot maps and try to find obvious clusters or strings. They might even try to formulate hypotheses to explain the patterns they see. But do not expect too much, because young children are likely to have some difficulty with the distinction between "even" and "random."

Middle-school students can begin to investigate the ideas of bias and balance. For example, a city may have an unbalanced pattern of income, with most of the wealthy people on one side of the city and most of the poor people in other directions from the center.

At **all levels**, students should learn that spatial-pattern analysis is a subjective and surprisingly sophisticated activity – the human eye and brain can make instant judgments that would take a computer many millions of calculations to duplicate. The goal is for students to develop an increasingly sophisticated "eye" for geographic patterns. This will eventually allow them to make comparative statements such as "this feature has a string pattern, but it is not as obvious as the strings of Israeli settlements on the West Bank" or "most of the trees in this area are in small clusters on the north sides of the hills."

10. Comparing Spatial Patterns — Are the Spatial Patterns Similar?

Pattern comparison

Are the spatial patterns similar?

In map-pattern analysis, students look for clusters, strings, donuts, and other shapes that might be caused by the presence of important point or line features, such as volcanoes, mineral deposits, highways, or navigable rivers. An obvious extension of this logic is to examine the map patterns of two different phenomena, to see if the phenomena might be related. For example, if all the cases of a particular pet disease occur in the same area that a specific kind of plant grows, it might be reasonable to suggest that the disease and the plant are related in some way. This logic cannot *prove* that the plant caused the disease, but it can definitely help to narrow the list of suspects.

Elementary-school students can examine the map patterns of some well-known combinations of features, such as malaria and anopheles mosquitoes; corn fields and flat, fertile soils; rainfall and population density in Africa, Australia, or the United States; recent volcanoes and black-sand beaches in Hawaii; or even Communist government and slow economic growth in 20th-century Europe.

Middle- and high-school students can learn how to use sampling and scattergrams to conduct more rigorous comparisons of map patterns. Suitable topics might be Civil War plantations and particular kinds of soil, road patterns and former colonial jurisdiction, and (horror of horrors, an immediately practical application of a school skill!) government contracts, political contributions, and local voting patterns in previous elections.

11. Determining the Exceptions to a Rule — Where Are the Places That Have More or Less of Something Than Expected?

Like the ideas of region and transition, the skills of map comparison and identification of exceptions use complementary forms of logic – they become especially powerful when used together. The basic idea behind identifying exceptions is simple: to examine maps of two features that seem to have similar patterns, and look for aspects of the pattern that are not similar and thus violate the "rule."

For example, you might compare maps of hilly land and family income in a city. These two features tend to have similar map patterns, because wealthy people often choose to live in the hillier parts of typical U.S. urban areas. (Sometimes, they even proclaim that choice through the use of place names such as Arlington Heights, Barrington Hills, Berkeley Heights, Beverly Hills, or Cherry Hills, to pick just a few from an atlas list of well-known upper-income American suburbs).

This relationship between topography and wealth is far from universal, however. Some wealthy residential areas occur on flat land. Meanwhile, some hilly areas have poor people living in them. Making maps that show the locations of such exceptions is a skill worth learning, because it is one of the best ways to identify topics that might justify further investigation.

Exceptions

Where are the places that do not fit the rule?

Elementary-school students might investigate the "rule" that a country rich in natural resources will tend to have prosperous people. In general, this is true, but there are countries (e.g., Nigeria, Ukraine) that are not wealthy even though they have rich farmland or valuable deposits of oil or other minerals, whereas some countries (e.g., Japan, Finland) have few resources but have achieved high standards of living.

Middle-school and **high-school** students might look for:

- Countries whose populations have longer (or shorter) life-spans than one might expect based on their per-capita income, or
- Neighborhoods that have lower (or higher) crime rates than one might expect based on their police expenditures, or
- Towns that have fewer (or more) flood damages than one might expect based on their elevation.

They can then display these results using two-way point symbols (e.g., plus and minus signs) or choropleth map shading. After careful inspection of their maps, students should be encouraged to suggest hypotheses for further inquiry. In short, this rather abstract form of spatial reasoning can have obvious citizenship values, because it can produce direct evidence about the effectiveness of various policies to combat crime, provide medical care, or protect from flooding.

12. Analyzing Changes in Pattern Through Time — How Do Things Spread?

Unless you have better information, it is reasonable to assume that something will continue to move in the future as it has in the past. If it doesn't, something else might have acted either as a barrier or as an avenue that allowed more rapid movement.

Pattern change

Spread or contraction of a process through time

Those simple statements are the basis for a wide range of applied geography skills, from daily weather forecasting and military strategizing to the investigation of ocean currents, disease, and suburban growth. The analysis of temporal change in spatial patterns is a very abstract form of spatial reasoning, yet people do it more or less unconsciously starting in early childhood.

Elementary-school students can examine a sequence of maps showing the position of a cold front at several earlier points and try to predict when it will arrive at a specified location (picking your community as that location should make it more interesting). Weather analysis has the added advantage of being self-correcting – tonight's Internet site or tomorrow's newspaper will have the map of what actually happened, to allow students to check their predictions and see how well they did.

Middle-school students can examine the historic spread of empires, language groups, or major epidemics such as the bubonic plague in Europe.

High-school students can also use the same logic to examine the spread of smoke from a factory, a new form of slang, or diseases such as West Nile Virus or SARS. (Those diseases were fairly new at the time this paragraph was written, but please feel free to substitute the one that is making headlines in the year you read this; I'm reasonably sure there will be one! If nothing else, those headlines underscore the relevance of this kind of spatial thinking.)

13. Devising Spatial Models — Are Places Linked by One or More Intermediate Processes?

Spatial model

Statement about influences across great distances

To make this list a "baker's dozen" (a dozen groups plus an extra one just in case we miscounted somewhere), I will add the most abstract and uniquely geographical of all aspects of spatial thinking. This is the process of making spatial models, which are scientific statements of how we think things might be related in space.

One good example is the phenomenon called El Niño, which has made the news quite often in recent years. El Niño is a sudden warming of ocean water off the west coast of South America. It appears to be the result of what is now called the Southern Oscillation, a cycle in ocean circulation that stretches all the way across the Pacific. As a result, an observation of sea-surface height near Indonesia can provide a reasonably accurate prediction of El Niño several months in the future.

At the same time, a strong El Niño in the Pacific Ocean seems to be connected to several climate events that occur months or even years later in other places. These include warm and wet conditions in the Carolinas, a reduced frequency of hurricanes in the Gulf of Mexico, and milder conditions in coastal Europe.

Geographers have become aware of many other long-distance connections between places. For example, a civil war in east Africa can result in an influx of migrants and refugees in places as far apart as Minneapolis and London. A drought in Brazil can raise the price of soybeans and corn in Iowa, which in turn affects the price of soft drinks in New York. A currency crisis in Thailand can affect interest rates in Europe.

And so on, and so on, and so on.

The main objective in teaching about spatial models is attitudinal. In elementary, middle, and high school, students should learn to appreciate that things happening in one part of the world can affect other places, often very far away. This concept is not hard to illustrate to an elementary-school student, and yet it has the potential to challenge future generations of postdoctoral researchers. And, in a wonderful closing of a disciplinary circle, the Internet now allows elementary-school students to participate in gathering information that might eventually help professional geographers refine their spatial models.

We are not going to provide any specific prescriptions for teaching this "extra" form of spatial thinking. The main objective, as noted above, is awareness of long-distance connections. The precise methods can be radically different in each school. Just remember: the idea of making and refining a spatial model basically incorporates all of the other skills of spatial cognition that have been described in this chapter.

So there you have it: brief descriptions of a baker's dozen skills of spatial thinking, with some suggested ways to teach these skills at different grade levels. This list should still be viewed as tentative, and any comments from readers would be welcome, either about the list, or about tactics that you have found to be effective in teaching these skills.

7

Three Kinds of Tests for Three Kinds of Meaning

"Is this going to be on the test?"

That might be the most frequently asked question in American classrooms.

Geography is more vulnerable than most other disciplines to that kind of "test mentality." By taking the world as its subject, geography lends itself to factual questions, such as:

Geographical trivia as opposed to a multistranded geography

"Where is the source of the Nile River?"

"What is the population of Singapore?"

"What is the elevation of the highest peak in Nepal?"

When people in the United States think of "geography" they often associate it with this kind of question. What might be responsible for this link? television quiz shows, computer programs, a popular parlor game called Trivial Pursuit, some newspaper polls that ask only factual questions like this.

The kind of geography this book seeks to engender, however, is much richer and more varied; it includes analytical skills, explanatory theories, and evaluative opinions, as well as concrete facts about places. Specific features in particular places gain meaning through their relationship to each other and with other places. To learn the "language" of relationships of this complexity, a student must deal with concrete images, abstract theories, and value judgments, all at the same time (as discussed in Chapter 3).

Educators in a number of disciplines have long acknowledged the existence of qualitatively different kinds of learning. Also, several **taxonomies** of learner outcomes have been devised to categorize different kinds of learning.

Taxonomy

Set of categories to help organize our concepts of the world

These taxonomies are especially important at the two extreme ends of the pedagogical process:

- Formulating objectives and goals prior to instruction
- Evaluating student performance after instruction

Those two phrases use the word "instruction" broadly, to include all kinds of teacher-guided learning activities, both in and out of class. The selection of a specific mode of instruction depends on the objectives, background, and aptitude of the students, as well as the skill of the teacher. Some kinds of facts, skills, and ideas are easier to teach with particular pedagogical methods. What approach to take is like choosing a specific tool to perform a specific task. To choose tools wisely, one has to know what outcome is sought.

Writing Behavioral Objectives and Learner Outcomes

Question: If course or class objectives are so important, why didn't this book deal with them sooner?

Answer: Because, to be honest, I am not sure it makes much sense to talk in purely abstract terms about the process formulating objectives. On the contrary, it seems plausible that a teacher has to have a fairly thorough grasp of the scope and philosophy of a subject before the process of writing course objectives can hope to be more than rote memorization of a model.

Like many other topics, the craft of teaching may well be something that is best learned inductively. Exposure to a wide variety of examples likely builds up to give a solid sense of what is important. For that reason, I structured this book so that readers would see a large number of examples of different kinds of geography before trying to wrestle with the question of writing objectives for a daily class, let alone a semester-long course. (If I am wrong in this decision, feel free to read this chapter first and then go back and read or reread the others!)

In contemplating the task of writing objectives for a geography class, one observation that seems hard to avoid is that geography is a vast subject. It is simply not possible to "cover" the whole world.

For this reason, teachers need clear objectives in order to decide what parts of the complex world to include in their daily lesson plans, handouts, activities, and readings.

Course objectives should span the entire range of learner outcomes. At one extreme are the basic grammatical details of map language (e.g., what symbols are conventionally used for a map of a common topic such as house types or storm probability?) Other strands include widely accepted geographic theories as well as potentially controversial opinions about spatial equity (e.g., is the geographic arrangement of medical clinics in this area fair to all groups of people who live here?).

One useful way to view the task of writing objectives is as a mandate on what items should be included in the four areas I outline below (several lists of suggestions are included on the CD to show the range of specificity that should be attempted). These four areas are basically the answers to four general questions:

What locations are important?

1. What placenames should be learned? What specific countries, cities, mountains, rivers, borders, and other features are so important that they should be included in the material to be presented and evaluated?

What images or other facts about places are important?

2. What facts about places should be learned? What specific clothing styles, crops, house types, landforms, religious denominations, factories, festivals, and so forth should students learn to associate with specific places? It is important to avoid the trap of vague objectives, such as: "the learner should be familiar with the pattern of natural vegetation in the United States." One problem with that statement is that the word "pattern" is far too vague. It is much better to say "the learner should be able to state whether the plants that occur in a specified part of the country are mostly broadleaf trees, needleleaf trees, grasses, desert shrubs, or chaparral." That degree of specificity is precisely what is needed to make an intelligent selection of modes of presentation and evaluation. Less detail only postpones the tough decisions until the night before class or the day you make up the test.

What geographic theories are useful?

3. What geographic theories should be learned? What explanatory concepts should students learn well enough that they can apply them to new data? At what level of sophistication should these theories be applied? Unlike 1 and 2, which may contain hundreds of items, the number of geographic theories should be quite small. The authors of the National Geography Standards spent thousands of hours refining the list of theories they provided, in order to have a manageable number of important ideas. Good texts and workbooks should also have a

clear way of identifying and summarizing the key theories they try to present.

4. What evaluative criteria should students use for judging the features of a place? What measures of safety, fairness, efficiency, or beauty should students be encouraged to apply to the features and patterns they encounter in different places? This is obviously a very sensitive area; but if we are not interested in seeing students learn how to make value-judgments knowledgeably, why are we teaching? Writing down a specific list of evaluation criteria is one way of making tough choices about what topics are worth discussing even in the face of possible criticism from parents or board members.

What evaluative criteria should students use?

The discussion so far has rested on a basic assumption: A detailed set of learner objectives does not add to a teacher's list of time-consuming tasks. Writing a detailed set of class objectives may take a certain amount of time early in the year, but it should save a considerable amount of time later, when preparing daily class materials or evaluation instruments.

Good objectives don't take time, they save it!

Dozens of teacher guides provide specific hints about writing behavioral objectives. For more on this subject, consult a general source such as Norman Gronlund's short book: *Stating Objectives for Classroom Instruction*.

Caveat. I have to be honest: I write detailed behavioral objectives for only one course: my course on geographic education. I do that because a slight majority of teachers still say that the process is worthwhile, and therefore I feel I should provide a model. Many other teachers (myself included, in most of my other classes) prefer to combine the tasks of assessment and objective-writing. For us, writing the final exam is the *first* step in planning a class; that assessment instrument then becomes our statement of objectives. I submit that this admission does not alter the main point of this chapter in any way – it just specifies a different form for the planning document. The point is that goals need to be stated very clearly, whether as a formal objective or as an assessment task.

Regardless of the form chosen by the teacher, a well-phrased learning goal contains a specific statement of four things:

Elements of a well-worded behavioral objective

1. A setting
2. A specific task to be performed
3. The expected level of performance
4. The criteria used for evaluation

Here are twelve examples of goals for geography students, chosen to represent a variety of grade levels, degrees of abstractness, and time requirements (whether they will take part of a day or the full term). The sequence on this list goes from concrete to abstract, which is *not* the same thing as grade level; elementary-, middle-, and high-school objectives are mixed together in this list.

Examples of well-worded geography objectives

Given a list of several "million-plus-population" cities and a world map with dots showing their locations, write the name of each city next to the correct location.

Given the latitude-longitude coordinates of two locations and a globe with a latitude-longitude grid and a bar scale, plot the locations and measure the distance between them. (You might even add an expected level of precision, such as "within three hundred miles of the correct distance.")

Given three photographs of typical buildings in a city (e.g., Japanese, British), arrange them in the order of date of construction, or distance from the town center, or apparent wealth of their owner, and explain why you put them in that order.

Given a topographic map, determine the elevation of two specified points, select an appropriate horizontal scale and amount of vertical exaggeration, and draw a side profile of the terrain between the points.

Given a set of (four) population pyramids, match them with the appropriate names from a list of (five or six) countries. (Making the list of choices larger than the number of pyramids makes the test more statistically valid by minimizing process-of-elimination logic.)

Given a dot map of population and the sites of (five) stores in a city, outline the market area of each store and estimate the population there. Then, recommend a site for a new store.

Given a description of the climate in a place, identify a climatic analog (a similar climate on another continent) and briefly describe some similarities and differences in the way people live in the two places.

Given a blank map and the name of a key population group, identify the area where the population group used to live, draw a line showing the general route of their migration to a new place, and describe the push or pull conditions that induced them to move.

Cite and briefly explain three different cases in which a human action that was appropriate in one environment (e.g., a forest) turned out to be dangerous, damaging, or otherwise inappropriate in another environment (e.g., a desert).

More examples of well-worded geography objectives

Given a news event in a particular location, find three thematic maps with information that might be useful in interpreting the event, write an essay that describes the background of the event, and use your maps to illustrate your essay.

Given a thematic map, describe the graphic vocabulary that is used on it (e.g., isolines, graduated symbols, choropleth shading, etc.) and discuss whether this particular choice of symbols is acceptable as a way of communicating the topic that is the subject of the map.

Given an environmental issue and a description of a particular interest group in the population affected by that issue, summarize the group's position. Design maps and other graphics that are technically accurate but also present the information in a way that might help persuade a jury that the position of the group is valid.

Behavioral objectives should use verbs that describe what students do

Note that these objectives do *not* use terms meant for the teacher, such as "cover this topic" or "present this material." Nor do they use vague, open-ended terms such as "understand" or "comprehend." Objectives that begin with those verbs don't show *how* to cover or understand the topics, to choose teaching tactics, conduct class discussions, or evaluate performance. By contrast, objectives that use an action verb such as "match," "compare," "list," or "write" can communicate the intended outcome of the learning much more clearly.

Some people say that broad ("cover" or "understand") objectives have a role in defining the scope of topics in a course. I say they are a waste of time, even for general planning – if you really know what you want students to be able to do, you can frame it in the form of a specific objective right away (even for the final exam!). And if you don't, a vague objective just postpones the time when you have to think about what you really want students to learn.

So, to repeat, a useful objective should include a setting, a task, an expected level of performance, and an implied (or maybe even a completely drafted) method of evaluation. Once the task and a level of performance are set, a teacher can use the written objective as a guide in selecting appropriate examples, readings, visual aids, group activities, and other strategies for presenting the material. In that way, students will receive a systematic exposure to the ideas, skills, locations, and place traits that have been deemed important in the list of objectives. The written objectives can then help in the design of appropriate evaluation instruments.

Designing Test Questions and Other Evaluation Devices

"So is this going to be on the test or not?"

It depends on what you call a test.

Designing assessments to fit different kinds of learning

Different kinds of learning require different modes of evaluation. Using a time-consuming essay question on a test rather than a multiple-choice question, for a factual objective, is overkill. It wastes time for the student and teacher, and probably doesn't provide more information than a well-designed multiple-choice question would.

By the same token, trying to evaluate students' understanding of some theoretical point with a true-false test is likely to be frustrating for both students and teacher. Some good students almost always read something into the question that the teacher did not foresee, and bad students have a 50/50 chance of getting even a well-designed question correct.

The key is to match the mode of evaluation with the cognitive strand of learning involved. (Read that again: according to one teacher who reviewed a previous edition, it's superimportant.)

In general, matching and multiple-choice questions work just fine for checking student knowledge of locations and the traits of places. This kind of knowledge does not demand elaborate modes of testing (especially if students are told in advance that they need this factual knowledge for theory and evaluation, but that we do not want to waste valuable time testing it in labor-intensive ways).

In large classes, multiple-choice questions are especially handy because it may be possible to download questions from an Internet site and devise multiple forms of the test by making relatively minor changes. For example:

A sample question with multiple forms

_______ Which of the following features are typical in the landscape of northern China?

A. rice fields

B. wheat fields

C. desert basins

D. tin mines

Substituting "southern" or "western" for "northern" makes A or C the correct answer instead of B. Creating several forms with these subtle alternatives makes a test virtually cheat-proof.

Item analysis

Statistical study of the difficulty and validity of individual test questions

Another advantage of multiple-choice tests is that a computer can perform item-analysis to check the difficulty and validity of the questions. If a test question is answered "correctly" more often by mediocre students than by good students, it is probably better to know that before relying on that test for information about student performance! Moreover, publicly acknowledging that a particular question was poorly designed (and therefore will not "count") can be reassuring to students, and in some cases it can also be an effective way to begin a discussion about a topic that students may not have mastered as well as desired.

Some people object to any use of matching or multiple-choice tests because they seem to focus too much on factual memory. One possible response to that objection is a blunt statement like: *Facts are an important part of learning, and of life; get used to it.*

A more diplomatic and ultimately more effective response is a carefully prepared multistranded list of objectives. This list should clearly specify the kind and amount of factual knowledge that you have decided is a necessary foundation for understanding the concepts you have chosen to teach, or for being able to choose and defend opinions about them.

This leads into another good defense for the practice of detailed objective writing: It forces the teacher to decide whether a particular kind of fact is really worth knowing. As noted earlier, it is all too easy to fall into the trap of teaching **categorillas**. Extreme statements can definitely help grab student attention. However, do students gain a balanced perspective by learning about places that have the longest suspension bridge, the best art museum, or the biggest ball of twine in the country? Having to justify the teaching of specific facts in terms of their utility in learning geographic skills or theories is a good discipline against picking facts just for their shock value or memorability.

Categorilla

Term coined for the largest, longest, deepest, highest, or otherwise most exceptional feature in a category

In short, a good set of course objectives can provide a solid foundation for the process of choosing examples and anecdotes for everyday class use. A teacher can also use that same criterion to evaluate textbooks and other course materials: a well-crafted text should have clear evidence that the primary objectives of each chapter were instrumental in guiding the selection of examples, illustrations, photographs, graphs, and other visual aids.

That integrated approach to class design, of course, demands that teachers do exactly what we want students to learn how to do: to think in multiple strands – images, explanatory theories, and value judgments – at the same time.

And that, in turn, requires teachers to write multiple objectives at the same time. In other words, teachers should be thinking about how to evaluate their nonfactual objectives even while they are making up test questions for their factual objectives.

Designing Evaluation Tools for Nonfactual Objectives

HOTS

Higher-order thinking skills (see Postscript 3-1)

Authentic assessment

Evaluating mastery of skills or concepts in ways that mimic their application in the real world

The list of learner outcomes for a geography class should include all of the cognitive strands that form the geographical rope. The other strands, however, are not as easy to evaluate by the same methods that can be used to test factual knowledge. Applications, analyses, syntheses, and other so-called **higher-order thinking skills** are more effectively tested with essay questions, projects, and other open-ended methods.

Here, the National Geography Standards are right in line with the **authentic assessment** movement. Their focus on analytical skills as well as factual knowledge implies the kind of evaluation in which students demonstrate the use of skills in realistic situations. The best way to test that ability is (wait for it!) to identify real situations and ask students to use the skills they have learned.

For example, a great way to evaluate the fundamental skills of map creation and analysis is to give students some data, ask them to put the data on a graph or a map, and then have them interpret the results. The key to effective evaluation of students' performance is to make a separate appraisal of each major step in the students' task: processing of data, selection of map symbols, analysis of patterns, and interpretation of the map. This realization, in turn, is one foundation for the current focus on student portfolios and the use of rubrics to assess mastery of a process.

> Caveat: Evaluation rubrics work when they carefully identify specific stages of student understanding of a concept or process. They fail, often spectacularly, when they become a rote formula, such as
>
> 3 points thorough understanding of the concept
>
> 2 points partial understanding of the concept
>
> 1 point incomplete understanding of the concept
>
> 0 points lack of understanding of the concept

Like behavioral objectives, learner outcomes, constructivist approaches, and many other buzzwords of professional education, evaluation rubrics are useful only when they are written well. A bad rubric is just as useless as an objective like "after doing this activity, students will understand the concept of ________." If you cannot specify what constitutes understanding, how can you decide if a student has it? And if you cannot specify what constitutes "partial understanding," how will you recognize it, let alone judge different students fairly?

In trying to assess a student's understanding of a topic, it helps to remember that there is a "gap" between the domain of concrete facts and the domains of theories and value judgments. Bridging that gap are some fundamental analytical skills, such as making maps, measuring distance, calculating density, and making fair comparisons. Skills like these are frequently tested with projects and lab exercises, but they can also be tested with multiple-choice tests.

Given the nuts-and-bolts nature of these skills, it is often good practice to treat the skills as means but not ends (that was our approach in Chapter 6). Students should learn to respect and even admire skills, but only as intermediate steps between ordinary sensory impressions and the kind of systematically gathered data that can be used to suggest or test hypotheses.

As noted earlier, there is a strong argument for always using real places as examples to teach skills. Devising an imaginary town or continent to teach a skill may place too much emphasis on the skill. Setting the lesson in a real place, by contrast, tends to convey the idea that the skill is worth knowing to the degree it contributes to our understanding of real places.

A way to shift emphasis from simply learning skills to valuing them as a means to an end is to give students a sample of skills tests in advance. In primary grades, a teacher could distribute a prop such as a ruler or compass and say, "tomorrow we are going to use this ruler to measure the distance between two places on a map. Here (Transparency 7A) is an example we will use for practice today." Then say something like:

Modular skill tests let you use similar forms for practice and the test

> "These are the skills we need to do ________, and this is how we'll know when we master them. I will give you a form just like this one on Friday, but the map will be different. You will have no trouble seeing the changes in the questions, because the words and numbers that are underlined are the only ones that might be changed, and they can be changed only to similar terms. For example, I might change the map coordinate 'D4' in question 1 to 'C2.' What you need to remember is that the test will be basically the same as this sample – it will have exactly the same kind of questions but in a different place! Therefore, you can practice with this sample skill test. You can even make up similar forms to test each other. If you have questions, ask me during study break."

Transparency 7A is an example of this kind of skill test. In time, as our software improves, questions like these could be transferred to a computer, which could use the answers to the first few questions to select additional questions that fit the exact level of individual students.

Mental maps can be tested by sketching or with matching questions

Mental maps are cerebral images of spatial relationships, which students should acquire for use in interpreting the world (as discussed in Chapter 3 and illustrated in Transparencies 3I to 3O). These maps are the subject of the second group of National Geography Standards and many state geography and social-studies standards, but they pose some unique assessment issues. In general, mental maps can be tested in two different ways: by asking students to sketch a map of a specific feature, or by asking them to match several maps with a list of the phenomena represented on the maps. Transparency 7B is an example; the teacher notes for this page show how to adjust the level of complexity for this kind of matching test.

Understanding of modern theories can be tested with essay questions

Student understanding of standard geographic theories is yet another kind of knowledge. Not surprisingly, it is best tested by another kind of test: a short-essay exam or a take-home project. Some caveats still apply, however: theories become much more meaningful if students have to apply them in the context of real-world situations. Transparency 7C is an example; it shows how students can be asked to link a historic process with present-day features.

Ability to formulate and defend opinions can be evaluated with essay tests, role simulations, or observation

In addition to tests of factual knowledge, skills, mental maps, and theoretical understanding, teachers should make an explicit attempt to evaluate the ability of students to marshal data in support of opinions or value judgments. Essay tests or term projects are time-tested ways of doing this, but we should not downplay the value of simply observing student participation in group and class discussions. This can often provide insight that complements what you get from other forms of evaluation. For example, Transparencies 7D and 7E are some instructions and evaluation criteria for one kind of community profile (like Transparency 3B). A teacher can easily extend the evaluation to include an assessment of the research strategies that students choose, the questions they ask, and the advice they give each other.

The ultimate test of geographic insight and competence is real-world performance. Can a student identify features "out there," formulate hypotheses about why they are located where they are, test those hypotheses, and make reasoned value judgments about the locations?

Evaluating performance in the field is difficult, even under ideal circumstances. For this reason, teachers should start by making sure that field trips and travel are seen as key components of a geography class, not as a respite from it.

This has implications for both the classroom and the field experience. Classroom activities should be justified in terms of how well they support the goal of geographic competence in the real world. Field experiences should be as carefully structured and theme-driven as the classroom activities, and they should be evaluated (Transparencies 7F and 7G).

Multiple forms of evaluation are insurance against misjudgment

In sum, different kinds of objectives require different means of evaluation. Moreover, using multiple evaluation tools is a form of insurance against misjudging students who simply do not perform as well on certain kinds of tests (more about that below).

There is more than a bit of truth to the idea that good teachers seek evaluative tools that provide numbers to support what we already "know" is the correct grade for individual students. That is why teachers should be ready to articulate their objections to standardized tests that do not span the entire range of desirable learner outcomes.

One can use standardized tests as a convenient third party to help motivate students. They play a useful role in helping teachers redefine the mood of classroom as "you and me against the test" rather than "me against you." But teachers should resist any attempt to have a standardized test be the sole measure of success. At the very least, they should insist on a formal mechanism to add teacher comments to the same page of the student record that includes the standardized test scores.

Before we leave this chapter on evaluation, do we want to open the can of worms labeled "ungraded"?

Evaluation without grades

Ungraded classes?

Not me. I am willing to concede that assessment of performance can have different bases – absolute, relative to other students, or relative to previous performance by the same student. But I am in the business of educating in order to change students' ability to perform. That raises a pointed question: how can I judge how well I am doing, unless I at least try to measure how well students can perform? Any rigorous attempt at measurement is bound to be thought of as a form of grading. It is as simple and as complicated as that.

Qualitatively different kinds of intelligence

Do people really have different learning styles?

There is another can of worms that is not so easy to keep safely closed. Some oft-quoted books about cognitive psychology (e.g., *Frames of Mind* by Howard Gardner) have suggested that different students may have different preferred modes of learning – verbal, spatial, numerical, kinesthetic, and so on. The topic has been trivialized by people who blithely use the term "intelligence" to describe different learning modes. They speak of "verbal intelligence," "kinesthetic intelligence," and so on. This reaches the point of absurdity when a student confronts a teacher with a jargon-rich

It is a sad day when students have to inform teachers of what they cannot do

self-descriptive statement like, "I've been diagnosed as a visual learner – you can't expect me to learn with all these words and numbers."

In one way, this new rhetoric is much like "the" five themes of geography: we have already covered the basic principles in previous chapters on disciplinary perspectives, cognitive strands, and modes of instruction. Still, it might be worthwhile to review one key idea from Chapter 1:

> Geography is not just a kind of knowledge, it is a qualitatively different perspective on the world.

The discipline of geography begins with the assumption that looking at locations, distances, and other spatial relationships can help people solve problems. To attack a geographic problem, we describe locations, gather sensory impressions, write verbal descriptions, perform quantitative analyses, display the results on maps, and occasionally use musical, artistic, or poetic analogies to interpret the results. In doing so, we employ many (if not all) of the "intelligences" that have been identified.

Nevertheless, geography is an easy subject to adapt to a classroom mode that explicitly tries to acknowledge different learning styles. A conscientious teacher with a well-conceived set of objectives can choose modes of instruction that cater to different learning styles. In so doing, we may help counter the temptation to pigeonhole people according to their self-evaluative statements (e.g., "I hate mathematics" or "I can't learn by role playing" or "I'm a visual learner – don't make me look at numbers").

I suppose if I were asked for a summary statement about learning styles, my answer would be that no individual is exclusively verbal or musical or whatever, and no one is hopelessly unintelligent in any of the learning modes. The best advice for someone who really wants to improve would probably be to strengthen the weak areas as well as play to the strengths.

Geography is a good subject for the kind of exploration involved in using different kinds of learning styles. If teachers consciously plan multimedia experiences, cooperative activities, field trips, simulations, and evaluation instruments that employ different learning styles, students will all be richer for it.

POSTSCRIPT 7-1

An Editorial about Standardized Tests*

If you want to use written standards and high-stakes tests as a way of evaluating students, teachers, and schools, then you have to be willing to invest the time and money that it takes to design truly superior standards and assessments.

*Documentary footnote: It is my observation (having participated in hearings or other aspects of the standards-writing process in 6 states, conducted teacher workshops in 23 states, and carefully read the state standards in geography for all 49 states that have them), that no state so far has been willing to spend what it takes to do this job well.

Indeed, for at least half of the states, the standards are distinctly inferior (you can view them on the accompanying CD). They seem to be hastily assembled products of well-intentioned but overworked and sometimes underqualified bureaucrats, committees, and even student interns.

I am hardly alone in thinking that this is not what is needed to ensure that No Child (Will Be) Left Behind.

To rescue this process, the assessments must be better!

8

A Multiwheeled Cart: Understanding Resistance to Educational Change

At the scale of a whole society, education is like a factory – students enter at one end and emerge at the other, changed and (we hope) improved by the experience.

At the scale of a single discipline such as geography, however, it might be better to view the educational enterprise as a cart with four wheels. Each of the wheels is a distinctively different kind of activity with its own specific role in the improvement of educational services (Transparency 8A):

Guild

Medieval organization in which apprentices work for a master until they learn enough to be accepted as full members of the guild

1. **Pre-service training**. Teaching is both a craft and an art. Learning to teach, therefore, has much in common with the ancient **guild** model of master and apprentice. In effect, teachers often start by teaching what (and how) they were taught. If they experience no geography in their teacher-training, how can they see its role or value?
2. **In-service training**. Only 5 to 10 percent of the teachers in a typical school district are replaced each year. This has two implications for the timely implementation of National (and state) Geography Standards and the other components of the No Child Left Behind Act (NCLB). First, some teachers of other subjects may be asked to teach geography, even though they have no prior training in the field. Second, some teachers of other subjects may be asked to mentor new teachers in geography. If these people have no access to in-service training, how can they develop the necessary skills?

Paragraphs adapted from the *Journal of Geography*, 1996

3. **Curricular position**. Each academic discipline that is listed as essential in the state standards needs a time-slot in the teaching schedule. This time-slot is what specifies how many hours should be devoted to a given subject at each grade level. Since there are always too many demands for time, geography teachers (like many others) have to negotiate for a position in the curriculum. Can their course proposals get a fair hearing from evaluators who may have little background in geography and may not see how the subject (as they understand it) has relevance to the real world of jobs, construction, crime, and health?

4. **Teaching materials**. Students need to work with maps, data, field observations, and other geographic content in order to learn geography. Most teachers have too many demands on their time to be able to create the materials they would need for effective geographic instruction. They look for support from textbooks, student activities, A/V materials, and computer software, as well as resource people and field guides.

Note: Transparency 8A does not put any of these four groups – education departments, teachers, administrators, or authors of teaching materials – in the driver's seat. This analogy is not about control!

The improvement cart needs all four wheels:

- *pre-service training*
- *in-service training*
- *place in curriculum*
- *teaching materials*

The value of the cart analogy lies in a simple realization: *the pedagogical cart needs all four wheels to run*. This gives people who dislike change (or fear it) a powerful tactic to delay or block action. When asked for assistance with any of the four activities, they can point to the relative flatness of other tires. For example:

Ask a publisher to print a high-school geography book. A likely response is that the market is too small because the subject is not required in very many states (i.e., the subject does not have a specific time allotment).

Ask an administrator to put a geography course in the curriculum. A likely response is that the proposal is impractical because few teachers are trained to teach this kind of course.

Ask an education department to offer courses in geography education. A likely response is that there is no need because there are few jobs for geography teachers.

Ask an in-service teacher to attend training in geography education. A likely response is that such training is of no use because the teacher's school does not offer a course in geography. (Moreover, there are no good new textbooks or materials available if one were to propose one, etc., etc.)

All of these responses have the same basic message: one can safely postpone repairing any given wheel of the cart, because the others are flat and the cart would not run anyway. This catch 22 has two results.

The first is that people can opt to direct their energy elsewhere. Teachers, administrators, and textbook authors all have plenty of demands on their time. Pumping a flat tire on the geography cart has been an unrewarding task for many years, and geography education has clearly suffered. Poll after poll has shown that American students have a level of geographic awareness that lags far behind their peers in other countries.

This has reached a point, finally, where it has caught the attention of mayors, governors, senators, and presidents, causing action on what had been ongoing efforts by a handful of dedicated proponents of geographic education. The result has been a flurry of activity, ranging from the symbolic (e.g., designation of a Geography Awareness Week) to substantive (e.g., listing among the core disciplines in the NCLB Act, inclusion in the New York State Regents' Test, and the addition of geography as a subject area in high-school Advanced Placement courses and tests).

This brings us to the second result of the four-wheel cart problem. People who have been working on individual tires are now being asked, far too often, to redirect their attention to the other wheels. Organizers of summer institutes have been asked to write teaching materials or assessments. Authors have been asked to conduct teacher-training workshops. Teachers of in-service programs have been asked to draft standards and create course proposals.

Some familiarity with all four wheels is beneficial

but . . .

too much switching among wheels is self-defeating

Such redirection of effort has been useful, perhaps even essential, when done in moderation. Authors who occasionally work with classroom teachers are less likely to churn out materials that teachers cannot use. Institute organizers who develop some of their own materials gain an appreciation for the challenges involved in producing books and activities for a wide range of users. And teachers who write materials and then teach them to each other almost always wind up broadened and recharged by the experience.

The redirection of effort is harmful, however, if it's carried past a modest level. Many competent and dedicated people have been pushed into roles where they are less effective; in these roles they may have unsatisfactory performance or accelerated burnout. Some are so busy switching from one tire to another that they do not have the time or energy to maintain a high level of competence at any one task.

A Better Goal, and How We Get There

Here is an all-too-common scenario: an author has just finished a presentation about a new classroom activity. During the question-and-answer period, a number of people ask pointed questions about teacher training and curriculum reform: "What are you doing to help retrain history and social-studies teachers to use these materials? What are you doing to present this material to education majors in college? What steps are you taking to get geography listed in the state curriculum?"

These are all valid concerns, but is a textbook or activity author necessarily the one to respond to them? If the questioners persist, and the speaker tries to answer, the result is likely to be unsatisfactory answers, followed by general commiseration about the sad state of education in America today. That is hardly a desirable outcome.

How can this scenario move toward a different conclusion? One good start would be to acknowledge that authors' expertise is in writing, and they came to answer questions about written materials. If that kind of question doesn't come up, their expertise is wasted.

Should we focus on questions we are not qualified to answer?

Or on those we are?

So, here is the same scenario, with a twist. The twist is that the author or moderator has a copy of the cart diagram (or a similar analogy), ready for the overhead projector. If someone asks what the author is doing to train teachers or convince administrators, the proper answer (in this scenario!) is, "not much. I don't need to. As this diagram shows, we need promoters, institute organizers, and pre-service instructors. I assume they're out there, pumping their tires as fast as they can. I am just trying to keep the materials corner of the cart up to their level. Any questions about these materials?"

If we can keep the attention of participants on the tires they know best, geography will do well. The four tires will get pumped up, more or less simultaneously. American students will have adequate time to work with good materials under the guidance of well-trained teachers. They will have a chance to learn the kind of geography that is useful for life:

At a local scale, students (future citizens, etc.) will learn how to arrange roads, cities, fields, clinics, and other things so that they are fair, safe, efficient, and beautiful.

At a global scale, students will learn about similarities and differences in climate, culture, and competitiveness.

And when we put the local and global scales together (like the two blades of a **scissors**!), the whole is more than the sum of the parts: fair, safe, efficient, and beautiful places tend also to be culturally tolerant, residentially desirable, and economically competitive.

The geographical scissors

Local and global, working together (see Chapter 2)

We will not reach that goal if we let the forces at work turn successful institute organizers into mediocre book authors, effective preservice teachers into frustrated curriculum advocates, and effective disciplinary spokespersons into beleaguered summer-institute administrators. When questions about the other tires come up, the proper response is to turn them back on the questioner. If you are pumping effectively on one tire and some people ask why you aren't working on another, tell them that you are doing your job and others are doing theirs.

The rest of this chapter is a brief outline of some ongoing efforts at each wheel of the cart. The order of presentation is immaterial; I arranged the four tires roughly in the chronological order in which someone studying to be a teacher might encounter them, as he or she begins a teaching career, then serves on a curriculum board, and finally helps write a student workbook.

Pre-Service Training

Pre-service training should begin with two subtly but profoundly different questions:

What does someone need to get a teaching job?

What does someone need to do a good job of teaching?

The answers to these questions should guide students in all phases of pre-service training: choosing courses, developing skills, and deciding what published materials to buy and what classroom materials to save.

To get a job interview

You should have a persuasive resume

So, what does someone need to get a teaching job? Here is a standard response: getting a teaching job involves putting together a resume that demonstrates focus, breadth, coherence, depth, adaptability, and other desirable traits. This book obviously is not the right place to do much more than repeat that prescription; taking a pre-service curriculum obviously will help to flesh that out.

This book can, however, provide assistance in answering the second question: What does someone need to do a good job of teaching geography?

We could start by acknowledging that the four tires of the geography cart are probably not going to be fully pumped up in the foreseeable future. To some extent, therefore, prospective teachers will have to be able to substitute some of their own time and expertise to offset the lack of proper support. In short, they will have to be able to train themselves, justify their courses to administrators, and make some of their own classroom materials.

The value of a teaching portfolio

Administrators realize this: they search a prospective teacher's resume for evidence that the person has these skills. One of the best ways to provide that evidence is to assemble a portfolio of teaching units, visual aids, and other materials that you have designed and produced yourself.

Making some high-quality overhead transparencies, computer presentations, or classroom handouts is one way for a pre-service teacher to show depth of understanding of a specific topic. It is also a way to display your mastery of the skills of page layout, graphics creation, and media use. It can demonstrate awareness of methods and topics that fit the National Standards and other pedagogical initiatives. Finally, it can demonstrate flexibility in acquiring and using new knowledge and skills (especially if the portfolio includes things like a personal website or electronic files of materials that can be adapted for other uses).

And, it gives you a head start on your first year of teaching!

How does a pre-service teacher trainee acquire such a portfolio? Here's one suggestion: treat each class project as an opportunity to add to your personal portfolio.

This mindset will influence the topic you pick for each class project. If one of your goals is to show breadth, it's probably not a good idea to do each project on the same general topic! Making at least one of your units from locally available materials is also useful, because it demonstrates a commitment to make what happens in the classroom relevant to the local community.

Some locally available sources of geographic information

Possible sources of raw data for a lesson with a local focus include:

- Telephone yellow pages, which list a wide variety of human activities that can be mapped and analyzed (Transparency 8B)
- Almanacs and census reports, which have information about towns, counties, states, and countries (Transparencies 8C and 8D)
- Topographic maps, which are detailed reference maps that can be transformed into teaching materials (Transparencies 8E and 8F)
- Soil surveys, which include detailed maps that can be simplified to show the suitability of particular sites for wildlife habitat and an assortment of human uses such as farming, building, and various forms of recreation (Transparency 8G)

- Newspapers and magazines, which present current events and a real-world context for places that can vividly illustrate geographic concepts (Transparency 8H)
- Environmental impact statements or planning documents for local projects, which (by law!) are supposed to make public disclosure of the kind of information that would allow an informed citizenry to make decisions about whether the benefits of a proposed project exceed its costs
- Road maps, tourist brochures, Chamber of Commerce profiles, and other sources of information about places of interest (Transparency 8I)
- And, of course, badly designed units in existing textbooks, which can always be improved!

All of these sources are available in both print form and online (see the CD, on local sources of geographic information). The key is to assemble these items into teaching units that feature all of the elements this book has already described as key components of geographic education:

Criteria for evaluating teaching units

1. A spatial perspective (Chapter 1),
2. An interplay of both local and global or national scales (the two blades of the analytical scissors of geography, as described in Chapter 2),
3. A well-balanced mix of concrete images, abstract theories, and value judgments (the three strands of meaning, as described in Chapter 3),
4. The preparation and analysis of regional maps that show the conditions at specific locations and their connections to other locations (the four cornerstones of Chapter 4),
5. An organizing theme or standard that will certify the student as having mastered one of the key ideas of geography (Chapter 5), and
6. A focus on introducing, expanding, or providing practice with a specific skill of spatial thinking (Chapter 6).

That is admittedly a tall order, but it is why geography teaching is both important and challenging.

In-Service Training

Once hired, a geography teacher has to figure out how to stay current in a changing world. That is another tall order, especially in light of typical constraints on time and money.

One possibility is to make efficient use of reading, television, travel, and recreation. In the interest of saving space, let me focus on just one of those. Here are half a dozen handy rules of thumb for geographically productive travel:

Some rules of thumb for geographically efficient travel

1. Take copious notes, if for no other reason than because they help you justify your income tax deduction for some of the costs of the trip. (A geography teacher who fails to deduct some travel costs is missing a real fringe benefit of the job!) Look for examples of features that you can describe to your students in an interesting way and that also illustrate an important geographical point (see the story about square ponds in Chapter 10 for an example).

2. Take photographs and record their locations. Even if you do not use your own photos in class, they can aid your memory of places you have visited. I find it interesting to compare my photos of a famous place with what I see in magazines and tourist leaflets. Without my own photos, I suspect my memories would inexorably shift toward those public-relations images of the place. In the same vein, retaking photos from the same places on a later trip can give a very valuable time perspective. (See Table 3-1 for more ideas about photography.)

"Adopting" a few places for repeated study is more rewarding than skimming many places

3. "Adopt" a few places and visit them repeatedly. Since traveling to see the whole world is an impossible dream, why not focus at least part of your time by picking a few places and getting to know them well? Choices can have different rationales. For example, I adopted Magnolia, Mississippi, just because its name seemed to evoke images of "Old Dixie." My wife's relatives lived in Nebraska near the Platte, a river that is both historically important and hydrologically similar to other Great Plains rivers. Over the last three decades, I have visited Magnolia half a dozen times; I try to read its newspaper for a few months every few years. I have never traveled more than a small distance along the Platte on any one trip, but I have often crossed it or followed along beside it for awhile. Occasionally, I stop and talk to someone along the river (or pick up some information from a county office or a town Chamber of Commerce). In time, this apparently casual strategy has resulted in a substantial collection of clippings, stories, and brochures. In short, Magnolia and the Platte have become benchmarks against which I can judge other things I read on the subjects of the South, the Great Plains, water resources, race relations, population trends, and many other topics. I am convinced that an equivalent amount of time spent reading or visiting a number of randomly chosen places would not be nearly as valuable.

Analyze something in a geographically systematic way

it is important for both the knowledge you get

and

the "body-language" you convey

4. Pick a subject to analyze systematically wherever you travel. For example, I check the price of adjustable pliers in hardware stores. I know, from personal observation, that seven-inch Vise-grip pliers range in price from $6.49 in a Wisconsin farm supply store to $12.95 in a small town Alabama hardware store. They are cheaper in Seoul, Korea, than in Manhattan, and over time their price does not go up as fast as inflation. In a similar way, I have observed that a Mississippi supermarket offered only one kind of mustard while a gourmet shop in Bethesda, Maryland, had at least 86. Systematic personal observations such as these have a ring of authenticity. They afford a standard benchmark for evaluating newspapers, television, and magazines – my mental "maps" of pliers and mustard are surprisingly easy to relate to national patterns of jobs and money. They clearly show that the real world is more cluttered than is shown by simple maps that appear in the mass media. In this way, a self-imposed routine of systematic observation helps keep a teacher aware of both the orderliness and the messiness of the real world. And (this is very important!) it sends a subliminal message that systematic observations have more analytical value than anecdotes. This is a message that many students desperately need to hear in an age of 20-second sound bites and talk-show "experts."

5. Make sketch-maps of interesting places. Like the systematic observations described above, these sketch maps help show students (by example!) that trying to figure out how things fit together in an area is important. Over the years, I have made sketch maps of large truck stops, houses in historic districts, birdsfoot trefoil plants in a pasture, parking fees in different parts of a city, and many other topics (admittedly not all equally fruitful!). I continue to do this because some intriguing geographic ideas emerge from serendipitous comparisons of this kind of map with traditional ones (Transparency 8J).

6. Buy a local newspaper or a "strange" magazine every once in awhile and scan it. If you're short of time and/or money, at least check what magazines are for sale in service stations in different places. For example, do automobile magazines outnumber cooking magazines in this area? You can get interesting perspective (and sometimes mappable data) from the presence of magazines and newspapers like the *American Rifleman*, *Art Forum*, *Modern Bride*, *Country Living*, *The Economist*, the *Farm Journal*, *Mother Jones*, *Paris Match*, *Sunset*, *Texas Monthly*, *the Utne Reader*, or *The Washington Post* (to name just a few in alphabetical order). The content of a local newspaper or specialized magazine can be especially useful because its subject matter is not selected by a national editor. Good regional novels or movies are also helpful – for a great blend of mystery and good

> geography, read any of various novels by James Lee Burke, Sue Grafton, Carl Hiaasen, Tony Hillerman, Margaret Maron, Sara Paretsky, Martin Cruz-Smith, and Margaret Truman (this list is hardly exhaustive!).

The items on this list of travel tips all underscore one basic idea: *Disciplined observation does not really take much more time than "ordinary" travel, but it has many tangible and intangible benefits.*

Two other ongoing strategies are worth mentioning. First, learn a new geography software program every year or two. That may sound like a chore, but each generation of students is using new and different technological learning aids. Doing the same has three plausible benefits: empathy on your part, respect for you on theirs, and an ever-improving ability to use a computer as a tool.

Second, take an occasional refresher course, in person or online.

> (I have to say that, since I'm a college teacher! But, frankly, this is one area in which I'd sincerely like to ensure my own obsolescence. If you have to skimp on any of these suggestions, a formal course is perhaps the least valuable item. I really think teachers would learn more geography in the long run by forcing themselves to take photos of trees, fields, barns, houses, stores, or schools wherever they travel, record the locations of the photos, and speculate about why they are the way they are in that particular place. Then, they can compare photos or descriptions of new places with their image collection.)

A course?

Sure, but not instead of your own disciplined observation

We probably should not leave the topic of in-service teacher training without noting, once again, that geography is profoundly interdisciplinary. It is worthwhile, therefore, to pay attention to what is happening in related disciplines such as earth science, ecology, economics, history, political science, social studies, even mathematics and English. Take a course, sign up for a weekend institute, scan a textbook, or talk to a colleague once in awhile.

Occasional forays into other fields can yield a huge increase in understanding with only a modest input of time. One big payoff comes when you have to work with people from other disciplines on a joint project. Knowledge of some corner of their domain can be very helpful in smoothing relationships and in finding ways to illustrate geographical themes with examples that have value in teaching other subjects.

Curricular Position

The key questions that should be asked by someone standing in front of this third wheel of the cart are: How many hours should be devoted to this particular topic? at what grade level?

Educators have been writing scope and sequence outlines at least since the times of Socrates and Sun Tzu. From such outlines we can see what a particular generation thinks is important to pass on to later generations.

Over the years, writing outlines has stimulated an endless cycle of numbingly repetitive debates on theory versus practice, fascism versus freedom, ethnocentrism versus tolerance, and challenge versus elitism. These debates are very important. Every teacher should come to reasoned conclusions about these issues. *But they should not dominate every discussion about curriculum.*

The Standards project involved more than 3 years of work by more than 350 people (including the author)

A very large group of dedicated people worked on the National Geography Standards project for several years to forge a consensus about what kind of geography should be taught, and in this book, we welcome that effort. Whether I agree with every aspect of the Standards is less important than that they give me the means to judge whether a particular set of teaching materials is faithful to the goals of the Standards.

Here, it is worthwhile to note that federal legislation has forced every state to write curricular standards. These state standards are useful for setting goals, but they are emphatically not to be used as outlines for classes or teaching units. In that respect, they are just like the five themes from Chapter 5 – they work best when they quietly pervade a curriculum rather than dictate its scope and sequence.

Themes and standards should pervade, not dictate

The difference between ideas that pervade curriculum and those that dictate it is subtle but profound. Geography classes and units that teach the Standards in a rote way will be boring and ineffective, because the human brain simply does not learn well in such a prescriptive way. (If you don't believe that, try reading the National Geography Standards from cover to cover in one sitting!)

On the other hand, geography classes and units that ignore the Standards can hardly claim to be a part of the discipline of geography. The middle ground is to read the Standards (a little at a time!), keep them in mind, and then design class outlines and materials to meet them in a pedagogically effective way. For a teacher, the aim is to be able to select materials and use them in a way that meets the Standards. One clue about how to do this can come from simple examination of available teaching materials (especially in some other disciplines as well as geography).

Teaching Materials

High-school history textbooks have caught a serious (if not fatal) disease called **OK-but-itis** (technically, the name of the disease is this-book-is-OK-but-it-doesn't-say-anything-about-topic-X-and-it-would-be-better-if-they-added-just-a-little-bit-about-that-itis).

OK-but-itis

A disease that leads inexorably to textbook obesity

Here is how the disease progresses. An author starts by writing 300 or 400 pages of coherent and fairly interesting prose. Then three dozen reviewers and editors get the manuscript. One by one they notice a few things missing, and they add a sentence here and a paragraph there. Then some state and local school boards review the book and make their own suggestions. No big deal, they all say, but the inevitable result is a massive tome of 900 or 1,100 or even 1,500 pages, which few students can carry comfortably, let alone read or comprehend.

Geography is genetically susceptible to the same disease; it resembles history in that it takes the entire world as its subject. It pays, therefore, to be able to recognize the early symptoms of OK-but-itis, because the later stages almost always include massive boredom and alienation on the part of students.

Let us start by acknowledging (shouting from the hilltops would be better!) that *the textbook does not have to cover every important topic.* Geographical themes can be taught with a wide variety of supplementary materials. The list of sources described above (under Pre-Service Training) is a good starting point.

A textbook cannot be the "expert" and still be concise and interesting

If the librarian or media resources person at your institution asks for suggestions for materials to purchase, you might consider nominating the following (italicized titles and/or trademarks are chosen for illustrative purposes only; see the CD, on local sources of geographic information, for a much wider selection):

- Several thematic atlases at different scales. Most major publishers produce a classroom atlas – look for one that has thematic as well as reference maps (see Chapter 2). Other affordable and effective

Other sources of information for a specific class topic

examples might include *Goode's World Atlas* (Rand McNally), the *Historical Atlas of the United States (National Geographic Society), and the State Atlas and county Soil Survey for your locality (if available).*

- *County and City Data Book, State and Metro Area Data Book*, and *Statistical Abstract of the United States (all can be bought from the U.S. Superintendent of Documents, and are also available online).*
- *United Nations Statistical Yearbook, CIA World Factbook*, other world almanacs. The information in these publications is available online, but for class purposes it is definitely better to have a printed copy, because you often want students to compare a number of countries, not just look up the data for one.
- Computer data-sets such as *PCWorld, PCGlobe,* or the *World Atlas.* Finding data in spreadsheet form can make it much easier to calculate ratios such as televisions per million people.
- A route-finding computer program such as *Automap, EasyMap,* or *DeLorme Map 'n' Go.*
- Classroom access to an online system with a good browser.

With these basic reference sources as one blade of a data analysis scissors, teachers can bring in topical ideas from newspapers or television and ask students to search the atlases and other data for context, background, and perspective.

In short, a good set of reference works can serve as half of the foundation for a whole variety of investigations. The other part of the foundation is a teacher's insight into what kinds of questions are geographically interesting.

A good test of a candidate question for a geography investigation is to ask if it:

- Features an interplay of global and local scales of analysis (as described in Chapter 2),
- Deals with all three strands of meaning (Chapter 3),
- Builds on the cornerstone ideas of geography (Chapter 4),
- Has a valid theme as an organizing tool (Chapter 5),
- Introduces or enhances one of the skills of spatial cognition that are the unique contribution of geography (Chapter 6).

These are precisely the kinds of questions that have been painstakingly developed in the National Geography Standards.

A final word of reassurance: the tasks described in this chapter might seem intimidating, but the whole point of the chapter is that no one person can do all of these jobs. A reasonable goal for one person is to have a basic appreciation of what is involved in each of these jobs, coupled with some ability to evaluate the results of someone else's work when you see it in textbooks, supplementary materials, state standards, and decisions on the local curriculum.

In short, getting the cart of geography education to go is also like a balancing act. On one hand, a teacher has to respect the efforts of those who choose to draft standards, write materials, create assessments, and other kinds of contributions. At the same time, the teacher should know enough to judge which standards are teachable, which textbooks and other materials are most faithful to the standards, and which classroom tactics are appropriate for students at the given grade level. To do this the balance between the extremes of do-everything and know-nothing must be maintained.

That search for balance is the topic of the next chapter.

9

Pairs of Tools, Working in Cooperation

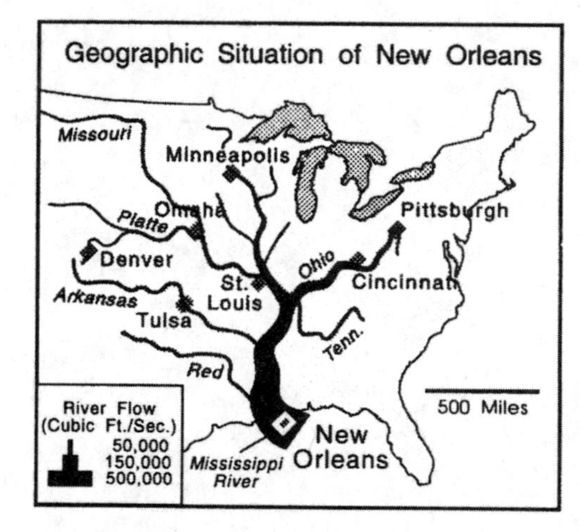

Look at the location of New Orleans, Louisiana. This old French and Creole city has a superb **situation**, using that word in its precise geographic sense, as a kind of summation of the connections a place has with other places. One reason why New Orleans has these great connections is because it is located near the mouth of the largest river system on the continent.

At this location, millions of tons of cargo are transferred between ocean ships and other modes of transportation, such as river barges, trains, trucks, or pipelines. All of that activity means jobs, not only for dockworkers and shipping clerks but also in banks, restaurants, hotels, and other businesses. Much of the New Orleans economy is thus a direct result of the favorable geographic situation of the city.

Situation (link)

Connections with other places

Site (place)

Conditions at a place

The geographic **site** of New Orleans, on the other hand, leaves much to be desired. Most of the city sits on mucky soil in a flood-prone swamp that was infested with snakes and mosquitoes before much of it was paved over. Huge pumps have to run most of the time to keep water from backing up in the sewers and streets. Even the graves in the cemeteries are aboveground, a cultural tradition that is especially apt in a place where a sealed casket might actually "float" to the surface in the waterlogged soil. To be blunt, the site of New Orleans would probably not be on anyone's list of preferred places for a city. People learned to cope with unfavorable *conditions* of New Orleans in order to take advantage of its superb *connections*.

To geographers, the concept of location is an interplay of two ideas:

Site and situation

Two blades of a geographical scissors (see Chapter

Site – Local conditions (the traits of a place, to use the language of "the five themes" from Chapter 5), and

Situation – Connections with other places (the movement of people, goods, and ideas to and from the location).

The ability to judge conditions and connections from maps is clearly a useful skill for planners and decision makers in business. This skill can also help ordinary people judge where they would like to live, work, or take a vacation (Transparencies 9A and 9B).

Site and situation (place and link, status and movement, local conditions and global connections) are like the opposing blades on a very useful pair of geographical scissors (as discussed in Chapters 2, 4, and 5). I suggest that they are one of a handful of pairs of seemingly unlike concepts that are pedagogically useful when used cooperatively. This chapter is a summary of some major ideas of geography, each one described as a seemingly opposed pair that is actually useful when used together.

Nature and Culture (Physical and Human Geography)

In evaluating places, students should keep in mind that what makes a location useful depends on the culture of the people there as well as the nature of the place. Here are three examples of this principle (often called the **cultural definition of resources**, as noted in the discussion of the concept of place in Chapter 4):

Cultural definition of resources

A feature is a resource only if people think it is

1. A rich deposit of iron ore is a resource only if people know how to use iron. If that kind of technology is not available, the presence of iron ore adds no value to a site. With an appropriate infrastructure of mines, roads, and steel mills, however, a rich deposit of iron ore can make people in an area quite wealthy.
2. A flood-prone field is useful for farming only if people grow crops (e.g., rice) that can tolerate flooding, or if they have the technology to build protective **levees** or drainage systems. A floodplain that has that kind of infrastructure, however, may rank among the most productive agricultural areas on earth.

Levee

Long dam to keep water out of an area

3. A great natural harbor has little value unless you need to move heavy products. Diamond producers, for example, are not likely to care that transportation by boat is much cheaper per ton than any other mode. Their product is so tiny and valuable that producers can send it by airmail!

In short, a geography student cannot just assume that a given natural resource will automatically encourage a particular human activity and produce a specific set of landscape features.

Determinism

The belief that conditions in specific places determine what people do there

The belief in that kind of causality is **determinism**. Geography erred by going too far in that direction in the early 1900s. Books written at the time featured many deterministic ideas, such as the notion that wealth and intelligence are a consequence primarily of favorable climate. These authors said that tropical climates tended to cause laziness and forgetfulness. Therefore (they said), it makes no sense to build schools or factories there.

This kind of extreme position will not last long when people are free to investigate the world. An honest and rigorous geographical analysis reveals many examples of people who overcame severe environmental problems, from heat to cold, flood to drought.

Although reaction against excessive determinism is valid, we should not go to the other extreme and assume that human technology can accomplish anything in any environment. As with most dichotomies, the answer lies in the middle. Nature and culture work together to define what is possible in a given place.

I know, it's possible to grow bananas at the North Pole, if you are willing to spend enough money for light, heat, and shelter. But would you really call that feasible in an economic or political sense? That is precisely the point. The interplay of natural and cultural traits is what gives a specific location its list of possibilities and challenges.

Understanding this interplay is important, which might be an argument against having separate introductory courses in physical and human geography. It would be better to have one course in which students use real-world data to make maps and then apply the maps to answering interesting questions such as

"What aspects of the terrain and culture in Bosnia made a guerrilla war there so difficult to stop?"

"What city should get a new professional baseball team if the major leagues decided to expand?"

"How many clinics are needed to provide service for an area?"

"What is a good location for a historical theme park?"

"What countries are likely to join South Korea and Taiwan on the list of 'Asian Tigers,' the countries that have experienced rapid industrialization and rising incomes?"

"Why do people in warm countries such as Mexico or Thailand tend to cook foods that are spicier than food from colder countries such as Sweden or Russia?"

"Why do people in Utah tend to vote for the Republican candidates for President?"

"What is the most efficient route for a new commuter rail line in this part of the city?"

"Should cities and suburbs merge into a single metropolitan government in order to solve urban problems?"

Answers to questions like these will automatically involve learning about locations, environmental conditions, cultural features, and connections between places. If the questions are carefully chosen, the result can be fairly complete "coverage" of major world regions, map skills, and geographical theories. (To see if coverage is complete, make a matrix such as Transparency 2J.)

Alternatively, one could organize a class with a focus on a few selected regions. In this case, a teacher can maintain the scissors action by using a specific economic, political, or environmental issue as an organizing theme in each region. The key is to have a specific set of "skill," "evaluation," and "appreciation" objectives in mind in each region. That will keep the course from drifting into being a mere catalog of the physical and cultural features of each region.

▪ ▪ ▪

Here are three examples of the kind of objective that works to keep physical and human geography together:

Given a list of imports and exports from a given country, make a **flowline map** of the data. Then, try to describe the conditions (environmental, economic, or whatever) that encourage people to produce those products. (You might add another related question: What kind of roads or other infrastructure did people have to build in each place in order to take advantage of the resources there?)

Flowline map

Map that shows volume of movement by varying the widths of arrows or lines

Given a location and a photograph of a house, describe three specific construction details of the house. Then, state whether (and why) you think each of those traits is appropriate in that particular location (recall Transparencies 3C and 3D?).

Given a latitude and longitude (or just a dot on a map), write a brief description of the climate at that location (e.g., southern California). Then, cite an example of a place on another continent with a similar climate (a **climatic analog**, such as Portugal or southwestern Australia). Finally, describe several similarities in the way people live in those regions. How are those lifestyle traits related to the climate?

Climatic analog

Place with a similar geographic position on another continent (e.g., Los Angeles and Casablanca; see Chapter 6)

Note that each question specifically asks a student to combine information about environment and culture in order to get an answer. Moreover, each question starts with a "simpler" subquestion. That allows beginning students to achieve some success even if they cannot answer the entire question correctly the first time around. One often has to learn to walk before trying to run, but it does not hurt to do both on the same good path! This principle is even more important as we consider other dichotomies, such as between theoretical and applied geography or between regional and topical geography.

Knowing and Doing (Theoretical and Applied Geography)

For some people, geographical knowledge is worth acquiring for its own sake. Others see it primarily as a means to other ends. For them, geographical knowledge is a path to a more efficient or humane society, a safer world, or perhaps just a larger income. To examine the interplay between theoretical and applied geography, let us look at five groups of geographers. Each group has its own professional organization that publishes a magazine or journal and does other things to promote itself.

1. Some people study other places for the sheer joy of knowing. For them, geography is an adventure. Their goal is to see, hear, smell, taste, and feel exotic places and cultures. Their main criterion for choosing places to study is its uniqueness. The more unusual a place is, the better they like it.

   National Geographic

   Exploring exotic places

   Dozens of television shows and magazines cater to the large demand for this kind of experience. The **National Geographic Society** is the most well-known embodiment of this perspective. With its head office in Washington, DC, this organization has sponsored thousands of expeditions, hundreds of films and television specials, as well as the magazine that bears its name. In recent years, the Society has started to do more formal education and even some theoretical research, which in previous decades would have been thought of as in the "domain" of other groups listed below.

2. People in the second group of geographers study other places in order to discover patterns and draw conclusions. The main goal is to understand and organize knowledge. (That urge may be at least as genetically strong in humankind as the urge to explore!) The reward for this study might be an elegant theory that ties many observations together into a neat package.

The professional organization called the American Geographical Society has this perspective. The journal of that group, the **Geographical Review**, consists mainly of articles about specific places and how they developed.

Geographical Review

Understanding other places

3. The third group on this list – the Association of American Geographers – tries to develop geographical theories. The primary focus of this professional group is research. To exchange ideas, the AAG has national and regional meetings with literally hundreds of sessions devoted to specific topics such as "Medical Care in South Asia," "Urban Transportation in East Africa," "Economic Restructuring of the English Midlands," or "The Biogeography of the Andes Mountains." The primary journals of this group are the **Annals of the AAG** and **The Professional Geographer**. Other research groups publish their own specialized journals, such as **Antipode** (the journal of radical geography), **Geographical Analysis** (for mathematically inclined geographers), **Economic Geography**, **The Southeastern Geographer**, and **Urban Geography**.

Annals of the AAG

Doing research in spatial theory

4. A fourth group of geographers is concerned with using geographical ideas to solve real-world problems. These "applied" geographers have basically the same range of interests as research geographers. Their studies tend to be much narrower, however, and usually have deadlines. These people often work in businesses, travel bureaus, or government agencies rather than universities. (Like many of the "divisions" discussed here, that distinction has a great deal of overlap with the other categories. Professors often act as consultants or do other kinds of applied geography. Agency people, meanwhile, sometimes contribute to theoretical journals.) Typical jobs for applied geographers include:

   - Forecasting traffic,
   - Analyzing the spread of diseases,
   - Deciding where to locate things such as new factories or sewage treatment plants,
   - Drawing borders around election districts to assure fair representation for minorities,
   - Evaluating the environmental impact of new settlements or subdivisions.

In recent years, applied geographers have started a number of new journals, including the Papers of the Regional Science Association and the Proceedings of the Urban and Regional Information Systems Association. (I think that's the name; hardly anybody says anything but the acronym **URISA**!)

URISA

Doing geography in the real world

Journal of Geography

Learning and teaching

5. The last group consists of educational geographers. They see geography both as a worthwhile subject in its own right and as a logical thread that can help teachers make connections among other subjects, such as art, history, economics, languages, mathematics, and social studies. The National Council for Geographic Education is the main organization for geographic educators in the United States. The NCGE publishes the **Journal of Geography** and many other teacher-oriented publications.

Like the distinction between regional and topical geography, the "splits" between theoretical and applied geography or between exploratory and educational geography are more conceptual than real. Any geographer who stayed too long doing one without the others would lose the creative tension that takes place when people approach an issue from different perspectives. After all, human need provides the strongest motivation for research and understanding. At the same time, a good theoretical discovery has the potential to make the work of an applied geographer both easier and more useful.

For that reason, this book has used examples from the journals of each organization on this list, but it has avoided using any of these terms until this near-the-end-of-the-book chapter. The preceding list of orientations, organizations, and journals is here primarily as a guide for those who are browsing through a library, trying to line up a guest speaker, or deciding which group(s) a new teacher should join (and joining at least one professional organization is part of being a professional, according to one teacher reviewer of an earlier edition of this book!). In those tasks, some guidance in avoiding "cognitive misfit" might be helpful!

Place and Process (Regional and Topical Geography)

The "split" between topical and regional geography may be responsible for the longest list of partly valid stereotypes in geography. Here's one of those stereotypes: regionally organized courses tend to become boring catalogs of features ("since today is Tuesday, we'll be studying Belgium"). Some people think of this tendency as inevitable and propose that geography courses should be organized topically, in which individual units would have titles such as population geography, economic geography, or political geography, rather than regional titles such as New England or South Pacific. A topical approach, however, is also subject to caricature, because in the wrong hands it easily becomes a series of process explanations illustrated with an occasional map. In short, the most effective geography uses both approaches, like the two blades of a scissors:

The analytical scissors of geography

Regional examples and topical ideas working together (see Chapter 2)

Use of regional examples in a course supplies the "puzzle" (the human-interest angle) that makes looking at the maps and analyzing spatial patterns seem worthwhile.

Topical ideas provide the broader view that helps put local facts into perspective.

As noted in Chapter 2, the contrast between regional and topical perspectives is deeper than simple definitions of the terms might suggest. Regional approaches tend to be more local in scale and descriptive in tone. They combine information from a variety of sources to describe what is happening in a specific place. By contrast, topical outlines are usually more global and analytical. The differences are evident in the kinds of maps that tend to be used with each approach.

This classification of maps was one of the core ideas of Chapter 2, but it is so important that a brief review here might be in order. It is useful, at least conceptually, to recognize two fundamentally different kinds of maps:

Review of reference and thematic maps

A **reference map** uses a variety of symbols to show the positions of different kinds of features that occur together in a specific area. This kind of map seldom can show a large area unless the map maker deliberately chooses to omit features. That kind of simplification, however, slowly changes a reference map into the other kind of map:

A **thematic map** uses a few conventional symbols to show the spatial pattern of a limited number of variables, often only one. The size of the area shown on a thematic map can range from a backyard to the entire globe. Most thematic map making, however, should cover a large enough area to show how the theme applies to specific places.

In practice, these two categories of maps usually overlap. A reference map usually has to omit some features to avoid being unreadable. Meanwhile, a purely thematic map would be hard to read without at least some reference information (Transparencies 9C and 9D).

Despite this overlap, the distinction is worth noting, because it highlights the two analytical perspectives that geographers use to focus on what is happening in a specific location.

The Main Point of This Chapter

Do not use halves of these dichotomies as unit titles in a class.

They work best when applied together, not one at a time. That might be hard to do if a textbook breaks the field down into specific topics or regions. Such a textbook organization shouldn't be too troublesome for a teacher who has a specific set of themes (or well-worded behavioral objectives, or assessment ideas). That kind of preparation can easily provide the conceptual cross-links that are needed:

Putting student experience against the textbook in order to create a conceptual scissors

A specific way of asking questions and finding answers

If the text is using general phrases and thematic maps, try to redirect student attention (once in awhile) by asking how the generalization applies in a specific place. For example, a lesson on urban geography can be improved by showing a few photos of distinctive urban neighborhoods, such as a cluster of high-rise apartment buildings. Then ask where that kind of building is likely to be found on the textbook's map of "a typical city." Or, provide a description of a person's activity or lifestyle and ask where it would fit on the map of land use.

If, on the other hand, the text spends most of its time describing specific places, try to redirect student attention (once in awhile) by asking how the places in the text fit into a broader picture. For example, show a transparency of a thematic map of rainfall. Then ask students to identify places on the map where a particular photograph in the text might be found. Ask how the information on the map helps explain what they see in the photograph.

This kind of redirection is especially useful if done with a newspaper or magazine as the "foil." Tie a newspaper account of a flood, for example, to a thematic map of airmass source regions. Or, tie a magazine map of the "nation's best hospitals" to a written description of the philanthropists and doctors who started well-known hospitals, such as the Sloan-Kettering Cancer Institute or the Mayo Clinic. In this way, students begin to realize that these famous places did not arise in a vacuum. Like most features in the landscape, they are products of the interplay of nature and culture in specific places.

Taking Advantage of the Teachable Moment

A detailed list of learner outcomes (e.g., the **National Geography Standards**) allows a teacher to take advantage of ephemeral student interest in particular topics. The key is to scan the Standards and classify them according to the natural "hooks" one might use to get students interested a particular bit of knowledge.

National Standards

Lists of learner outcomes prepared by teams from various academic disciplines (see Postscript 5-1)

Some topics, for example, are well suited to be handled as a response to current events. Different teachers may have different ideas about which topics fit that criterion, depending on their personal experience as well as the nature of the class. Some teachers, for example, find the Standard on plate tectonics much easier to "cover" in the context of heightened student interest just after a major earthquake (Transparency 9E). But if a class spends all its time studying this year's headline event, students might be unprepared for next year's.

The Standard on flows of investment, by contrast, probably fits better as part of a structured discussion of world patterns of production, consumption, energy use, and other aspects of the world economy.

> Caveat: That assertion about the difficulty of teaching about the world economy in response to news events is true, unless your community has the economic misfortune (but pedagogical good fortune!) of a shutdown of a nearby major factory during the class.

There is an obvious danger in either extreme: a completely ad-hoc class "organization" or a rigid sequence of topics. For that reason, let us just call this another pedagogical dichotomy for which the correct answer is usually "both." A teacher might look over a list of desired learner outcomes for a course and classify them into three rough lists:

1. Topics that seem easy to teach (by that particular teacher!) in response to daily news events;
2. Topics that seem best to teach in a prescribed sequence;
3. Topics that could go either way, depending on events, student mood, time of day, day of week, month of year, teacher health, cloud cover, amount of time before or since vacation, and so on.

There are many different ways of combining the skills of geography – finding information, making maps, looking for patterns, comparing places, evaluating infrastructure, and so on – with the list of places and theories that will be "covered" in a class. Teachers should insist on a seeing a matrix, table, or list of those cross-connections when evaluating a book or activity packet. Documentary evidence of this kind of careful forethought is a good clue to the quality, flexibility, and usability of a book!

If the author of a textbook has provided a table or matrix that shows what skills and ideas are taught in the context of what places, the teacher using the book can choose to substitute other skills or places according to need. A teacher of an advanced class can use the same principle to assemble a course out of raw materials from several sources (e.g., the ones listed near the end of Chapter 6 or on the CD, local sources of geographic information).

In short, many of the opposing blades of geography teaching can be interchanged by substituting different places, data sets, maps, or skills for ones specified in the text or workbook.

Teachable moment

Time when students have been "primed" for a specific topic by events that occur outside the classroom

Individual teachers should feel free to choose topics they can comfortably put on their own mental list of subjects to reserve for discussion what some call the **teachable moments**. The key is to treat these topics in the same way as everyday topics. In other words, focus on an interplay of site and situation, place and link, region and topic, theory and application, physical and human features, local and global scales – all the word-pairs that help to capture the essential geographical perspective.

If the perspective stays firm and identifiable, occasional excursions away from the "regular" course outline provide a sense of immediacy and practical value. That, in turn, makes it easier to continue with the planned sequence of geographical stories on the "ordinary" days.

So what is a geographical story? That is the topic for the final chapter in this book. There is, however, one additional "both-and" dichotomy that is important for a geography teacher.

Freedom, the Tao, and Other Existential Dilemmas

Geography is "about" a mundane thing – the spatial arrangement of hills, roads, houses, video stores, murals, and other landscape features. Like leaves on a tree, however, those visible features reflect the nature and vitality of the roots. Those roots are the perceptions, beliefs, and aspirations of the people who live and work in a place.

If cultural "roots" of a place change, the landscape "leaves" in that place will also change. Societies with different ideas and values have created very different landscapes, even in similar environments. That is how landscapes form and change.

It also works the other way around: If the leaves change, the roots will respond. Enforcing a rule that rewards or punishes some kinds of behavior will in due time influence the beliefs of the people affected by the rule. That is how cultures persist and change.

Like the other dichotomies in this chapter, the pedagogical answer to a question about freedom and structure is "both, in balance." From the "jen/t'ien" of Confucius to the "becoming/being" of Luther and Kierkegaard, most human philosophies have converged on an unfathomable mystery: *individual human beings achieve their fullest freedom when they help build (and thus partially lose themselves in) a mutually sought larger project.*

For example, as a research geographer, I long ago concluded that it would be pointless to measure the chemistry of soil in different places if that investigation did not intersect with the issue of social justice. On the surface, I am interested in specific technical questions about soil traits and erosion. That's true, but it is not the whole story. I focus on those narrow issues precisely because they can help us address broader questions about the interplay of environment and mind.

Here are a few examples of those broad questions: would my German ancestors have learned such a strong ethic of saving if they had grown up in an environment where fruit was ripe for the picking every month of the year? What is the role of children's stories such as the Three Little Pigs or the Grasshopper and the Ant in passing this kind of ethic to the next generation? To what degree should we view the much discussed "German" or "Japanese" economic attitudes as the result of growing up in an environment where the soil freezes part of the year (and families who do not think carefully about saving might freeze or starve before spring)?

These are intriguing questions, and answering them probes some of the dichotomies we have been discussing: perception and "reality," nature and culture, physical and human geography. We probably all agree that people often behave on the basis of what they perceive, rather than what is "really" there. And what they perceive depends (often in complex ways) on what they learned in childhood.

A person's "predisposition to perceive" in particular ways, therefore, has been and still is affected by what is "out there."

So, as a research geographer, I study topics such as soil chemistry because I think they have an influence on perception, ethics, and cultural beliefs. I am persuaded that we cannot achieve social justice if we are ignorant of geographic differences in soil chemistry. I come to that conclusion for many reasons, some easy to explain and some that seem quite obscure on first glance.

Let me offer just one example. It is a medical fact that a child whose blood has certain chemical imbalances will probably not develop "normal" brain capacity. It seems to me that trying to prevent that personal calamity is an ethical imperative, especially if it is related to some environmental condition that is beyond a child's control (e.g., parental income, food-purity laws, or even soil chemistry!).

Applying that kind of logic has helped solve some medical mysteries. Records of several childhood nervous disorders, for example, showed them to be more prevalent in urban areas with old houses and a lot of auto traffic. That observation led researchers to measure the level of specific chemicals in the soil and in the blood of young children. Once a medical link between health problems and the amount of lead in a person's bloodstream was found, politicians voted to ban lead in paint and gasoline.

In other words, it took a cooperative effort by many disciplines, each looking at the world through its unique perspective, to solve the problem. An educator tabulated test scores. A geographer made maps of lead in house paint and playground dirt. A doctor studied symptoms in children. A chemist tried to find an anti-knock chemical that could be used instead of lead in gasoline. All of these people could legitimately say that the goal of their work was to help prevent brain damage in children. Moreover, all could say that they made their contributions because of (not in spite of) the fact that they were working within the constraints of an academic discipline. It is another example of the mutually agreed upon self-restraint that is an essential part of true freedom.

Paradigm

Way of looking at the world; way of asking questions

(Wait a minute! Isn't it also true that an academic discipline can become a straitjacket that limits thought? Of course; nearly every discipline can point to times in its history when it ran into a conceptual rut, and then someone came up with a new **paradigm**. The classic example, often cited in books on the history of science, is the "scientific revolution" that occurred when Galileo and Copernicus formulated a new sun-centered view to replace the incomplete and inadequate earth-centered model of the universe.

So, new paradigms may occasionally arise: but in a time of considerable discussion about new paradigms in many disciplines, I also suggest that the number of truly new paradigms is far, far fewer than the number of people who would dearly love to be extolled by generations of admiring followers as originators of new paradigms. In the case of geography, I think we have more than enough good paradigms at the moment – what we need are some more practitioners out there doing good research and teaching! Before I get too far into that particular sermon, however, perhaps I should just admit that although I have plenty of opinions about philosophy, this is supposed to be a book about teaching; I'd better just conclude this chapter and go on.)

So, I am a geographer who measures chemicals in the soil in order to understand more about how people think. Like the medieval craftsman who was digging a hole when someone asked him what he was doing, my answer is, I dig holes in the ground, in particular places for my own reasons, but I am also (at exactly the same time) helping to build a cathedral.

It is that unity of ground and spirit that is so satisfying in the discipline of geography.

(No doubt a similar metaphysical unity exists in other well-founded human endeavors, but this book is about geography!)

The unity of geography is the topic of the next chapter.

POSTSCRIPT 9-1

About Computer Programs in Geography

First, a confession. I need computers, and I like computers. I am typing these words on a computer. All of the transparency masters were drawn with a computer. Another computer I use helps me store and manipulate census data. My wife teaches cartography (map making) and has several computers. A computer goes in our car when we travel. There is even a computer connected to my synthesizer when I'm in the mood for making music.

Some geography computer programs are very good at grabbing student interest. Here are some thoughts on a few of them:

Where in the World is Carmen Sandiego? was a pioneering program that helped put geography back in the public eye. If not balanced with sound geographic theory, however, this kind of quiz program can make geography a trivial pursuit. Blank games such as online *Jeopardy*, which allow teachers to suggest their own questions, have more promise.

Travel planners such as *Automap, Map 'n' Go, Mapquest, Tripmaker,* and others are valuable in planning a trip. But their view of geography is narrow and incomplete, with a strong focus on tourist attractions and categorillas such as the longest suspension bridge and the largest ball of twine.

Sim City, Sim Earth, and their clones can do a good job of emphasizing that decisions have consequences. If these imaginary places are not balanced with facts about real places, however, they can imply that the world is an evil, unpredictable place that is out to get you.

The Oregon Trail and several related programs show how a computer can link history, geography, and decision making in an intriguing way. Students need guidance in trying to transfer these decision models to the real world.

Puzzlemaker, Quizlab, and many similar programs can help teachers make handouts and test forms. The quality of the result depends mainly on the teacher.

A variety of computer atlases, encyclopedias, mapping programs, and websites such as *www.cia.gov* and *www.prb.org* have become standard references for many teachers. Teachers' almost unanimous opinion, however, is that students need guidance in using these references, or their projects may consist of collections of unrelated facts.

The Internet and its many strands are great ways to connect a class with information or peers in other places.

Here's the last item on this incomplete list: Both novice and experienced teachers should make it a practice to consult teacher sites such as *www.about.com*, as well as the sites of professional organizations such as *www.aag.org*, *www.ncge.org*, or *www.ngs.org*.

In short, computers can be a great teaching resource (something I'd also say about cameras, sound recorders, projectors, etc.). Someone, however, must always be monitoring the balance between local and global, nature and culture, perception and "reality," freedom and responsibility. As modes of data storage and delivery become more sophisticated, the teacher's role as diagnostician and "balance fixer" becomes more vital.

If indeed these new technologies have the potential to engender more powerful learning, then it is even more important that someone be watching to keep things in balance. That is precisely what live teachers have to do, conscientiously and skillfully. That means knowing about the computer programs that students are using, understanding their objectives, their trade-offs, and the symptoms of typical student responses to them.

Here's a question: If people can sue a computer maker because they claim to have been injured by a keyboard, shouldn't someone be thinking long and hard about the possible side-effects of powerful educational technologies?

What do you think?

10

Building a Cathedral

What is appropriate for a person to do here (in this place)?

At the scale of an individual, the practical value of geographical knowledge boils down to the ability to answer a single question:

"What is it appropriate to do in this place?"

Site and situation

Two parts of the concept of location (see Chapter 2)

In a more elaborate form, the question can use several different verbs that span quite a range of options: "What kinds of things are people required, expected, encouraged, allowed, tolerated, discouraged, forbidden, or unable to do here?" The answer may depend on any of the factors that help make a particular place what it is – in other words, on any of the **site** conditions or **situational** connections that have an influence on the nature of the place.

At the scale of a community, a worthwhile geographic goal is to locate such things as roads, stores, clinics, schools, and so on in such a way that the result is not only beneficial for individuals, but is also safe, fair, efficient, and beautiful for others. At the national scale, a reasonable goal is to arrange borders and trade so that the effects are efficient and fair. At any scale (community, state, national, or global), the result of answering the basic geographical question is more than just the sum of the choices made by individuals, because those choices can also have effects on other places, both near and far away.

When a decision is complicated or may have effects on many people, one person might not want to make the decision alone. For this reason, decisions about where to locate things are often a group effort of some kind, whether it is in the form of a town meeting, a marketing cooperative, a municipal planning department, a legislative hearing, a treaty, or an international trade agreement. The main goal of these meetings is to ensure that everyone's opinion gets a fair hearing and everyone's rights are respected.

When we study how people actually make choices, however, we soon run into something that has been noted in a number of disciplines. Economists call it "the paradox of the aggregate." Ecologists call it "the **tragedy of the commons**." Other disciplines have their own names for the same basic idea: *something that is good for one individual may be bad if too many do it.*

Tragedy of the commons

What is appropriate for one individual may not be so for many people

For example,

If there is only one cow in a pasture, it can eat all it wants. If there are too many cows, however, the pasture is likely to be overgrazed. Inedible weeds may take over, and all the cows suffer.

If there is only one billboard on a road, it draws a lot of attention and can convey its message. If there are too many signs, the individual billboards get lost in the clutter and fail to communicate.

If there is only one cabin on a lake, it can be a quiet retreat, with a great view and clean water. If there are too many cabins, the result is likely to be a lot of noise, a cluttered view, and polluted water.

In short, there is a proper amount of anything for a given place.

And what is good for one person may be bad if too many do it.

▪ ▪ ▪

This is the basis for a useful form of applied geography – deciding about the number and locations for stores, roads, factories, and other structures. Owners of a chain of video-rental stores, for example, can study maps in order to decide how many new stores are needed and where they should be located (Transparencies 10A and 10B).

At a simplistic level, one might just look for gaps in the pattern of existing stores. A more advanced kind of **location analysis** might consider other conditions such as population density or per capita income (that's one blade of the geographical scissors described in Chapter 2). A still more elaborate study would include an analysis of connections, travel time, and congestion (that's the other blade).

Location analysis

Process of deciding how many things to put in a place and where to put them

We could continue adding variables to this list to show how geographical analysis is an open-ended process, with ever more sophisticated levels of inquiry. This book, however, is about teaching geography, not doing it. Why are we taking this detour into a discussion of the paradox of the aggregate?

Why? Because the paradox of the aggregate also applies to teaching: There is a proper amount of any given approach to teaching. Things that may be very effective if done once in awhile can become ineffective if done too often.

You know the buzzwords: checklist, learner outcome, scaffolding, Socratic approach, Binko method, storyboard, individualization, case study, simulation, CAI, cooperative learning, constructivism, and so forth. And let's not forget "standards based." Whatever the fad, it probably has a persuasive guru with plenty of amazing personal experience being wildly successful using the approach all over the world.

The need for variety in a classroom

Like a single cabin on a lake, almost any good teaching tactic works well for awhile. Classroom teachers, however, know that different students respond to different approaches on different days. Variety is necessary, if for no other reason than to make today into something other than a monotonously predictable continuation of yesterday. At the same time, most students need to be, to some degree, on familiar and predictable ground so that they can develop a sense of mastery.

This balance between predictability and variety is complicated by the fact that the desirable level of predictability varies from one age group to another, one student to another, one day to another, one term to another.

▪ ▪ ▪

How do teachers find a happy medium along this continuum from monotony to excessive variety? That is the challenge of teaching. It is why teachers should view educational gurus with a healthy mix of appreciation and skepticism. It is true that educational research does provide interesting insights. It is also true that teachers should always be experimenting with better ways of getting ideas across. In the last analysis, however, there is no single method that works for all students all the time.

The best approaches are the ones that get students to learn to the best of their ability. That means tailoring approaches to the kind of students that a specific teacher encounters on a given day.

The selection of teaching approaches thus depends on the abilities, background, and preferences of an individual teacher. The choice also depends on the nature of the students. It depends on the time and place. Finally, it depends on how well various approaches can contribute to helping students meet specific learner outcomes.

The purpose of this book was to explore a range of options and to fit them into a coherent perspective on the unique nature of geographical teaching. That exploration can be summarized in one final question, one that has a deceptively simple answer:

Q: What should we use as unit titles in a geography class?

A: Whatever lets us get our themes across most effectively.

Throughout this book, we have advised against using **geographical themes** or skills as unit titles. This is because geographic themes, skills, and approaches resemble the working of a scissors. They are like blades that cooperate to cut through the complexity of the world and isolate useful principles.

Geographical theme

Idea that permeates a discussion and helps define what is important (see Chapter 5)

This can be illustrated with one final case study. To show how the analytical blades of geography can work almost anywhere, I propose to focus attention on what might seem like one of the most boring landscapes in the United States. (If nothing else, it gives me one last chance to underscore the idea that geography is about all places, not just the unusual and spectacular.)

Let us, then, take a drive across the Maumee plain, a numbingly flat expanse of cropland in northwestern Ohio. The tall corn suggests that this land is exceptionally fertile. Meanwhile, there are a large number of long, straight, deep ditches along the road. To a knowledgeable observer, these ditches imply that the fertility of the soil is not easy to exploit. People here have done a lot of work to deal with the conditions in this place, particularly the land's apparent tendency to remain unworkably wet for long periods of time. Someone who knows about the annual cycle of temperature and precipitation would rightly speculate that the worst period is probably right after the spring snowmelt, before the summer heat can evaporate the water.

(That short descriptive paragraph already notes several facts about the physical and human geography of the region: its flat terrain, fertile soil, snowy winters, and accessibility. After all, who would bother to dig deep drainage ditches if the soil were infertile or the fields too far from potential markets? Moreover, each of these facts about this **region** is basically a **condition** that is a result of the **movement** of airmasses, water, tractors, and grain trucks from one **location** to another – all four themes, working together.)

Core themes of geography:

location (position)
conditions (site, place)
connections (movement)
region
(see Chapters 4 and 5)

The farmhouses in this area are smaller and not as old as those on the hillier land both east and west of the lake plain. The frontier of European settlement swept over this area in the early 1800s (Transparency 3M). Actual occupation of the lake plain, however, was delayed until some German immigrants arrived in the middle to late 1800s. These people from the north European plain already had the skills, community organization, and work ethic needed to dig the ditches and install drainage tiles across several hundred square miles of wet clay soil. By that time, however, the network of roads and railroads had matured to the point that families did not view their farmstead as the center of their lives. As a result, the houses are more modest than the tall Federalist and Italianate "mansions" one sees on farms in other parts of Ohio and Indiana.

All of this is background for a puzzle that is the core of this case study:

Why build a diving board on an ugly little square pond?

As you drive across the lake plain, you might notice a small pond or lake near some of the railroad overpasses or highway interchanges. Look closely at one of the ponds. It is probably square or rectangular, not a very natural-looking shape. You might also see some mobile homes or sheds near the shore, maybe even a boat dock, a beach, a water ski ramp, or a diving board.

We can use the puzzle of these features to show how all of the analytical blades of geography come into play. Why would people build boat docks, beaches, and even water ski ramps on artificial ponds not much bigger than a football field? Looking for the answer to this question makes you think about many different consequences of the flatness of the terrain.

What are some of those logical consequences of flatness?

Thematic map

Shows the pattern of a specific thing in a large area (see Chapter 2)

1. In spring, the entire area looks like a shallow lake, before the water from melting snow can drain away or evaporate. This simple sentence actually draws on information gathered from **thematic maps** of the terrain, precipitation, and temperature. In other words, having mental maps of the climate and the former location of Ice Age lakes can help with the interpretation of a local view from a car window in northwestern Ohio. (See Transparencies 3I, 3K, and 10C. Better yet, lay the transparencies on top of each other to make the relationships clearer).

Image, theory, value

Multiple strands of meaning (see Chapter 3)

2. By late summer, the area looks just like a huge flat farm field. One can reasonably assume that it has all of the exciting recreational opportunities associated with huge flat fields. That is a geographic fact, but it is also a **value judgment** for many people.

Infrastructure

Roads, powerlines, and other features built to facilitate links between places (see Chapter 4)

3. Interstate highways are supposed to go over or under other roads, rather than meet them at a stoplight or uncontrolled intersection. The purpose of this rather expensive kind of **infrastructure** is to promote rapid movement of people and goods. That helps connect places with the world economy,

which in turn can make land near the highway exits more useful and valuable.

4. Digging a tunnel under another road is not wise in a place where spring floods cover the flat land. Road underpasses are likely to fill with water unless they are pumped out. Moreover, there is really nowhere to pump excess water on this flat land.
5. A bridge is the answer. To get enough dirt for the approaches to the bridge, engineers have to "borrow" from nearby fields. Since those fields are pool-table flat, the removal of dirt inevitably leaves a pit.
6. The pit fills with water. That also is hardly a surprise, because when it rains, water cannot flow across the flat land, and it cannot drain downward through the sticky clay soil. It has nowhere else to go.
7. Finally, even a small area of open water is better than none in an area that has few recreational options. For that reason, people make use of the "resource" created by the highway engineers. To do this, they build boat docks, water ski ramps, diving boards, artificial beaches, and even vacation homes on the tiny ponds. You can see those features from a car window. They are also visible on **reference maps** of local areas (Transparency 10D).

Reference map

Shows the locations of a variety of things in a specific area (see Chapter 2)

These small rectangular ponds thus can be viewed as the landscape consequences of reasonable human responses to the environmental conditions in a particular location.

These ponds also provide one last illustration of another basic geographic principle: *the same words can mean different things in different regions*. Despite the ingenuity of the people who built parking areas, changing rooms, diving boards, picnic shelters with vending machines, and even water ski ramps on the ponds, there is no way that the word "beach" can mean exactly the same thing here as it does in Daytona, Florida.

The same word can mean different things in different places (see Chapter 3)

People use the little Ohio ponds for swimming and sunbathing only because Daytona is too far away for easy commuting after school or work. That is yet another geographical principle in action – *the cost of overcoming distance has an obvious effect on behavior.* In this case, distance exerts a severe financial penalty on anyone who would try to visit a Florida beach too often from Ohio.

In short, geography matters. It has an influence on what people can do in a place, and on where they choose to do particular things. If we can illustrate that principle in the "boring" flat lake plain of northwest Ohio, we can apply it in practically any environment on earth.

People who have a good stock of knowledge about other places also have a solid basis to judge proposals for doing specific things in the place where they live. As we said in Chapter 1, geography is the art and science of knowing why something we know, for sure, in one location may be wrong in another.

The flip side of that coin is that geography is the art and science that can help us understand why people do what they do in other places. This understanding can then guide us in choosing the best of those ideas and applying them to help solve problems where we are. Or, even if we conclude that we cannot use an idea, we can at least appreciate how other people solved some of the problems they encounter where they live.

One more thing: you do not have to pay too much attention to graffiti in northwest Ohio. Painted marks on bridges do not carry nearly as many ominous meanings in rural Ohio as they do in parts of Los Angeles. That is not a bad trade-off for someone growing up in a place where the word "beach" means fifty feet of muddy sand on a tiny square pond next to a heavily traveled highway. The simple fact is that places are different, and different kinds of behavior are appropriate in different places.

The study of geography is the way in which human beings organize and communicate this information to each other.

▪ ▪ ▪

Here's one last "rationale-sentence" for geography – if you look at what a geographer says about a place, such as northwest Ohio, you might conclude that geography is the science of the perfectly obvious. So be it – the whole point is that something that is perfectly obvious in LA is not necessarily so in northwest Ohio.

APPENDIX

Some Facts Every Geography Student Should Know for Perspective

Latitudes of some U.S. places and international equivalents

__ - your home town

55 – southern Alaska - Moscow (also northern Ireland, southern tip of Chile)

49 – Montana/Canada border - Paris (northern Mongolia, south of New Zealand)

40 – Philadelphia - Beijing (North Korea, middle of Spain, southern Italy)

35 – north of Los Angeles - Tokyo (Spain, Tehran, Buenos Aires)

30 – New Orleans - New Delhi (Kuwait, middle of South Africa)

25 – tip of Florida - Taiwan (central Egypt, middle of Australia)

22 – Honolulu - Rio de Janeiro (Mecca, Hong Kong)

Low-high temperatures, winter and summer, and annual precipitation in inches

your home town – (____ to ____ in January, ____ to ____ in July, ____ inches)

East – Miami (59–75, 76–89, 56 inches), New York (26–37, 69–84, 42 inches)

Center – Dallas (36–55, 75–94, 37 inches), Minneapolis (3–21, 63–83, 28 inches)

Interior – Phoenix (40–66, 80–105, 8 inches), Denver 16–43, 59–88, 15 inches)

West – Los Angeles (47–67, 65–84, 15 inches), Seattle (35–45, 55–75, 37 inches)

Half a dozen Distances

two prominent local features that are one mile apart – ________ and ________

two prominent local features that are 100 miles apart – ________ and ________

distance across your home state – ________

New York to San Francisco – a bit less than 3,000 miles

Canada border to Texas tip – a bit more than 1,500 miles

around the world – about 25,000 miles

Half a dozen Areas

your home state - ________

a prominent local feature that is one square mile in area - ________

Connecticut - about 5 thousand square miles

Alabama - about 50 thousand square miles

Texas - a bit more than 1/4 million square miles

the United States - about 3 1/2 million square miles

Half a dozen Populations

your home town - ________

your home state - ________

the largest urban area in your state - ________

the United States - 290 million

the world in 2004 - 6,300 million

the world in 1900 - 1,400 million

A dozen Amounts

the GNP of the United States - about 10 million million dollars

the U.S. Federal Budget - about 2.2 million million dollars

the military budget - about 500 billion dollars

annual interest on the national debt - more than 300 billion dollars

average income per person in the United States - more than 36 thousand dollars

average income per person in your home state - ________

the official poverty line for a family of four - about 18 thousand dollars

annual budget of your entire school - ________

value of three-bedroom house in your home town - ________

cost of a typical pickup truck - about 25 thousand dollars

cost of a front-line fighter plane - 6 million dollars

tuition and fees at a good private college - 17 thousand dollars per year

Half a dozen Rates

unemployment during the worst of the Depression - more than 20 percent

unemployment at present - about 5 percent (this can change quickly)

inflation at present - about 3 percent (this can change quickly)

the prime rate - about 5 percent (this can change quickly)

population growth rate and doubling time, United States - 0.6 percent, double in 120 years

population growth rate and doubling time, world - 1.4 percent, double in 50 years

A dozen Percentages

agriculture as percent of national income - less than 2 percent

percentage of cropland in the United States - about 19 percent

percentage of cropland in your home state - ______
manufacturing as percent of home state income- ______
manufacturing as percent of national product - about 15 percent
government as percent of national product - about 13 percent
people on military payroll in home state - ______
people on military payroll in nation - about 2.5 million, 2 percent of total
foreign-born as percent of the total population - about 11.5 percent
female income as percent of male in several occupations - ______
southern wealth - 32 percent of nation before the Civil War
southern wealth - 12 percent of nation after the Civil War

Half a dozen Trends and Miscellanies

a good corn yield - 140 bushels per acre
average commuting time in your home town - ______
average life expectancy in 1776 and 2000 - 38 and 74 years
average income of someone between the ages of 25 and 34 in 2000
without a high-school degree - $23,000
with a high-school degree - $33,000
with a college degree - $59,000

Many of these figures came from the *Statistical Abstract of the United States* (available from the Superintendent of Documents, Washington, DC, 202-783-3238) and from standard almanacs - it is a good exercise for students to update these figures, many of which are probably out of date by the time you get this book!

Illustrations (Transparency Masters)

As noted in the text, the illustrations in this book have been designed as copier-ready transparency masters, so that a teacher could use them immediately in an elementary- or secondary-school classroom. The "caption" for each graphic is an outline of a classroom activity that could be based on it. In most cases, these activities were designed to stand alone (no one should need prerequisites or other materials). Deep inside, however, we do not believe that a geography class should consist of a bunch of stand-alone activities (much more about that later). For that reason, we sometimes suggest ways in which the Transparencies can be combined to reinforce each other, perhaps even over many months.

The transparencies are included in this printed text for convenience – if you have a copier that is capable of enlarging an image by about 25%, you can make transparencies directly from this book. The accompanying CD, however, has full-size .pdf graphic files, which you can print on any computer that can handle the accompanying software. It also has a folder that has all of the captions as text files, which you can load into any word processor and edit for your own notes or for classroom use.

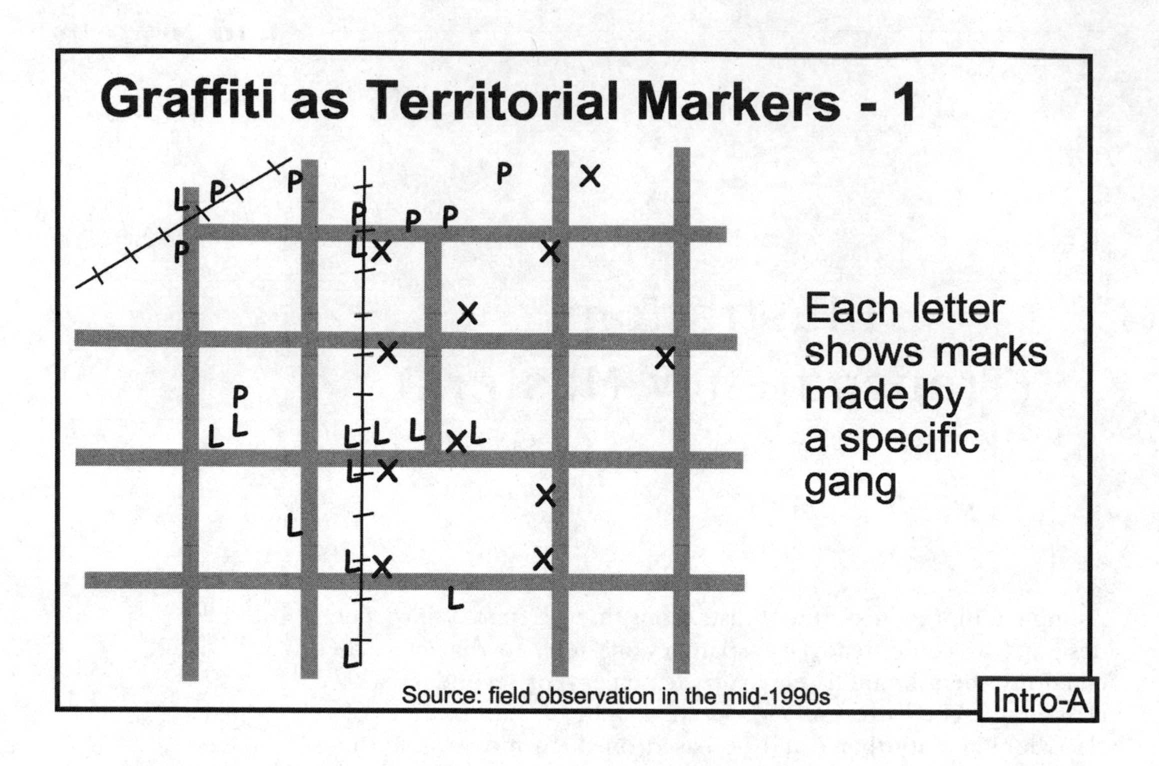

Graffiti as Territorial Markers - 2

Marking territory -- "this is ours"

Disputing territory -- "no, it's not"

Warning -- "no trespassing"

Threat -- "we know where you live"

Intro-B

Teacher's Guide for Transparencies Intro-A and Intro-B

A point-symbol map is a simple and straightforward kind of map. To make it, a feature in the real world is identified and a symbol is put on the map to show its location.

In this transparency, the feature is a particular kind of graffiti painted on a wall or bridge abutment. The map is adapted from one I saw in a Los Angeles police station in the mid-1990s (letters and some details are changed).

At that time in that particular part of Los Angeles, graffiti often were territorial markers – they indicated which gang claimed control of an area. The graffiti "code" had three parts:

1. A member of one group "signs its name" to a wall or bridge by painting a symbol or some letters that the gang used as its "tag."
2. A member of a group writes over another group's tag with a symbol to indicate disagreement. In this part of Los Angeles and at the time of the creation of this particular map, one gang drew an X over a rival gang's tag if they wanted to show that they denied the other gang's claim to the area. An asterisk goes one step further, announcing that anyone who strays over the line risks being beaten or even killed. An infinity sign was the strongest statement – it basically extends the threat to include families of rival gang members.
3. A special letter code returns the challenge (saying, basically, "same to you"). An intriguing example was the sign in front of a police station in East Los Angeles. This sign had an engraved "C/s" in one corner; that was local gang code for "same to you." It was, presumably, written by a cop and warned gangs not to deface the police sign because, as one cop said to me, "I suppose you could say we're the biggest and baddest gang around."

Meaningful signs were often hidden in a mass of what could fairly be described as run-of-the-mill graffiti. Over 90 percent of the graffiti had no special significance. This is a way graffiti are like a microcosm of the real world; some landscape features are very revealing but most have no special message.

That, by the way, is a good reason to study geography: to learn what landscape features carry messages that are worth knowing.

Activity: Have each student make a point-symbol map. The topic can be practically anything – fire hydrants, pizza restaurants, pickup trucks, eagle sightings, dust bunnies under a bed, peas on a plate, and so on. Start with a base map, a sketch that has some key features, such as major roads, buildings, walls, the edge of the plate. Then add point symbols (dots, X's, circles, triangles, etc.) in the proper locations to represent the feature(s) you want to map.

Then follow up with a simple test of whether the point-symbol map is well made: Have another student read the map, go to a place indicated on it, and check whether the feature the map says is there is really there. As a test, this can be remarkably difficult, because a graphic "language" that makes sense to a map maker does not always communicate a clear message to a map reader.

That is why the National Geography Standards and most state geography or social studies standards have a large section on map skills. Map making and map reading are essential skills that help us understand places, but students have to "learn the code" in order to use the skills.

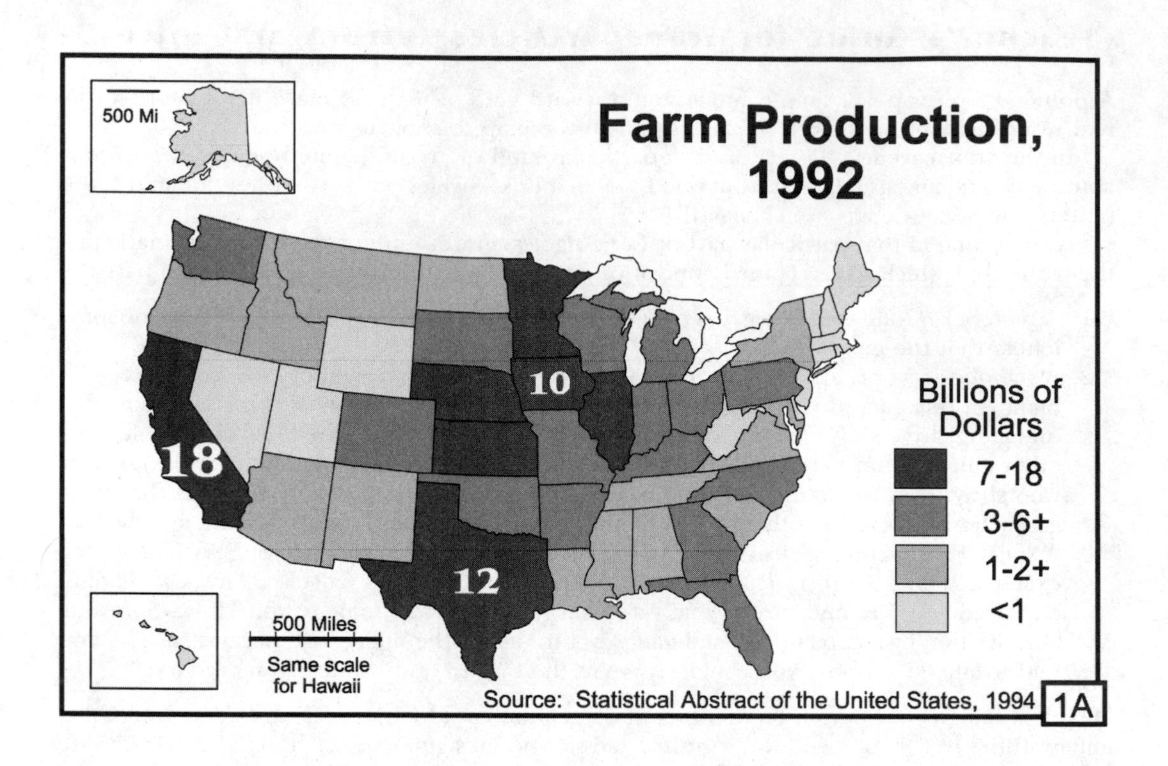

500 Mi
Farm Production, 1992
18
10
12
Billions of Dollars
7-18
3-6+
1-2+
<1
500 Miles
Same scale for Hawaii
Source: Statistical Abstract of the United States, 1994
1A

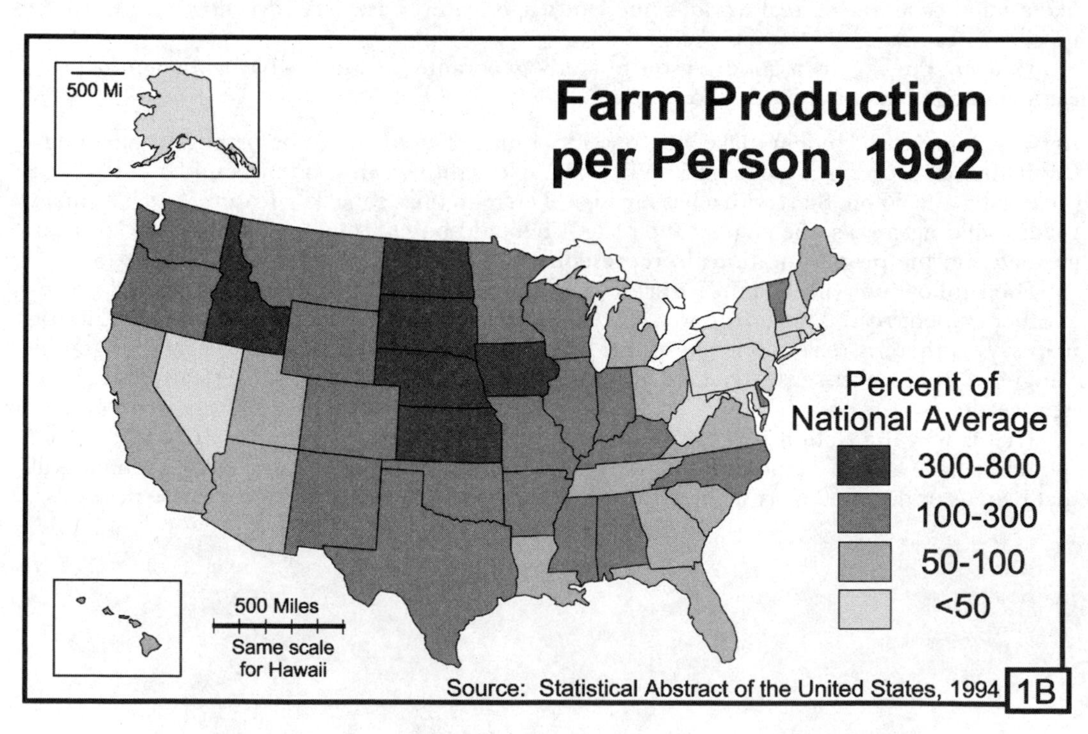

500 Mi
Farm Production per Person, 1992
Percent of National Average
300-800
100-300
50-100
<50
500 Miles
Same scale for Hawaii
Source: Statistical Abstract of the United States, 1994
1B

Teacher's Guide for Transparencies 1A and 1B

Transparency 1A is a bad example (and 1B is a good example) of a choropleth (CORE-oh-pleth) map.

The choropleth map "language" is used to show information that is collected and reported for political units, such as countries, states, counties, or election districts. Examples of such information are census data, crime rates, percentages of people who voted for a particular candidate.

Reporting information by political areas does not tell us about the internal arrangement of an area. To avoid map readers' being misled, therefore, cartographers have adopted three conventions about how to make choropleth maps:

1. Use choropleth maps for ratio data (numbers computed by dividing two counts, such as the amount of production and number of people, in order to get a ratio: production per person). That is what Transparency 1B does right. Map 1A uses choropleth symbols to show counts. This can be misleading, because states with smaller population might actually have greater production than a high-population state, but their lighter color implies otherwise. It would better to use scaled symbols (e.g., circles of different sizes) to show counts, because readers can visually compare the size of a graduated symbol with the size of a state to get a more accurate picture.
2. Pay attention to category boundaries. We put numbers on Map 1A to show actual production in the top three states. How different would the map seem if the darkest category had only states with more than 10 billion dollars of farm products? This question illustrates an important point: a map maker has plenty of opportunities to make choices that are not "wrong" but still alter what the map conveys. That realization is a useful learner outcome.
3. Colors should follow a logical sequence. The conventional rule is to represent low values with light colors and higher values with progressively darker colors.

Activity: Have students find choropleth maps in magazines and newspapers. Did the authors of the maps follow these three basic rules? You'd be surprised how often they don't. If you find a particularly good (bad!) example, cover the legend and ask students which areas have the highest values. Discussion should lead to two conclusions:

- A map with poor category choices and/or a nonintuitive color sequence may accurately portray each individual area, BUT
- A frequent need to consult the legend makes the reader work harder to see the larger pattern.

Since the clear depiction of spatial patterns is a major goal in making maps, students should wonder whether someone who chooses awkward categories or a bad color sequence is just ignorant or is deliberately trying to mislead.

Activity: Give students a table of numbers for an area with about 20 sub-areas, and have them make choropleth maps with different numbers of categories and boundaries between categories.

For a how-to manual with interesting data, get *Thematic Maps: Visualizing Patterns*, by Carol Gersmehl, prepared for the Tennessee Geographic Alliance summer institutes and used in many other states. I freely admit that I "borrowed" many ideas from this teacher-tested manual. I also realize that the one-page discussion here is sketchy, but remember, this page has just two of 85 example transparencies; keep reading!

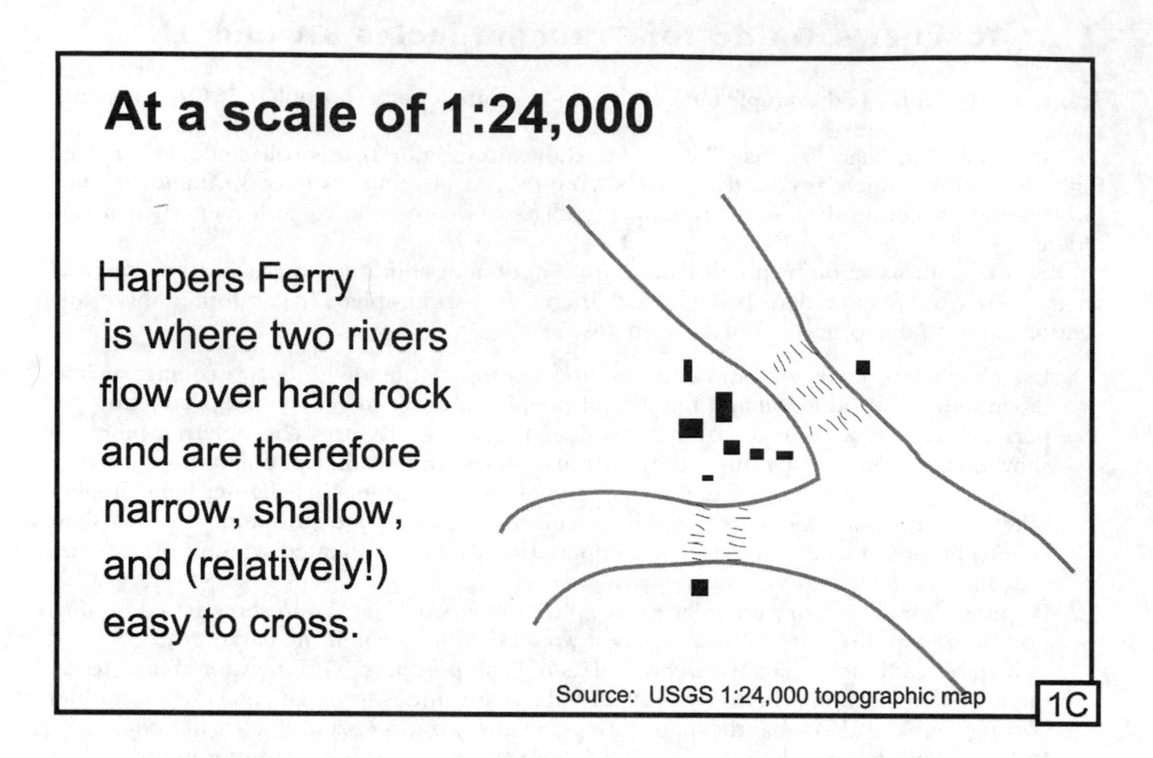
At a scale of 1:24,000
Harpers Ferry
is where two rivers
flow over hard rock
and are therefore
narrow, shallow,
and (relatively!)
easy to cross.
Source: USGS 1:24,000 topographic map
1C

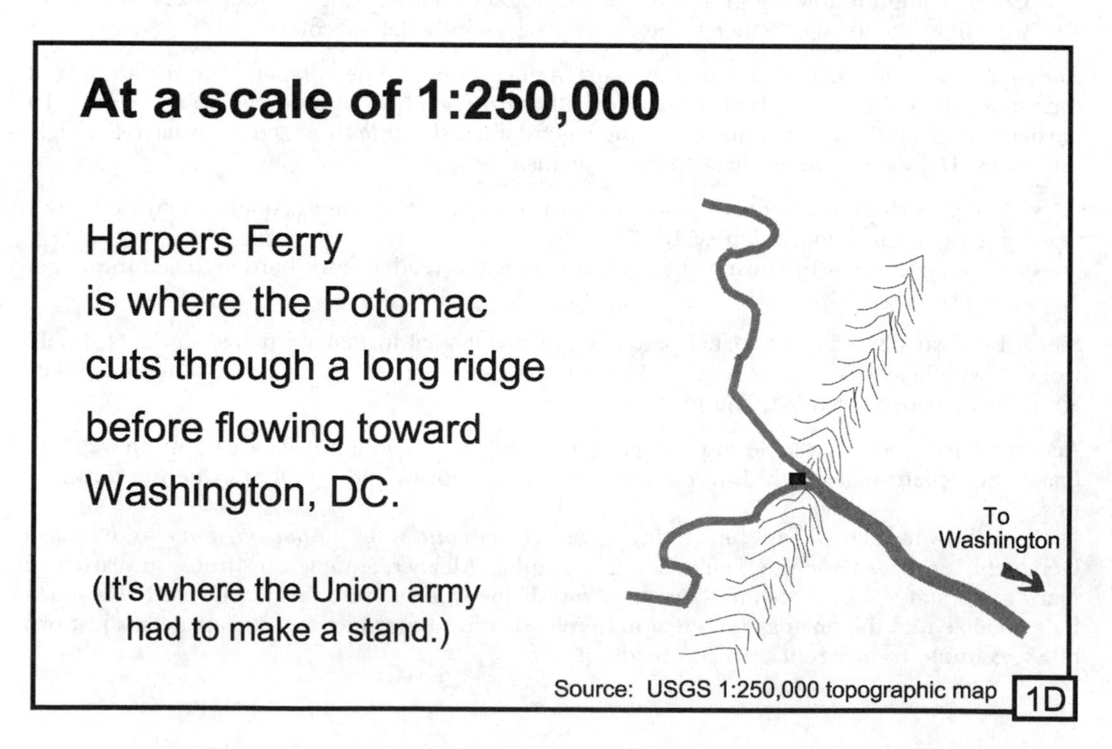
At a scale of 1:250,000
Harpers Ferry
is where the Potomac
cuts through a long ridge
before flowing toward
Washington, DC.
(It's where the Union army
had to make a stand.)
To
Washington
Source: USGS 1:250,000 topographic map
1D

Teacher's Guide for Transparency 1C

This map is a greatly simplified 1:24,000 topographic map. Topographic maps like this are available for most parts of the United States; go to *www.topozone.com* or write to the Map Distribution Center, U.S. Geological Survey, Federal Center, Denver, CO 80225, for a (free!) index map of your state, which you can use to select maps for purchase.

Activity: Put a topographic map of your local area on a bulletin board – this can prompt lively discussion of subjects being covered in class and subjects that students generate on their own. Hopefully, such a map will show familiar details in a new light.

A key to successful use of topographic maps is to know the advantages and limitations of each scale. Maps with a scale of 1:24,000 (the fraction means that an inch on the map is 24,000 inches or 2,000 feet in the real world) have a wealth of fine detail but cannot show more than a few dozen square miles at a time. If you have a lot of display space, tile several of these maps together to show a larger area. A subject the size of Atlanta, Georgia, and its suburbs, for instance, would take up more space than most teachers probably have available.

The answer to this dilemma is to find other maps at different scales. The U.S. Geological Survey also makes available maps at scales of 1:50,000; 1:100,000; 1:250,000; and 1:500,000.

Teacher's Guide for Transparency 1D

At a scale of 1:250,000, a topographic map is useful for planning how to get from one place to another. Maps a this scale show major roads, rivers, terrain features, and forest regions. As you can see on this transparency, however, a 1:250,000 map cannot show individual houses or details such as the rapids in the river. In other words, someone looking at a map of Harpers Ferry at a scale of 1:250,000 will miss precisely what seemed most strategically important at a scale of 1:24,000, namely the presence of shallow water that makes the river easy to cross.

This raises an intriguing and very important point: much of what we "know" about the world depends on the scale at which we view it. It is therefore wise to examine places at several scales.

Activity: Have students gather as many different maps as they can find for a specific area. Look at maps in newspapers, county and city offices, Chambers of Commerce, travel brochures (e.g., in rest stops and plazas on Interstate Highways), and road maps.

Then, post the maps (or copies of them) in scale order, with the most detailed maps at one end and maps that cover an entire state or country on the other. Ask students to examine the maps and tabulate what kinds of information appear at each scale. The flip side of that coin is a good test question: "What scale map should you ask for if you want to find out about the pattern of houses in Roseville? the distance from Moscow to Berlin? and so on?"

Extra: The DeLorme Company (*www.delorme.com*) is in the process of issuing books of topographic maps for each state; about 40 states are complete. Check a local bookstore if you live or travel in those states. Those books are great resources!

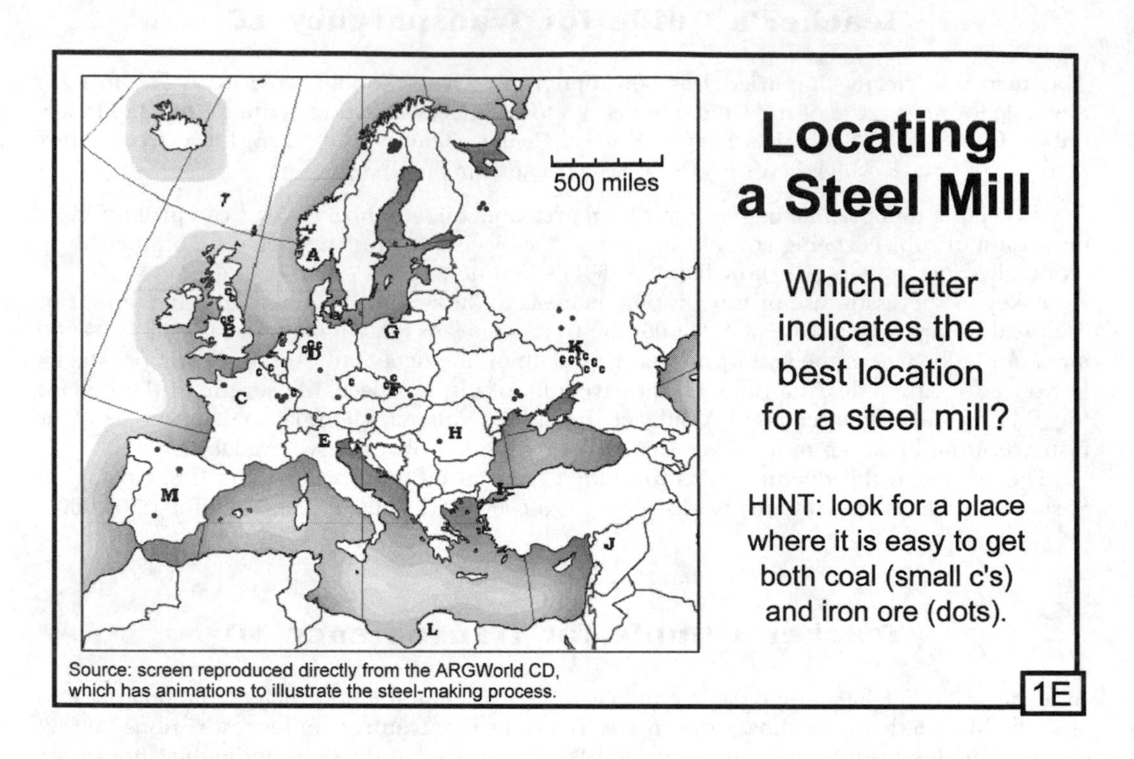
Locating a Steel Mill
Which letter indicates the best location for a steel mill?
HINT: look for a place where it is easy to get both coal (small c's) and iron ore (dots).
500 miles
A
B
C
D
E
F
G
H
I
J
K
L
M
Source: screen reproduced directly from the ARGWorld CD, which has animations to illustrate the steel-making process.
1E

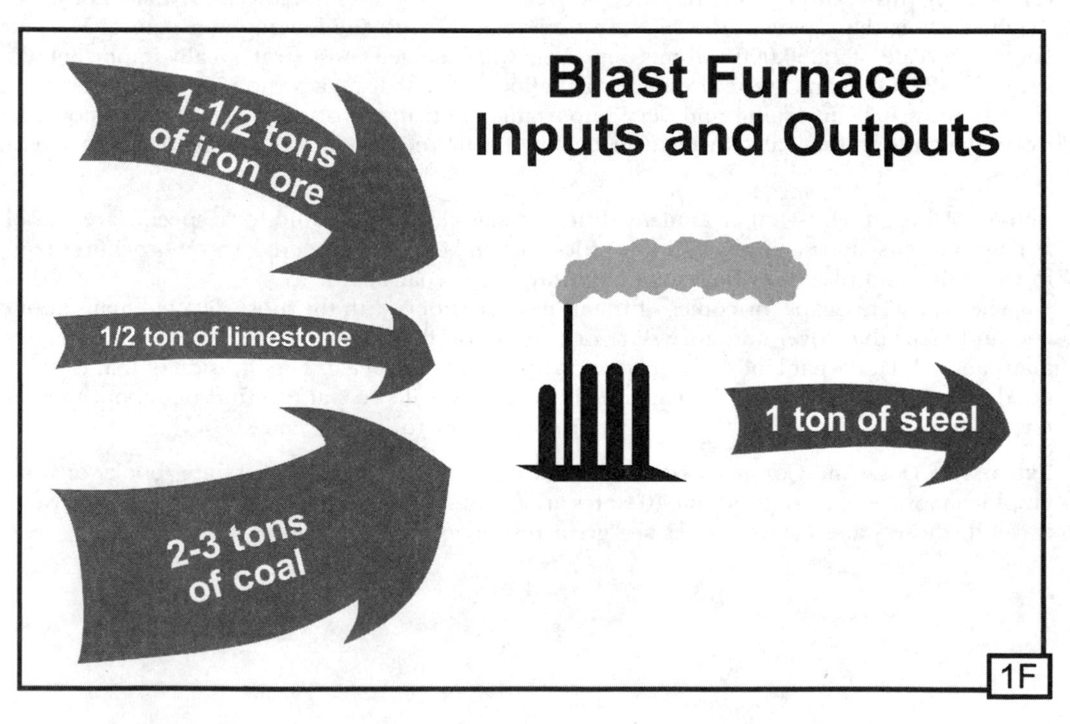
Blast Furnace Inputs and Outputs
1-1/2 tons of iron ore
1/2 ton of limestone
2-3 tons of coal
1 ton of steel
1F

Teacher's Guide for Transparency 1E

Iron and steel mills were big business, particularly in the late 1800s, like textiles in the early 1800s, autos in the mid-1900s, and computers in the late 1900s. Each of these industries employed, and continue to employ, millions of workers. Moreover, production in these industries required the purchase of many other kinds of products, and they provided "raw" materials for a great many other goods and services. As a result, during their peak each of these industries had a great influence on patterns of wealth and population.

If you want to understand the world economy of the late 1800s, therefore, you have to look at where iron and steel were made.

This map shows part of Europe. It is deliberately drawn with very faint lines for national borders, because in fact these borders changed several times. Moreover, many of the international tensions that led to wars in the 20th century involved control of key transportation routes and the raw materials for iron and steel mills.

Activity: Show the transparency and ask which lettered places would be good locations for steel mills. The names are not important for this activity, but in order from A to I, the places are a mountain village in Norway; Birmingham, England; Paris, France; Essen-Dusseldorf (the Ruhr Valley of Germany); Verona, Italy; the western end of the Carpathian Mountains; the port city of Gdansk, Poland; Budapest, Hungary; Istanbul; Aleppo, Syria; the Donetsk area of Ukraine; Banghazi, Libya; and Madrid, Spain.

Of those, D was and is the largest steel center; it is located close to major coal deposits and Is directly downstream from the largest iron ore mines in Europe. Places B and K are also important. To make this Activity more analytical, read on.

Teacher's Guide for Transparency 1F

How to locate a steel mill is a decision involving trade-offs. The recipe for steel is simple: mix coal, iron ore, and limestone. Cook well, making sure you don't burn yourself! Then ship the steel to the customers. Unfortunately, the raw materials and buyers are usually in different places. As a result, people building a steel mill must pick a compromise location.

This transparency graphically shows that a blast furnace needs more coal than ore, and therefore it usually makes sense to put it close to coal mines. Moreover, since it is cheaper to transport heavy material in boats than by other means, it usually makes sense to put mills on major rivers or in ports. (Think of the locations of famous steel mill areas in the United States: Pittsburgh, Cleveland, Detroit, Gary, Chicago, and Baltimore are all on navigable water, and they all are much closer to sources of coal than to iron ore deposits, which are in northern Michigan, Minnesota, and Brazil.)

Activity: Use this diagram to aid a discussion of Transparency 1E. Then, if desired, have students try to draw a similar diagram showing the inputs and outputs of another industry.

> Caution: Steel production is a "heavy" industry; the raw materials and final product weigh a great deal and are expensive to transport. As a result, people try to locate factories between the sources of raw materials and the market. Some industries have raw materials and products that are light and easy to transport, so that factories can be located almost anywhere.

Activity: Have students try to classify specified industries as heavy or light; for example, oil refining (H), computer assembly (L), copper smelting (H), diamond cutting (L), grain milling (H), and sewing shirts or backpacks (sort of in between). Heavy industries are usually located close to raw materials. Most industries in the light and in-between categories tend to be located closer to other inputs such as skilled labor, capital, clean water, and so on.

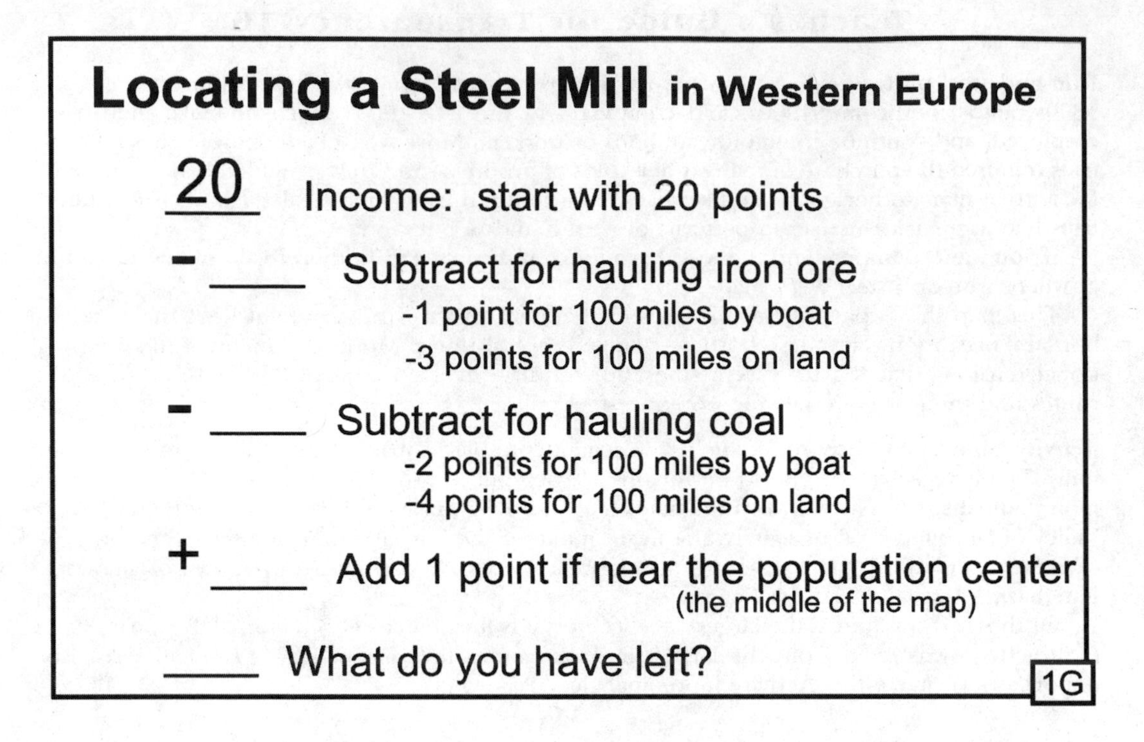
Locating a Steel Mill in Western Europe
20 Income: start with 20 points
- ____ Subtract for hauling iron ore
-1 point for 100 miles by boat
-3 points for 100 miles on land
- ____ Subtract for hauling coal
-2 points for 100 miles by boat
-4 points for 100 miles on land
+ ____ Add 1 point if near the population center
(the middle of the map)
____ What do you have left?
1G

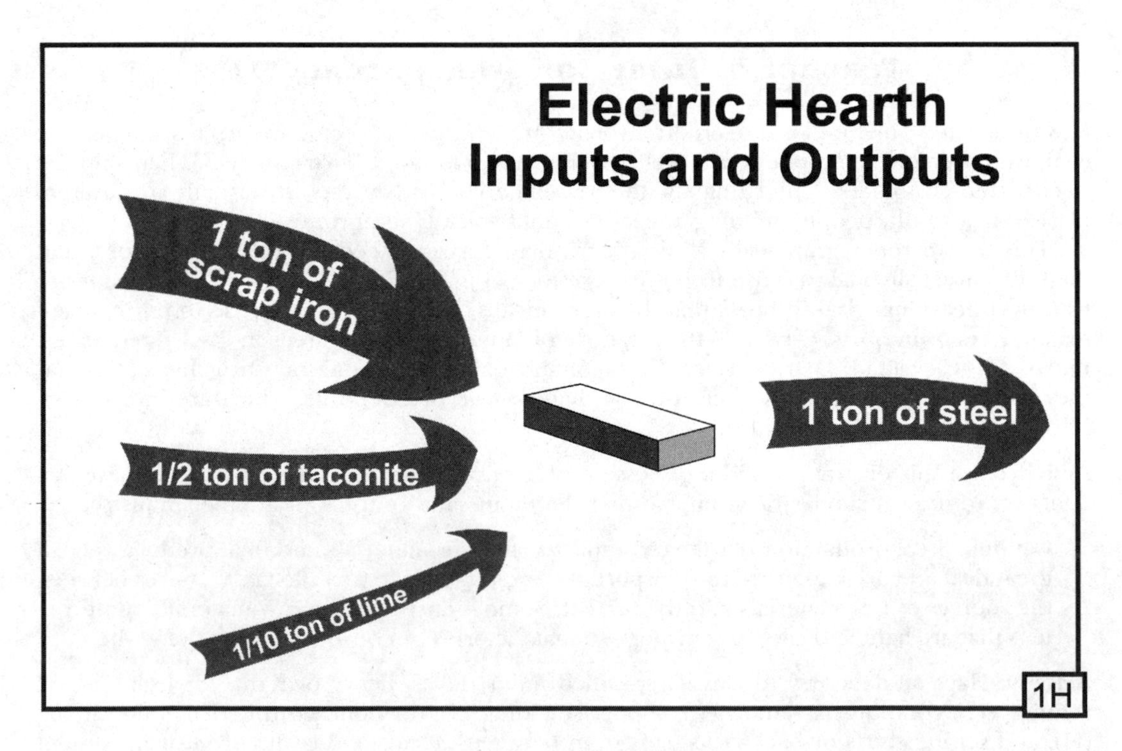
Electric Hearth
Inputs and Outputs
1 ton of scrap iron
1/2 ton of taconite
1/10 ton of lime
1 ton of steel
1H

Teacher's Guide for Transparency 1G

A key to finding a good location for production in a heavy industry is to minimize the cost of transporting raw materials. Assigning "penalties" to potential sites is a way to make analysis of this factor more rigorous than simply "eyeballing" a map.

Activity: Use the map in Transparency 1E, but with the following calculations. Assign groups or individual students to figure out the penalties for transporting materials to each lettered site:

1. Estimate the distance.
2. Note whether the route goes over land or by boat.
3. Calculate the transportation penalty.

If this seems too easy for your students, you can add complexity in several ways. Add penalties for crossing mountains or national borders. Add penalties for areas with severe air and water pollution, such as the Ruhr or middle Elbe. Have them do research on mining and add penalties for using low quality coal or iron ore.

Teacher's Guide for Transparency 1H

With Transparency 1F, this diagram helps introduce some really knotty policy questions:

What should people do when changing technology has made their town a less desirable location for a particular kind of industry?

What should people do when doing "what they always have done" no longer makes them able to compete in the international economy?

Phrasing a question in a particular way tends to restrict the range of options that might be considered. This is important for students to learn, because the two questions above really address different aspects of the same issue. Emphasize that a decreasing ability to compete may not be anyone's "fault." The human tendency is to seek a scapegoat – to blame factory owners, labor unions, recent immigrants, or government policy for something that may be nothing more (and nothing less!) than a geographic consequence of a change in technology.

It may be useful to put the question into a larger perspective: a change in technology has made steel production more efficient and steel less expensive. Consumers all around a particular country have gained, though producers in a few areas have been hurt. A humane society should take this geographic fact into account in designing policies for dealing with unemployment and migration.

Activity: Assign roles (e.g., mayor, labor leader, minister, storekeeper, etc.) for students to play in simulating a town that loses its major employer.

Transparencies 1E–1H can help a teacher show why a steel mill using older technology in one place may not be able to compete with a mill using new electric furnaces in a location well suited to that technology. Or, if desired, modify the simulation above to fit some other economic activity, such as a textile mill, a meat-packing plant, a drive-in theater, or a small video store threatened by a new superstore. Have students discuss the threatened business's options: moving away, retraining for other work, rebuilding, and so on. Have them prepare maps to illustrate the options. The key question is: Can the proposed new activity work in the current environment? This is *the* big question of economic geography; it can take years to learn how to answer it well. Four transparencies aren't enough, but we might have learned enough to say whether a society is in trouble if its citizens graduate from school and still see no need to think about the locational consequences of technological change!

Land per Person

	Million acres	Million people	Acres per person
Argentina	686	38	18
Bangladesh	36	131	0.3
Canada	2465	31	80
United States	2263	280	8
Mexico	488	101	___
Germany	88	82	___
Japan	93	126	___

Source: CIA Factbook, 2003

1I

Cropland per Person

	Million acres	Million people	Acres per person
Argentina	62	38	1.6
Bangladesh	22	131	0.2
Canada	123	31	4.0
United States	453	280	1.6
Mexico	59	101	___
Germany	28	82	___
Japan	10	126	___

Source: CIA Factbook, 2003

1J

Teacher's Guide for Transparency 1I

The basic idea of "per" data is simple: a quantity of one tangible item is divided by the number of individuals, or units, it is spread among.

It must be admitted, however, that the real-world use of per data can be confusing, especially if we are not sure what is being divided by what. This raises a difficult question: how can students accurately understand a newspaper or television story if they cannot figure out how the tables or maps of per data were made? It helps, then, in teaching to use per data frequently, so that students gradually improve their ability to interpret the results.

Activity: Project this Transparency and have students fill in the blanks by dividing the population number into the total land area number. Since both numbers are expressed in millions, the millions "cancel out" and the result is the number of acres per person. Work through some of the examples that have already been calculated – the answers for the blanks are 5.3 for Mexico, 1.1 for Germany, and 0.7 for Japan.

Once students get the idea, you can extend the discussion. For example, is the number of acres per person the most useful per number, or would something else be more meaningful? It might be noted that people cannot grow food in deserts or mountains. It might therefore be useful to know the proportion of the land that food could be grown on before performing the calculation. A transition to Transparency 1J would be to write the following numbers next to the country names: 9, 62, 5, 20, 12, 32, 11. These numbers indicate the percentage of land area that can be used for crops (right, that's another per number!). Multiplying Argentina's 686 million acres of land by 9 percent (9/100) gives a result of 62 million acres of cropland. The answers to this calculation are in the first column of Transparency 1J.

Teacher's Guide for Transparency 1J

Repeat the recipe for "per" data: "take the quantity you want to measure and divide it by the number of units." Agricultural land, for example, is meaningful in terms of how many people must be fed. A relevant per measure, therefore, is the number of acres per person (the number of acres of cropland divided by the number of people).

Activity: Project this Transparency and have students calculate the number of cropland acres per person (0.6 for Mexico, 0.3 for Germany, less than 0.1 for Japan; underscore that the amount of cropland per person in Japan is less than 1/20th as much as in the United States).

These numbers, as Postscript 1-1 says, are one reason Americans have not had to be as concerned as people in other countries about what I am calling "values" issues in geography, such as questions about the efficiency and equity of the use of land and the distribution of infrastructure. Compared with those of other industrial countries, residents of the United States have always had plenty of cropland for food and plenty of "empty" land for buildings, roads, parks, waste disposal, and other uses. As population increases, however, land issues become more important, which is one reason for the present upsurge of interest in geography in American schools.

It is important to keep in mind that all land is not equally productive. This observation leads to another kind of per data: yield per acre. The average yield of wheat per acre of cold Canadian soil is much less than the corn yield in Iowa, and both are low compared with that of tropical Bangladesh or southern China, where farmers can get several crops of rice each year. Almanacs or encyclopedias (or *www.cia.gov*) can provide data for calculating such ratios as food production per person, calories per pound of food, calories per person, protein per person, and so on. Such calculations will give students a good idea of how different countries compare.

The Geographical Scissors - 1

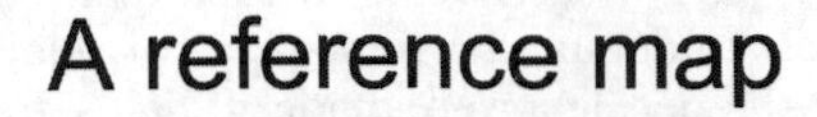

A reference map

the local puzzle
that shows how features
on thematic maps
fit together

A thematic map

the broader view
that puts what we see
on a reference map
into perspective

2A

The Geographical Scissors - 2

A reference map

uses a variety of symbols
to show how different things
fit together in a small area

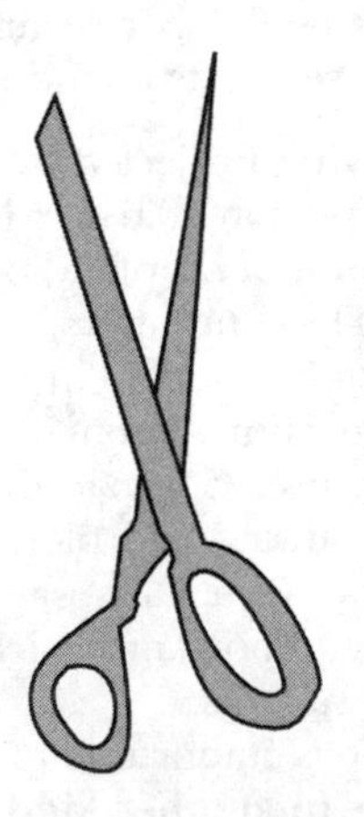

A thematic map

uses conventional symbols
to show how one specific thing
is arranged in a large area

2B

Teacher's Guide for Transparency 2A

I debated in writing this book whether to put the scissors diagram at the beginning or end of the Transparencies chapter. At the beginning, it would serve as a reminder of the scissors metaphor, one of the most crucial distinctions in this book. At the end, it would summarize that main point.

I decided not to put it in either place, but I still wonder if placing it that way would have helped convey the material in this chapter.

Activity: Have students approach the dual nature of the scissors metaphor by on the one hand thinking of examples of two modes of analysis working together (e.g., regional and thematic), and then on the other hand thinking of a single geographic subject and how to analyze it according to one of the scissors-pairs of perspectives. Another option is to think of other kinds of tools to use as metaphors: drill, chainsaw, hammer. (I would actually be interested in hearing what works well in classes, and why.)

Chapter 5 covers deductive and inductive reasoning and the use of themes in teaching. Here, let us just say that a teacher could use any of the diagrams in this book as a framework for discussion. There are no hard and fast rules about what works best in a given situation. In fact, most of the metaphors in this book have been used in any number of ways.

Teacher's Guide for Transparency 2B

It is often useful to show the same diagram several times during a class period, in order to review or amplify a point. Better yet, use several variations of an important diagram at different times in the discussion. Repetition with slight variations helps students abstract the most important concepts and eliminate unnecessary detail.

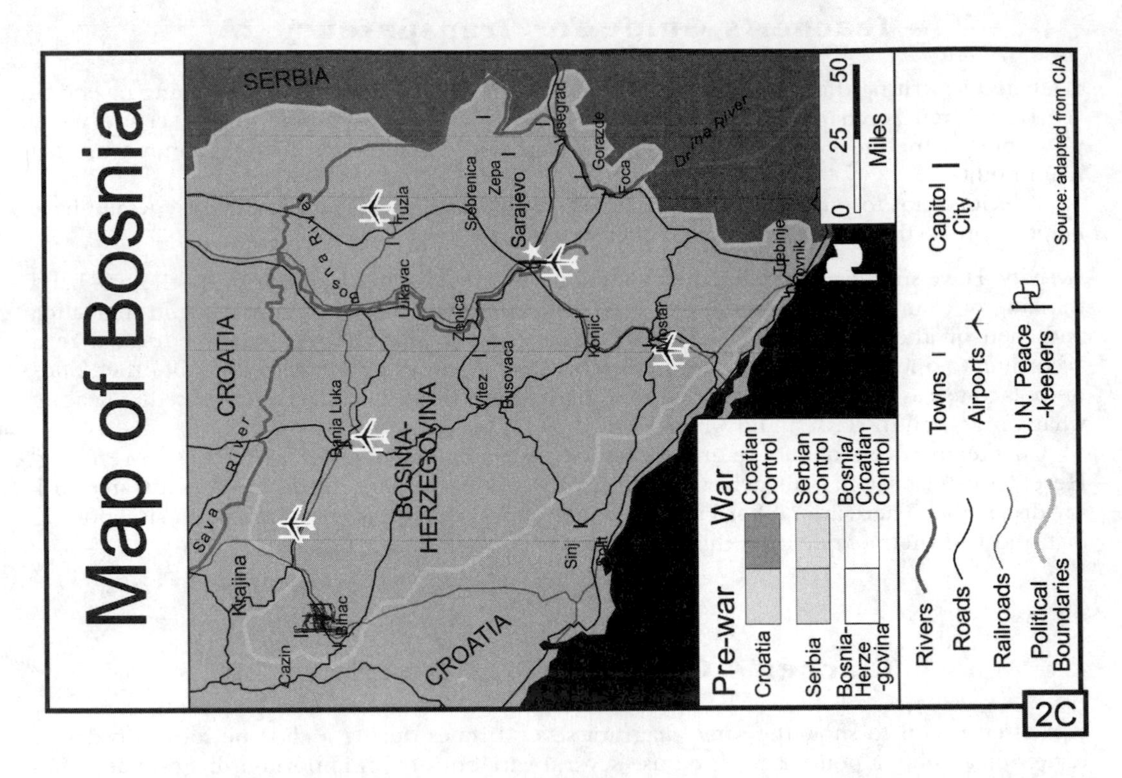

Map of Bosnia
SERBIA
CROATIA
BOSNIA-HERZEGOVINA
CROATIA
Sava River
Bosna River
Drina River
Krajina
Cazin
Bihac
Banja Luka
Lukavac
Tuzla
Srebrenica
Zepa
Sarajevo
Visegrad
Gorazde
Foca
Zenica
Vitez
Busovaca
Konjic
Mostar
Trebinje
Sinj
Split
0 25 50
Miles
Pre-war
War
Croatia
Croatian Control
Serbia
Serbian Control
Bosnia-Herze-govina
Bosnia/Croatian Control
Rivers
Roads
Railroads
Political Boundaries
Towns
Airports
U.N. Peace-keepers
Capitol City
Source: adapted from CIA
2C

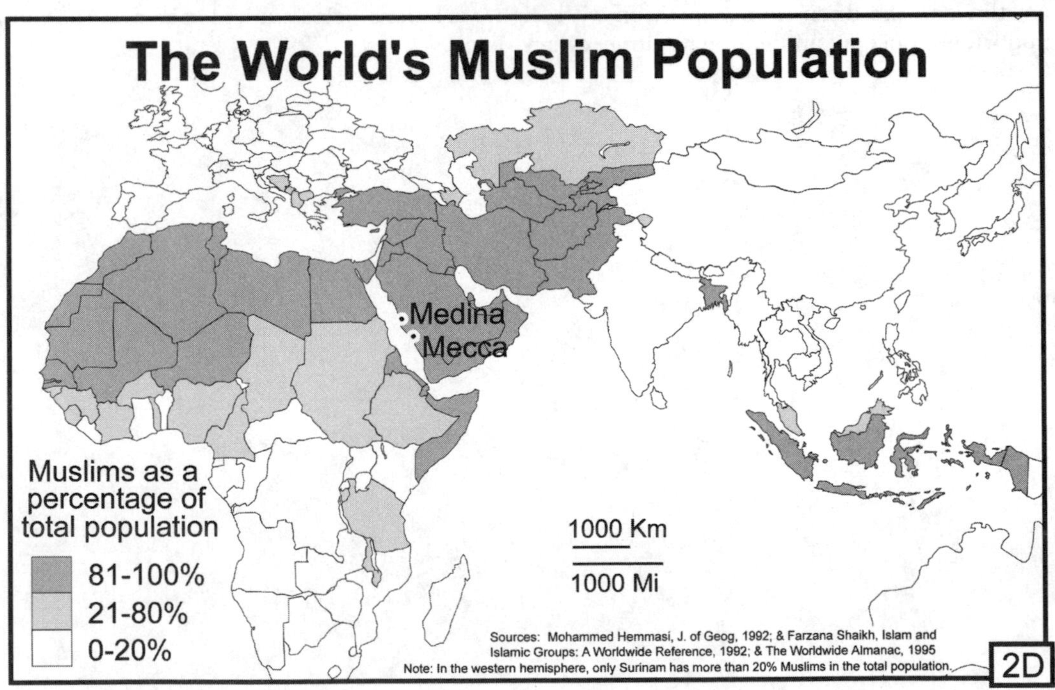

The World's Muslim Population
Medina
Mecca
Muslims as a percentage of total population
81-100%
21-80%
0-20%
1000 Km
1000 Mi
Sources: Mohammed Hemmasi, J. of Geog, 1992; & Farzana Shaikh, Islam and Islamic Groups: A Worldwide Reference, 1992; & The Worldwide Almanac, 1995
Note: In the western hemisphere, only Surinam has more than 20% Muslims in the total population.
2D

Teacher's Guide for Transparency 2C

A reference map shows the locations and spatial relationships of several different kinds of things on the area being mapped. To accommodate the many different kinds of information it contains, a reference map might have a fairly large legend; the legend must show the symbols used to depict features like mountains, rivers, swamps, cities, roads, and so on.

The designer of a reference map can make the map reader's task easier by selecting symbols that are easy to remember (e.g., wiggly lines for rivers, fuzzy patterns for forests). Color certainly helps, and there are natural associations with some colors (e.g., green with trees, blue with water, gray with borders or roads).

Activity: Post or project some reference maps from atlases, newspapers, or magazines. Then cover the legend and ask students what they think specific symbols might mean. Have them discuss why they associate particular meanings with particular colors, patterns, or other symbols. Summarize by pointing out that a good map maker usually tries to choose symbols that are intuitively clear to the intended readers of the map or that have become conventionally accepted ways of symbolizing particular features.

Teacher's Guide for Transparency 2D

A thematic map tries to show the spatial pattern of only a few things, sometimes only one. The designer of a thematic map can make the map reader's job easier by using established conventions for showing each idea - dots for cities, shaded areas for forests or property ownership, isolines for elevation or temperature, choropleth gradations or hierarchies for ratios such as crime rate or population density, and so forth. In the Transparencies that are thematic maps in this book I've tried to include a variety, to illustrate the many different approaches and symbols that can be used for different tasks.

Activity: To get students started on the path to evaluating maps well and using them well, find a number of reference and thematic maps in newspapers, magazines, or online sites. Post or project them, and have students try to classify the maps into four categories:

- Primarily reference maps, such as USGS topographic maps or maps of the whole country, such as often appear at the beginning of newspaper, encyclopedia, and magazine articles.
- Primarily thematic maps, such as dot maps of population, choropleth maps of ethnic origin, isoline maps of temperature.
- Mixtures of the two, as in the case of a map that emphasizes a single topic such as the locations of nuclear power plants but that also may show roads, cities, rivers, mountains, and other reference information.
- Badly designed maps, with unnecessary information or symbols that are unclear.

Since this is a rather basic skill, it is probably better to choose clear examples in which the category is fairly easy to distinguish (in other words, don't start the discussion with too many examples in the last category!). Many of the later Transparencies in this chapter are useful as illustrations of appropriate map symbols to express particular ideas. Cartographic literacy is not as easy to learn as some people seem to think (why else would we see so many juicy examples of terrible map design in newspapers and magazines?)

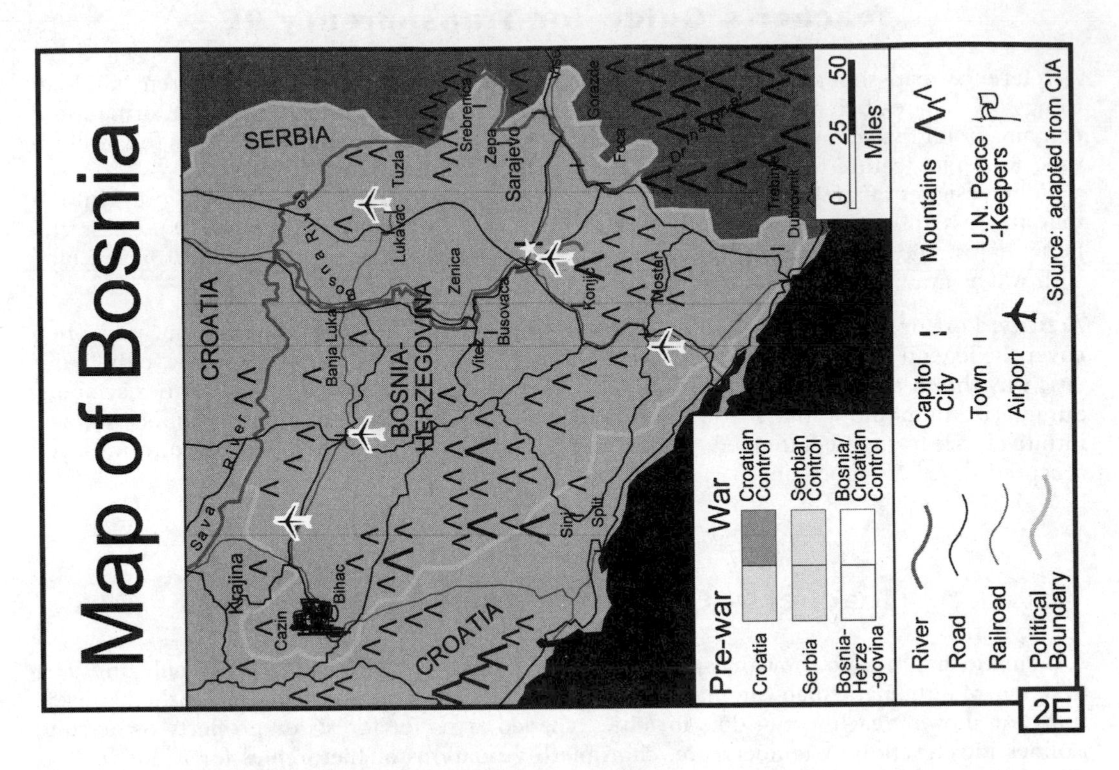

Map of Bosnia
SERBIA
CROATIA
BOSNIA-HERZEGOVINA
CROATIA
Sava River
Bosna River
Drina River
Krajina
Cazin
Bihac
Banja Luka
Tuzla
Lukavac
Zenica
Vitez
Busovaca
Srebrenica
Zepa
Sarajevo
Gorazde
Foca
Konjic
Mostar
Trebinje
Dubrovnik
Sinj
Split
Pre-war
War
Croatia
Croatian Control
Serbia
Serbian Control
Bosnia-Herze-govina
Bosnia/Croatian Control
River
Road
Railroad
Political Boundary
Capitol City
Town
Airport
Mountains
U.N. Peace-Keepers
0 25 50
Miles
Source: adapted from CIA
2E

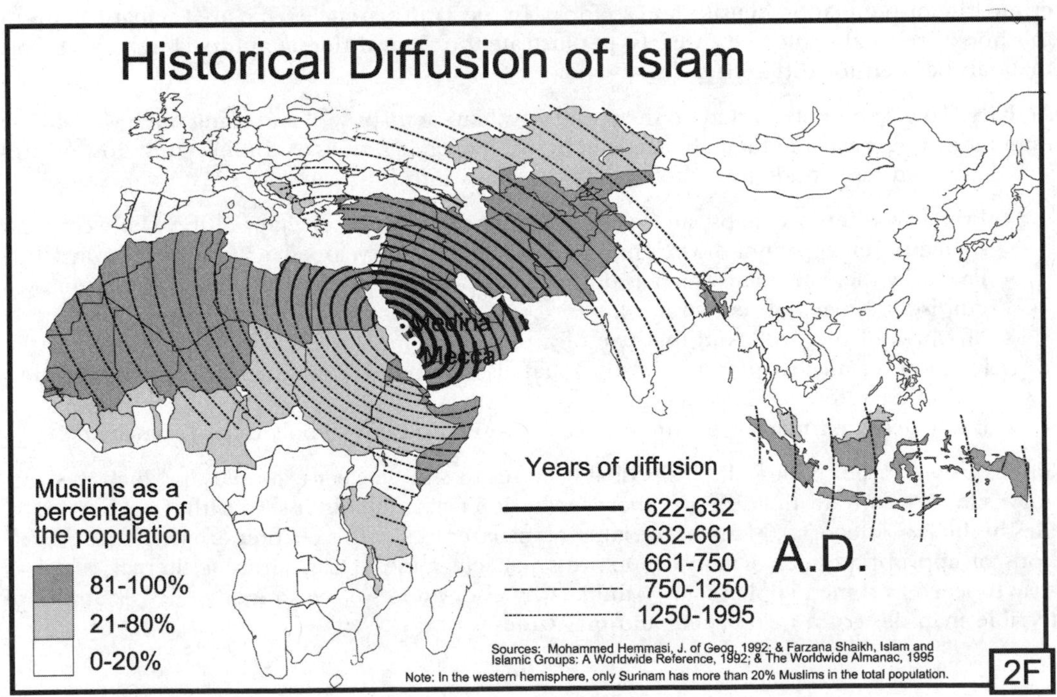

Historical Diffusion of Islam
Medina
Mecca
Muslims as a percentage of the population
81-100%
21-80%
0-20%
Years of diffusion
622-632
632-661
661-750
750-1250
1250-1995
A.D.
Sources: Mohammed Hemmasi, J. of Geog. 1992; & Farzana Shaikh, Islam and Islamic Groups: A Worldwide Reference, 1992; & The Worldwide Almanac, 1995
Note: In the western hemisphere, only Surinam has more than 20% Muslims in the total population.
2F

Teacher's Guide for Transparency 2E

By nature, it is difficult to discern in reference maps the spatial patterns of specific features. The complexity of the legend and the multitude of symbols can make patterns of individual features hard to isolate. This Transparency is a more complex version of Transparency 2C, and 2F is a more complex version of 2D.

Activity: Ask students to trace the outline of a country, state, or other area shown on a reference map, and then try to transfer a single feature from the reference map to their new outline version of it. For example, one student could try to copy all cities in their approximate locations; another could copy rivers, mountain areas, and so forth.

This is called "extracting a theme" from a reference map. In effect, one is taking information from a reference map and making a thematic map of it. The ability to copy things in their correct relative locations is an acquired skill that is useful for many other purposes. This is the kind of exercise students should practice a number of times, in a number of different contexts.

Teacher's Guide for Transparency 2F

This map deals with the same topic as Transparency 2D, but it uses different symbolic "language" to make its point. Many of the skills the National Geography Standards focus on involve selecting appropriate symbols for specific kinds of information and then analyzing the pattern.

Activity: Find thematic maps that use appropriate symbols:

- Isolines for climatic data, such as temperature, growing season (Transparency 3J) or dry areas (4J).
- Shading for surface features, such as glaciers (3L) or malaria (4V).
- Choropleth maps of ratio ("per") data in political areas, such as farm production per person (1B) or welfare payments per family (4Y).
- Dots for discrete features, such as plantations (4X), Palestinian settlements (9C), or video rental stores (10A).

Cover the title and legend of the map. Then post or project the map and ask students to describe the geographic pattern they see on the map:

- Does the feature cover a large or small part of the map?
- Is the feature evenly spread across the map or is it localized (primarily on one side of the area)?
- Is the feature shown on the map dispersed across the map? or does it occur in a compact cluster? a long string? many separate clusters, strings, pairs, trios, other shapes?
- Is the general shape of the pattern round or elongated? if elongated, in what direction? is the alignment of individual pairs, trios, or strings consistent throughout the map, or are they oriented in different directions?

The goal at this time is pattern recognition and vocabulary building, not accurate taxonomy. We are just starting to explore the uses of pattern analysis (more in Chapter 6). If the discussion leads to questions about the topics shown on the maps and maybe even the causes of the patterns, so much the better. That's what a geography course is about!

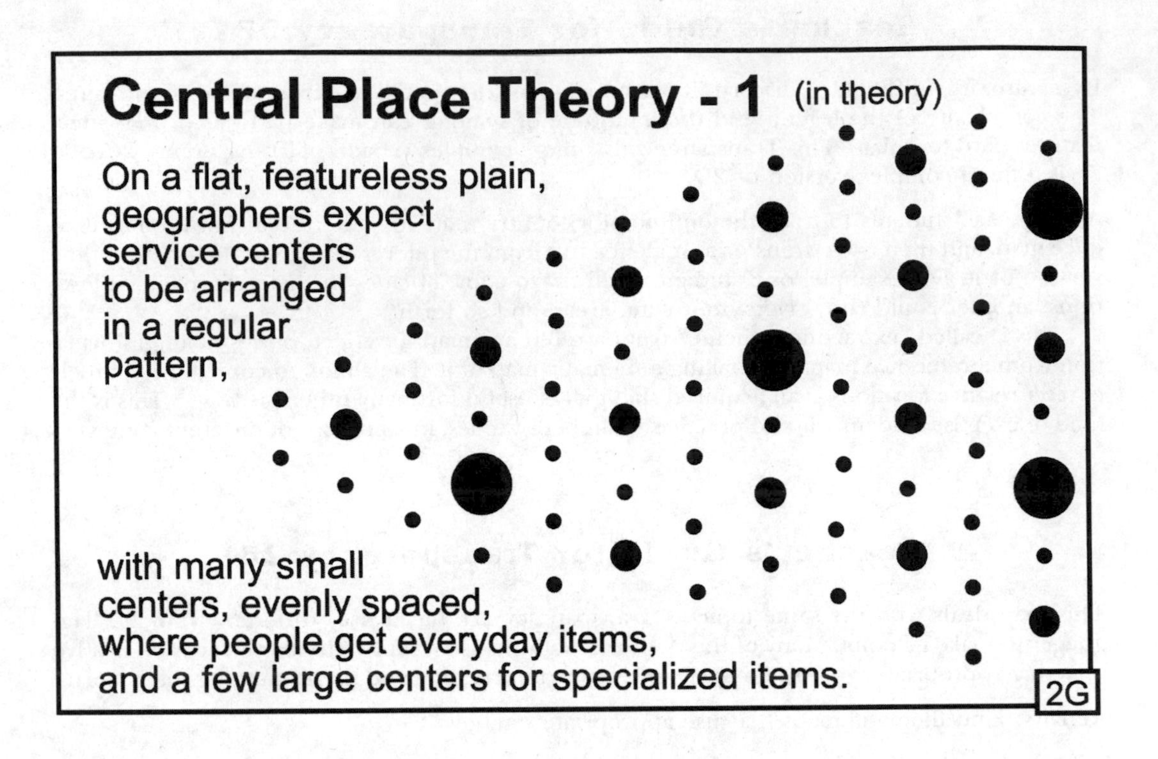
Central Place Theory - 1
(in theory)
On a flat, featureless plain, geographers expect service centers to be arranged in a regular pattern,
with many small centers, evenly spaced, where people get everyday items, and a few large centers for specialized items.
2G

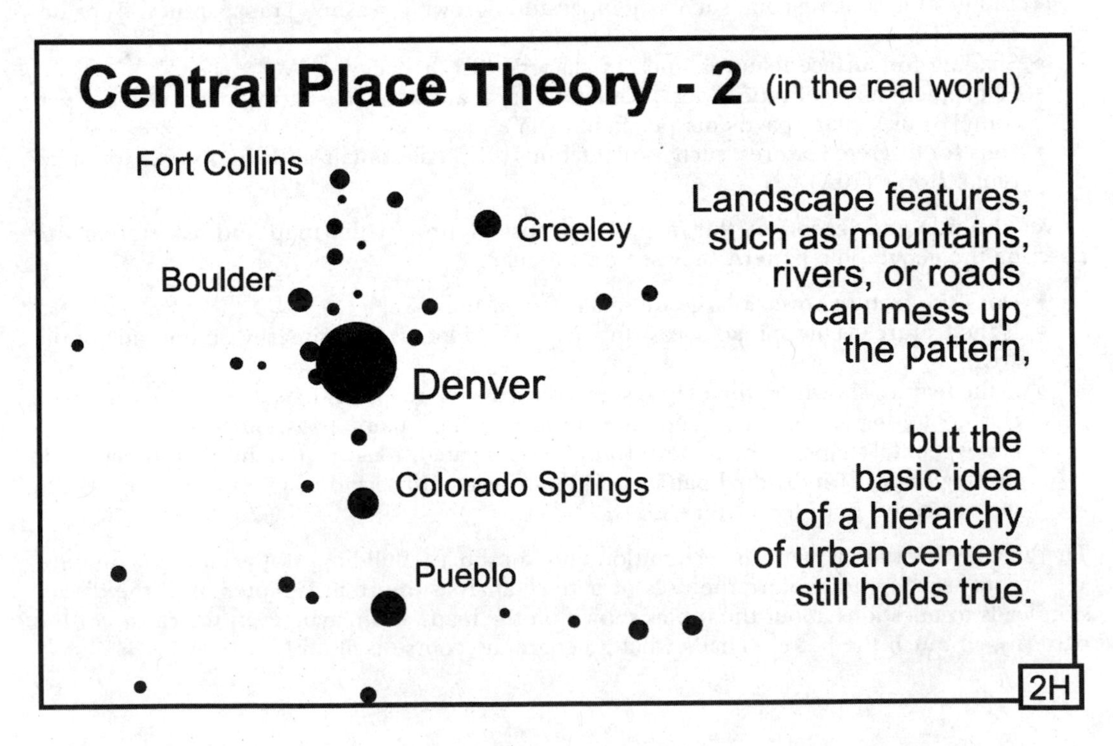
Central Place Theory - 2
(in the real world)
Fort Collins
Greeley
Boulder
Denver
Colorado Springs
Pueblo
Landscape features, such as mountains, rivers, or roads can mess up the pattern,
but the basic idea of a hierarchy of urban centers still holds true.
2H

Teacher's Guide for Transparency 2G

Part of geography is about the forces that make a particular place what it is. Many of these forces act from the outside – things such as solar energy, ocean currents, immigration, trade, or political influence.

Some of the forces that shape a place, however, are from inside. These are consequences of the everyday actions of the people who live there. For example, if people in an area buy food, then we expect to see food stores or markets of some kind there. The size and spacing of stores is a result of the way people typically buy things – how often they shop, how far they are willing to travel, and so forth.

Geographers often use an imaginary "flat, featureless plain" to explore the processes that occur when people behave "typically." (This is not a fictional place - it is an attempt to see what a real place would be like if it did not have so many complications.) A teacher can do the same kind of thing in a classroom.

Activity: Write a list like this on the board: new car, rental video, baseball game, loaf of bread, wedding dress, tank of gasoline. Project the Transparency, and ask students where on the flat plain they would expect to find stores selling the items on the list. The question may be phrased in several different ways:

- Would people sell (name a product from the list) in every small town or corner shopping area, or only in big cities or major malls?
- How many places sell new cars, as compared to gasoline?
- How far do people typically seem to be willing to travel in order to rent a video, to attend a baseball game, or go to college?
- How many grocery stores could one soft-drink bottling company serve?
- How many cancer hospitals would there be, as compared to dentists' offices?

What do all these questions have in common? They tease at the fact that some products and services are needed by a significant portion of the people, and frequently (e.g., gas stations or video rental stores), and others are inherently specialized and needed less often (e.g., cancer clinics or stores that sell wedding dresses). Of the former catergory there are many instances, close to customers. The latter are likely only to be in a few places, usually in high-traffic centers where there are enough people to support the providers.

Teacher's Guide for Transparency 2H

Putting the abstract idea embodied in Transparency 2G into the real world is the next step.

Activity: Have students look at maps of Colorado in an atlas.

- What accounts for the north-to-south string of fairly large cities, from Fort Collins to Pueblo? (They are all right at the eastern edge of the Rocky Mountains.)
- How about the smaller string that runs east from Pueblo? (They are along the Arkansas River [pronouced ARKansas here, ArKANsas in Kansas!].)
- Why is Denver the biggest city? (Perhaps it started earliest? Maybe because it is near the center of the settled area? Maybe its river is the biggest? That is important in a dry region!)

Activity: Have students make similar maps of other areas, in the United States or in other countries. What seems to influence the patterns of cities?

Two Ways to Organize a Geography Class

Regionally - from place to place

e.g. Equatorial Africa
The Sahel
The Sahara
Egypt
Arabian Peninsula
. . .

Topically -- from idea to idea

e.g. Climate
Agriculture
Industry
Urbanization
. . .

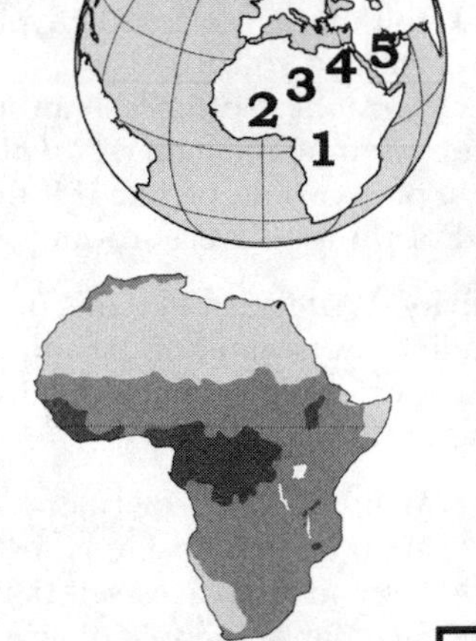

Gersmehl and Rohde, 1994; Global Change Database.

2I

An Organizing Matrix

because the world is too big to study everything in it

Within each regional unit, we will focus on a few specific topics. (or vice versa)

	Climatic Cycles	Regional History	Land Ownership	Economic Control
Megalopolis				X
Appalachia			X	X
Old South		X		
Great Plains	X		X	
. . .				

2J

Teacher's Guide for Transparency 2I

This Transparency might be useful early in a geography course.

Activity: Explain different ways of organizing the subjects covered in a course, and then ask students whether the textbook is organized topically, regionally, or as a combination of the two.

Talking about course organization is way of reminding students of the purpose of learning and doing geography. Here are two related questions: "Why are things like cities and malls located where they are?" and "Where should people locate things like malls, clinics, and political borders in order to be efficient and fair?" The first question implies a regional organization of a class; the second points to a more topical emphasis.

If you have already discussed climate a little, show this Transparency and ask students to think of a reason why, if using a regional organization, it might be a good idea to start at the equator. The goal is to get students to see that the issues that need to be conveyed about a subject can be arrived in a number of different ways.

This Transparency might be even more useful in discussions with parents, administrators, and colleagues. It was created as a presentation for the 1992 ARGUS Steering Committee meeting. That meeting was supposed to address editorial details, but basic questions about textbook organization kept coming up, even though they presumably had been "settled" at earlier meetings. The truth of the matter is that the organizational framework of a geography class does not matter nearly as much as the content.

Teacher's Guide for Transparency 2J

After a course framework has been set, we must decide how to fit into the major theories and regions into it. One useful tactic is to make a matrix like the one in this this Transparency, which was adapted from part of the outline of GIGI (Geographical Inquiry into Global Issues), shows a well-structured high school geography course in the form of a matrix. It can be seen at a glance whether coverage of topics and regions is sufficiently thorough. This kind of careful planning is necessary, because, as the Transparency says, "the world is too big to study everything."

Awareness of the vastness of the world does not give us a license to teach whatever we want. Society still expects a geography class to "cover" the major world regions (or at least equip students to learn about them on their own).

The prescription is simple: insist on seeing a matrix of this kind from every textbook publisher. In time (with luck!), they'll get the hint: teachers expect authors and publishers to show evidence of having thought about the relationships between topics and regions. The alternative is the kind of cheap educational schlock that often floods the market after a major initiative (like the No Child Left Behind Act). Teachers deserve better.

Activity: Students can learn from the effort of trying to figure out which region is a good setting in which to discuss a particular topic. An interesting way of introducing (or summarizing) this discussion would be to ask "absurd" questions and checking student reaction:

- Would it make sense to discuss farming in a unit on Antarctica?
- How about a book that "covers" steel production in a chapter on Florida?

If students see the humor in such suggestions, that is a good step in the right direction! They are showing a grasp of the idea that the world has some "spatial logic" – that there are reasons why things are where they are.

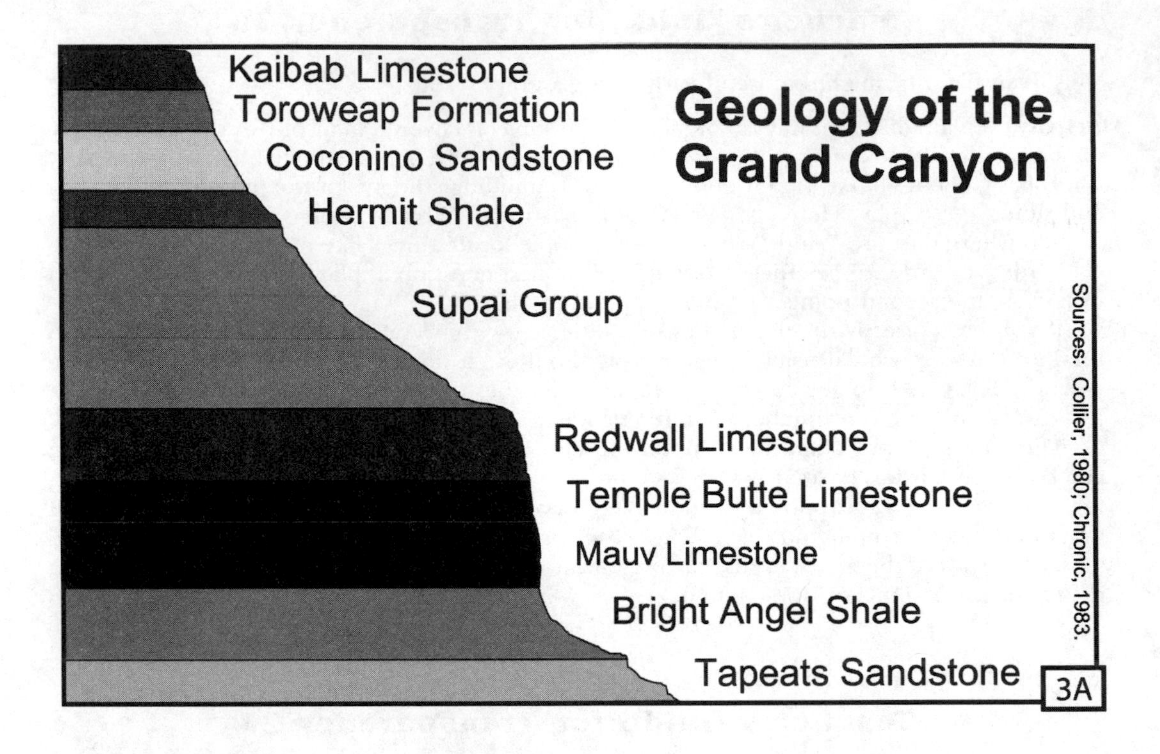
Kaibab Limestone
Toroweap Formation
Coconino Sandstone
Hermit Shale
Geology of the Grand Canyon
Supai Group
Redwall Limestone
Temple Butte Limestone
Mauv Limestone
Bright Angel Shale
Tapeats Sandstone
Sources: Collier, 1980; Chronic, 1983.
3A

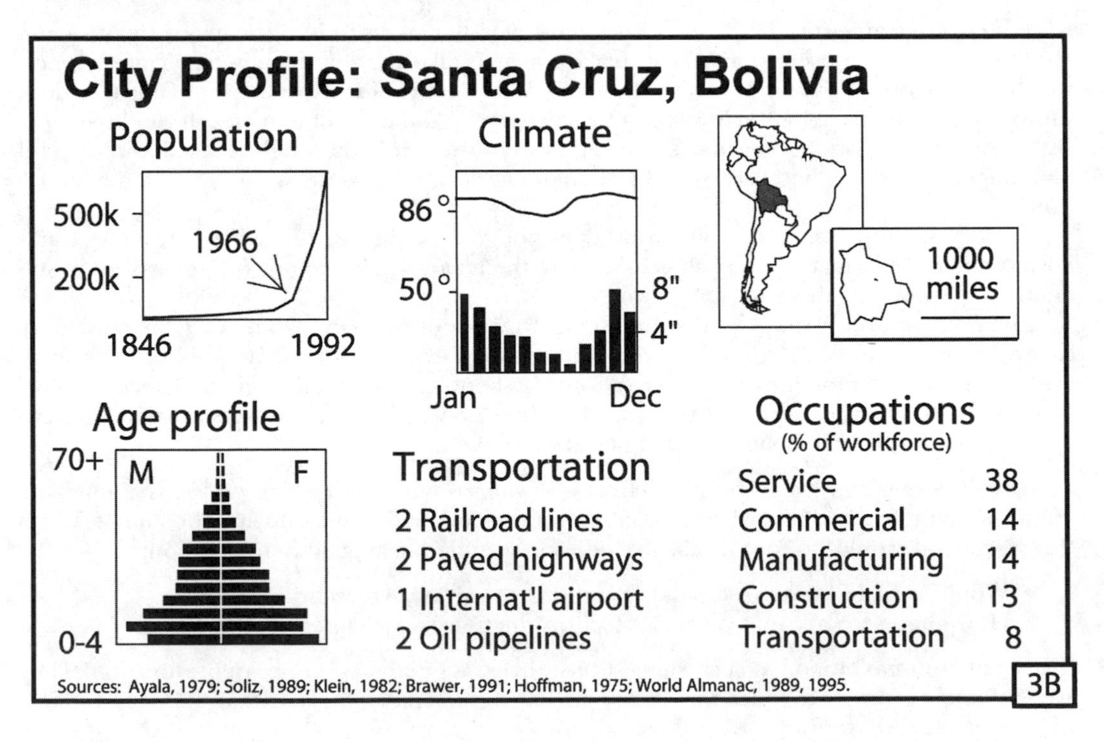
City Profile: Santa Cruz, Bolivia
Population
500k
200k
1966
1846
1992
Climate
86°
50°
8"
4"
Jan
Dec
1000 miles
Age profile
70+
0-4
M
F
Transportation
2 Railroad lines
2 Paved highways
1 Internat'l airport
2 Oil pipelines
Occupations
(% of workforce)
Service 38
Commercial 14
Manufacturing 14
Construction 13
Transportation 8
Sources: Ayala, 1979; Soliz, 1989; Klein, 1982; Brawer, 1991; Hoffman, 1975; World Almanac, 1989, 1995.
3B

Teacher's Guide for Transparency 3A

The Grand Canyon is a classic example of the effects of differential erosion. Differential erosion is admittedly a term for junior high or later, but it refers to the fact that different kinds of rock have different degrees of resistance to the destructive effects of wind and rain. The concept is important: it helps explain the locations of many things, including canals (Transparency 4N), sizeable Southern plantations (Transparency 4X), and Harpers Ferry (Transparencies 1C and 6F). Topics like this should be discussed early in children's geographic education, so the vocabulary should be simplified for early grades. But how?

One way is to refer to the rocks that resist erosion and form cliffs as "hard" and the easily eroded rocks as "soft." This is less than ideal, however, because the terms "hard" and "soft" have another very specific meaning in geology. They refer to the scratchability of a mineral. Diamond is hard because it can scratch any other mineral; calcite is soft because it can be scratched by most other minerals.

Activity: Bring samples of minerals into the classroom, and have students arrange them in order of scratchability, from the hardest down to the softest.

Here is why terminology can be problem: calcite is a soft mineral, but the calcite-based Redwall Limestone is a strong rock. It is therefore a cliff-making layer in the Grand Canyon. Quartz, by contrast, is a hard mineral. It can scratch tempered steel, but a sandstone made of weakly cemented quartz is not a resistant rock.

This potential for confusion is enough to make some university teachers wish that all educators had to take something like the doctors' Hippocratic Oath: "above all, do no harm." Most topics can be grasped at early grade levels, if packaged properly. The "packaging," however, must be done knowledgeably, so that children in early grades do not acquire erroneous ideas that must be "unlearned" later.

For the record, then: the Grand Canyon has some strong and some weak rocks, which account for the cliffs and gentle slopes in its profile. It also has some hard and some soft rocks, but hard rocks are not necessarily strong, nor are soft rocks necessarily weak.

Teacher's Guide for Transparency 3B

A visit to (or from) the local Chamber of Commerce can illustrate an important form of applied geography. Most Chambers of Commerce have prepared packets of information to give to people who might be interested in moving to a community or investing in it. These materials might be called community profiles, annual reports, or relocation packets (because the main consumers of the information may be factory owners, office builders, vacationers, or people who are considering moving to the area).

Activity: Have students make one-page profiles of places and post them for comparison. Doing this early in a term, without much guidance, and then toward the end when the students have more experience can be gratifying in showing students they can now "do geography." This Transparency could be set up as an example to be improved on, rather than emulated (see also Transparencies 7D and 7E). Say something like, "this is the old version, and we want to replace it with something that does a better job of attracting attention and presenting our community."

As with any short profile, the Santa Cruz example is selective in the information it shows. The population graph shows rapid growth after 1966, but it does not say that the growth was due to an oil boom. Only by noticing the 2 pipelines in the transportation section might one guess the role oil played in the local economy. The profile also fails to note farming in the surrounding area (let alone a thriving cocaine industry!). Bottom line: since it is impossible to show everything, profile designers should target their efforfs at a specific audience by anticipating questions the audience would want to see answered.

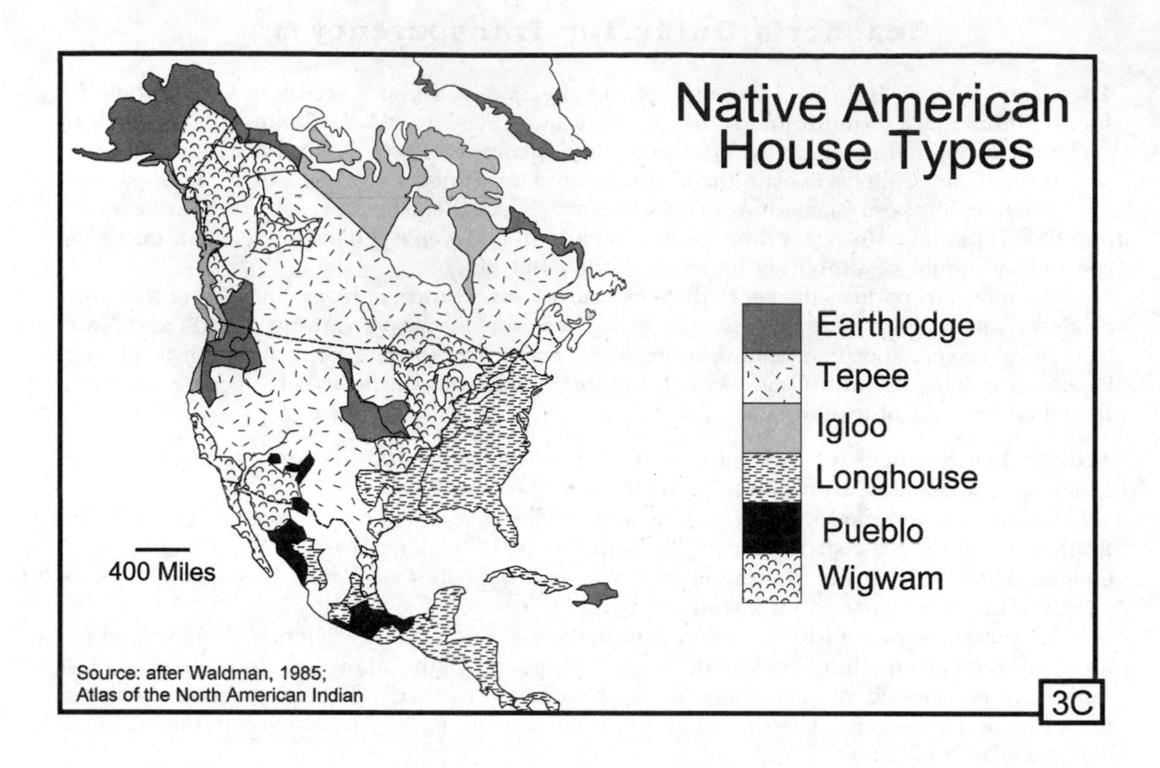
Native American House Types
Earthlodge
Tepee
Igloo
Longhouse
Pueblo
Wigwam
400 Miles
Source: after Waldman, 1985;
Atlas of the North American Indian
3C

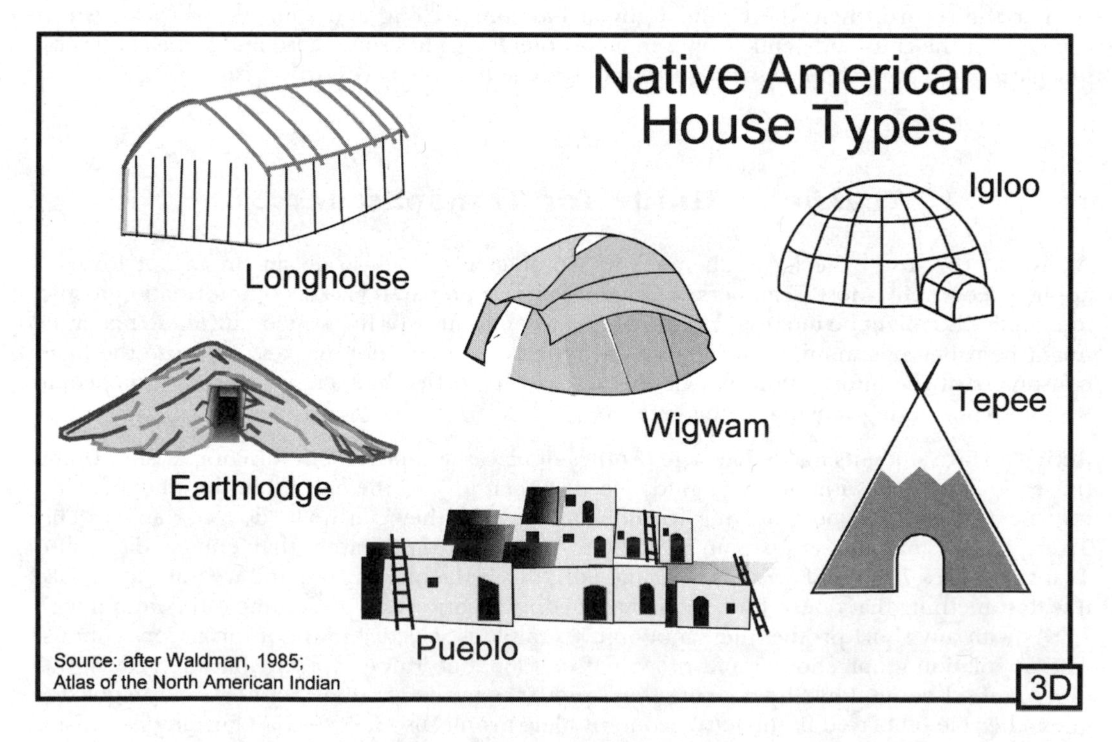
Native American House Types
Igloo
Longhouse
Wigwam
Tepee
Earthlodge
Pueblo
Source: after Waldman, 1985;
Atlas of the North American Indian
3D

Teacher's Guide for Transparency 3C

This is an example of a bounded-area map. A map maker identified places where a particular feature was common, drew a line around those places, and shaded the area. A bounded-area map is visually similar to but not quite the same as an isoline map (Transparencies 3J, 3K, and 4U) or a choropleth map (Transparencies 1A, 1B, and 4Y).

Isoline and choropleth maps are used for quantitative information, such as temperature or population density, which is measured on a scale from low to high. Bounded-area maps, by contrast, are used for qualitative information, things that are different in kind but not in degree. For example, the Bantu language is different from Swahili.

This basic difference in the nature of data should be reflected in the legend and the choice of colors. Colors on an isoline or choropleth map should go in a logical sequence (e.g., from light to dark), because the data go from low to high. Colors on a bounded-area map should not imply a sequence, because on category is not "lower" or "higher" than another.

Activity: Have students gather maps and classify them according to "graphic vocabulary." Don't expect mastery in one class, or even a year. The language of maps, like other languages, has many levels of meaning. Mastery begins with the realization that people invented different symbols (like bounded areas and isolines) to express different ideas.

Teacher's Guide for Transparency 3D

A house is a tangible expression of someone's dreams and personal history. It tells something about available technology, family income, and aesthetic ideas. In short, a house is a landscape feature that provides an interesting commentary on many other aspects of the local environment and culture. Part of this Activity is simple vocabulary building – learning words to use in describing a house.

Activity: Students should write a short verbal description of one of the houses on this Transparency, but they shouldn't label it or give other clues about which house is being described. The students then trade descriptions and try to identify which house is being described. If this seems too mundane, have them imagine that they are writers trying to describe the setting for a mystery. Descriptive writing is a skill that can be sharpened. Moreover, a person skilled in descriptive writing usually is good at evaluating other written work. This is why we hear so much about writing across the curriculum!

Activity: Have students sketch one of the houses and label key features. Like any activity designed for vocabulary building, this can have different levels of complexity in different grades. For example, a beginning student may notice only the conical shape of a wigwam or tipi. A more advanced student might notice the skin cover and framework of poles, which allow the house to be taken down and moved more easily than other house types. Still more advanced students might notice the air deflector that keeps wind from blowing into the smoke hole near the top. When they have labeled their sketches, have them write an essay describing how those features fit local environmental conditions.

Activity: Have students match the houses in the Transparency with possible locations on a map. This matching starts by assuming that people build houses to fit the conditions in their environment. For example, people who live in cold forests might build houses of wood, with steeply pitched roofs to shed snow. People in hot deserts, by contrast, have little wood and no reason to fear snow, so they often build flat-roofed houses out of adobe (dried mud) or stone.

The Gravity Model - simple version

Your attraction to a place depends on its size and its distance.

Attraction to this place is measured.

This place is twice as far, so it is half as attractive.

This place is twice as big, so it is twice as attractive.

3E

The Gravity Model - complex version

Things that tend to increase traffic:
- larger destination
- larger origin
- family ties
- economic links
- political connections

Things that tend to decrease traffic:
- greater distance
- rugged terrain
- bad roads
- closer places to get what you want

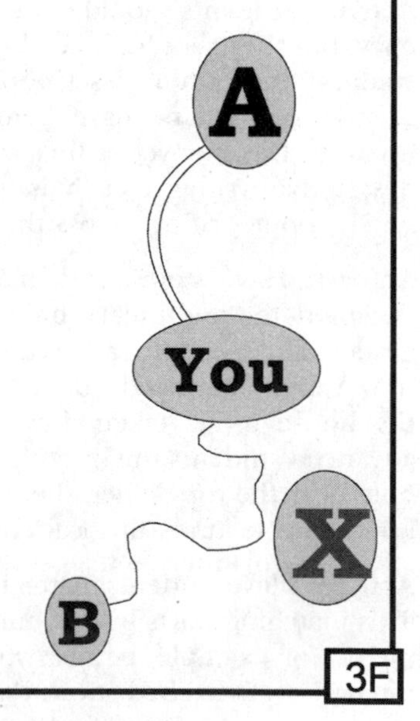

3F

Teacher's Guide for Transparency 3E

The gravity model is a theory that allows people to use measurements of traffic on one road to predict what the traffic on another road would be. Ask students if they have seen the rubber cables that engineers stretch across roads in order to count cars. Using such a measuring device is costly – it is cheaper to measure traffic on a few roads and use that data to predict the traffic on the rest of the roads being studied.

Like any theory, the gravity model starts with some assumptions. The main assumption is that large numbers of people behave in a predictable way (even though individuals might act independently). Suppose we measure the traffic on the road from "you" to one town (the one marked 4, which could mean 400 cars per hour or 4,000 phone calls per day or some other measure of traffic).

The gravity model says traffic is *directly* related to population. It therefore predicts twice as much traffic to a town that is twice as big. The model also says that traffic is *inversely* related to distance; it predicts half as much traffic to a town that is twice as far away. Question: What traffic would you predict to the question-mark town? Answer: It is twice as big, and twice as far, so the traffic is likely to be 4 times 2 (because it's twice as big) divided by 2 (because it's twice as far), or about the same as to the measured town.

What about a place 6 times as big and 1/2 as far away? 4 times 6 divided by 1/2, for a whopping 48. (Lookee, it's a math across the curriculum activity!)

Activity: Put this in the real world. Pick a well-known town near the school. Then pick some other towns with sizes and distances that are even multiples or fractions of that town (e.g., 1/3 as big, or 4 times as far away). Have students measure the distances on a road map. Tell them that engineers counted 1,000 cars per day on a road to ______; what traffic would the students predict for the other towns?

Who cares about this kind of information? The police, highway engineers, bicyclists, anyone who runs a gas station, restaurant, store, or theater; in short, lots of people.

Teacher's Guide for Transparency 3F

In the real world, people often have to adjust the results of gravity model analysis to make them more realistic. These adjustments take into account such things as congestion, road condition, alternative opportunities, or various political or economic connections. For example, one could express distance in terms of cost or minutes rather than simply in miles.

This kind of adjustment makes traffic forecasting partly an art as well as a science. It is a necessary prerequisite, however, for anyone who wants to build good infrastructure (whether we are talking about canals, roads, powerlines, telephone wires, bicycle paths, or fiber-optic cables for the internet). Not surprisingly, traffic prediction has become a big career opportunity for geographers. Moreover, citizens who want to participate in decisions about where to locate stores, highways, and so on should know how traffic is predicted.

In 1989, for example, I helped a neighborhood group use the gravity model to predict traffic on a new freeway connection the city was planning to build. A junior-high student did some of the calculations, and a retired woman presented the results at a public hearing. City officials had to admit that the extra traffic would probably cause excessive noise and danger to children. They agreed to pay the moving expenses for any people who wanted to move to other houses. Isn't that what the phrase "geography for life" is all about?

(Of course, people in another place might use the gravity model to argue in favor of a new road connection. This is an important point to make about spatial models: the model just predicts traffic; it does not say what people should want.)

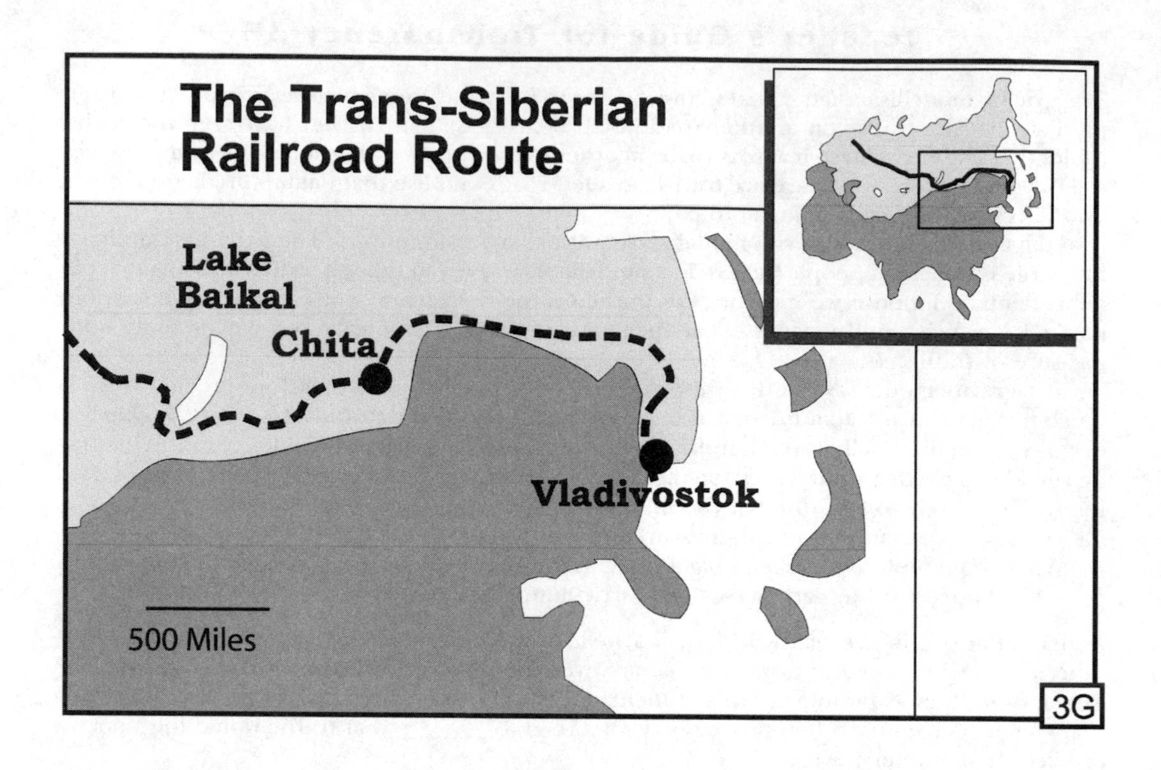
The Trans-Siberian Railroad Route
Lake Baikal
Chita
Vladivostok
500 Miles
3G

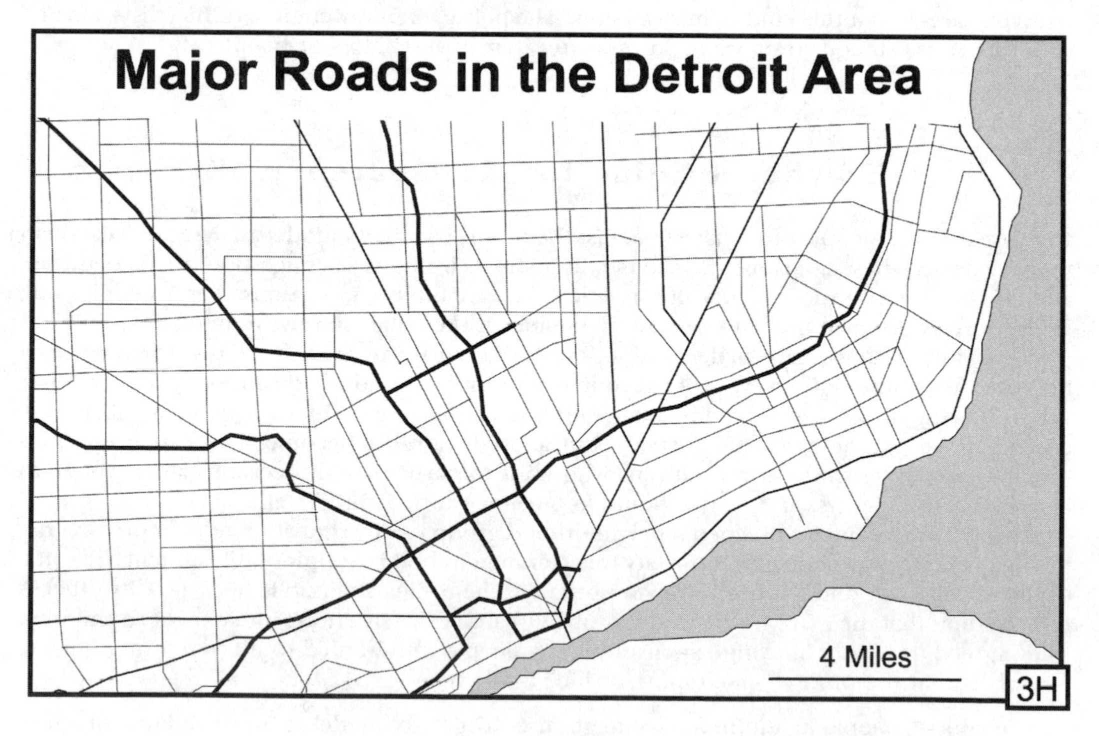
Major Roads in the Detroit Area
4 Miles
3H

Teacher's Guide for Transparency 3G

The famous Trans-Siberian Railroad links the European part of the former Soviet Union with the port and military base of Vladivostok on the east coast of Asia.

The western end of this extraordinary railroad makes a great deal of economic sense. It starts in the densely populated area near Moscow. The railroad then goes through a grain growing region, a skinny strip of fertile land between the deserts around the Aral Sea and the cold forests of Western Siberia. It passes through mineral-rich areas near Novosibirsk and curves south of Lake Baikal. Then, rather than go directly southeast through the well-populated area around Harbin, the railroad makes a big detour northeast, then east through essentially uninhabited area, then south, and finally southwest.

Activity: Measure the distance along a straight line from Chita to Vladivostok and then along the railroad. The actual route of the railroad is about 600 miles longer than a straight line. This is the "penalty" the Russians were willing to pay in order to stay entirely within their own country: every trip on this railroad takes an extra day and thousands of gallons of fuel. Moreover, the area north of the Amur River has few people. An emergency along that lonely stretch can be a big problem, because there are so few roads or villages there.

Activity: Examine an atlas and try to find other places where railroads or roads seem to go far out of their way in order to stay inside one state or country.

Teacher's Guide for Transparency 3H

Detroit was originally a French town. The downtown streets follow the French "longlots." In this system of land division, someone "owns" a certain distance along a river or road. Property lines extend directly away from the river.

Since access to the river was important for transportation, people would divide their land among their heirs by giving each son a share of the "frontage." The results are long, skinny farms oriented at right angles to the river. This pattern is still evident in many places near Detroit. In the Raisinville area, for example, many of the "farms" are a few hundred feet wide and half a mile or more long.

Activity: Look at the map of downtown Detroit. Use one color to trace the roads that appear to follow the French longlot rule (they go directly away from the river or parallel to it). Then use another color to mark roads that follow the Public Land Survey (they are one mile apart and run roughly north-south or east-west). Use a third color for radial (bicycle wheel spoke) roads that run in different directions away from downtown. Finally, use another color to put circles around corners that you think are likely to have complicated stoplights and possible traffic jams. (Intersections with five, six, and more inputs/outputs are harder to control with traffic lights than is a simple four-way stop.)

Activity: Look at street maps of mid-size cities such as Dallas, Denver, Indianapolis, Minneapolis, New Orleans, Oklahoma City, Pittsburgh, St. Louis, and Salt Lake City. Which cities seem to have traces of old longlot settlement? (M, N, P, and SL). What other things seem to influence the street patterns of some cities? (Denver and Pittsburgh have mountains; Salt Lake City has both mountains and a big lake).

Activity: Look at street patterns of cities in other parts of the world. Can you see evidence of former land uses (an old fort wall, for example [Edinburgh, Quebec City], a colonial survey [Saigon], or land left over from the palaces of former rulers [Rome, Tokyo])? What influence do these relics of former land use have on present-day traffic patterns?

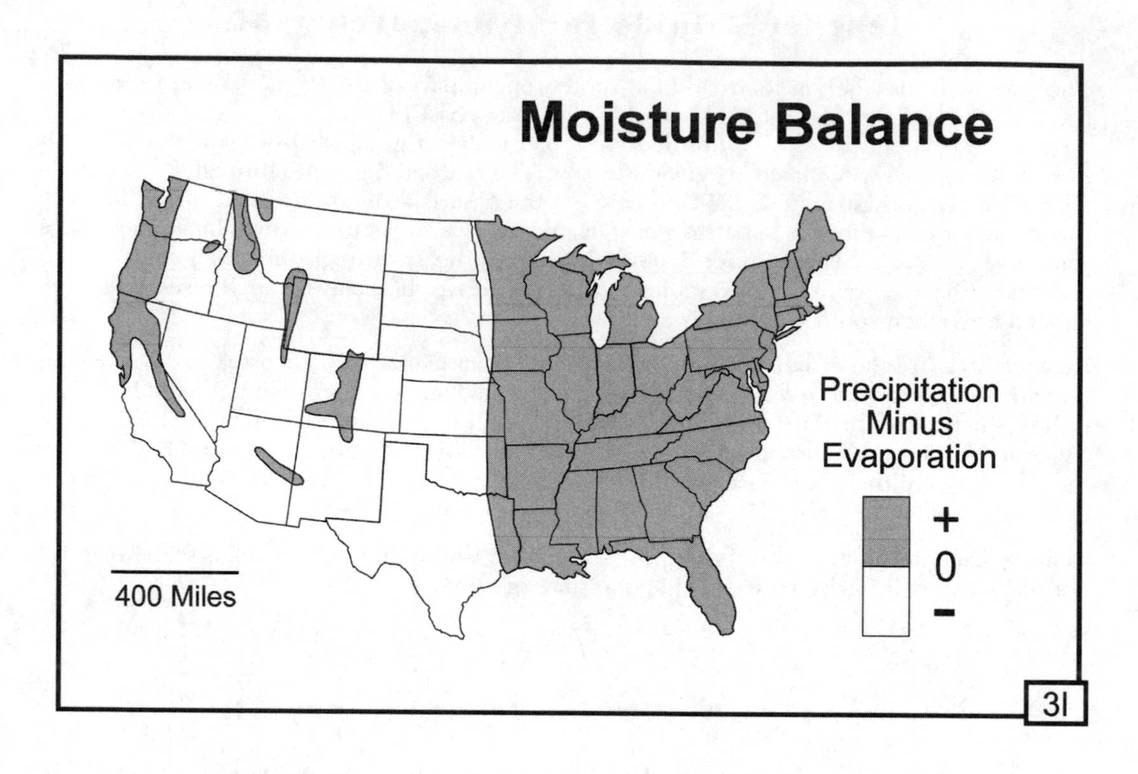
Moisture Balance
Precipitation
Minus
Evaporation
+
0
-
400 Miles
3I

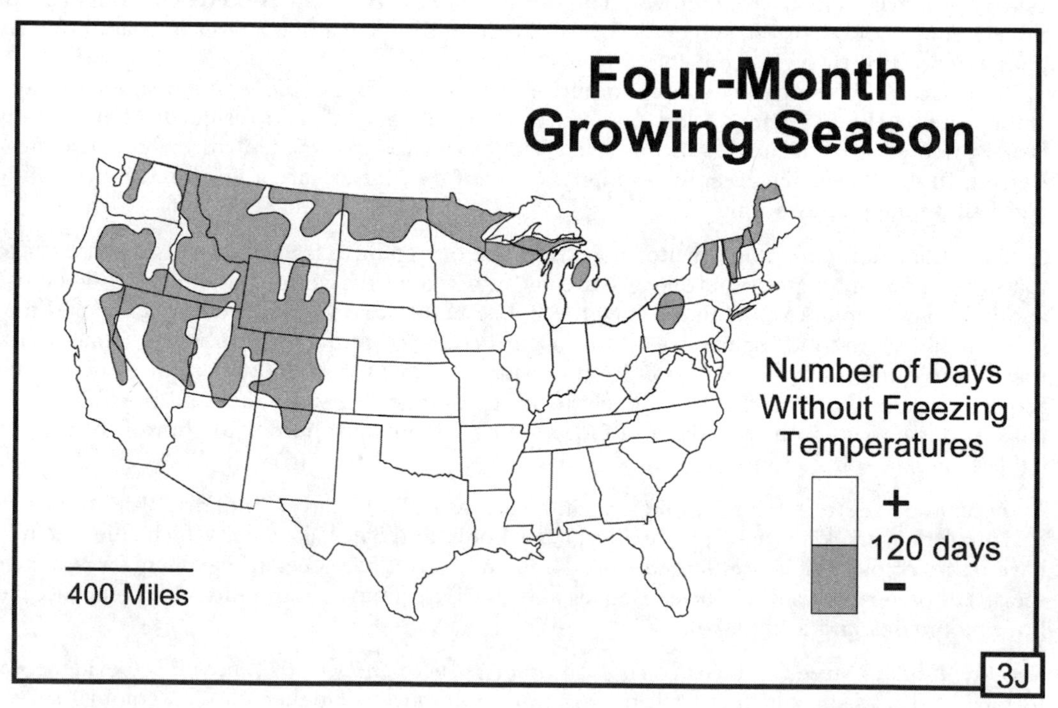
Four-Month
Growing Season
Number of Days
Without Freezing
Temperatures
+
120 days
-
400 Miles
3J

Teacher's Guide for Transparency 3I

The lines on this map have an important environmental message. On the gray side of the line, rain and snow supply more water than plants that grow there can use.

Activity: Color and name the states that have a moisture surplus (more water than plants need).

Activity: Explain the basic idea of the map and then ask students to make a list of things that surplus water "does" in the environment. For example, it grows forests. It makes rivers and lakes. It washes dust and smoke out of the air. It occasionally floods basements. It leaches nutrients from the soil, so that farmers need to fertilize their fields. It erodes land and makes gullies.

Activity: Show a picture of a covered wagon and ask students to list things that might be in it. Keep pressing until the list extends beyond household items and includes tools needed to make a living or build a house (e.g., clothing, rifle, gunpowder, harness, plow, corn seeds, axe, saw, hammer, nails, thread, cooking pot, matches or flint, well bucket, etc.) Then, project the Transparency and point out that areas with moisture deficit cannot support forests. Finally, ask students how many of the tools on their list would work if there were no trees around for wood.

In an influential book called *The Great Plains*, W. P. Webb (a Texas historian) said that European-American culture was not originally equipped to deal with a treeless environment. As a result, the westward movement of the settlement frontier slowed in the mid-1800s (except for people who leapfrogged the Plains and went on to places like Utah, Oregon, and California - as shown in Transparency 3M). It took time for people to invent (or find) tools that could make life in treeless areas easier. This list of new "tools" includes the revolver, sodcutter, steel plow, wheat thresher, longhorn cow, barbed wire, windmill, drilled well, and so on. Even so, much moisture-deficit land was never claimed and is still "owned" by the Federal government (Transparency 3O).

Activity: Ask students what they would take with them if they were settlers moving to (name a place and a time). Discuss the environmental appropriateness of the items on the lists they make.

Teacher's Guide for Transparency 3J

The 4-month frost-free season is an even more formidable barrier to human settlement than the moisture-deficit line. Hardly any highly productive crop can mature when the growing season is much shorter than 4 months. Only a few cultures, such as the American Eskimo and Inuit, the Lapps of Scandinavia, or the Yakut people of eastern Siberia have learned ways of living in lands with such short summers. Their total population probably was never more than that of a medium-size city such as Baltimore or Seattle.

Activity: Have students look at a world map or atlas and count cities that are north of about 60 degrees north latitude (maybe put pins in a map?). The few that exist are those lucky places with warm ocean currents, as well as a few mining areas, a few fishing villages, a few prisons and military bases, but in general this part of the world is uninhabited.

Activity: Have students think about ways to "solve" the problems of dryness or coldness (irrigation from wells or canals, heated greenhouses, seasonal migration). Is it easier to overcome a lack of water or a short growing season? Right - several cultures learned how to use irrigation in order to survive in a dry land and even build big cities (such as Cairo, Baghdad, Phoenix, Karachi, or Tashkent), but a short frost-free season is still a major impediment.

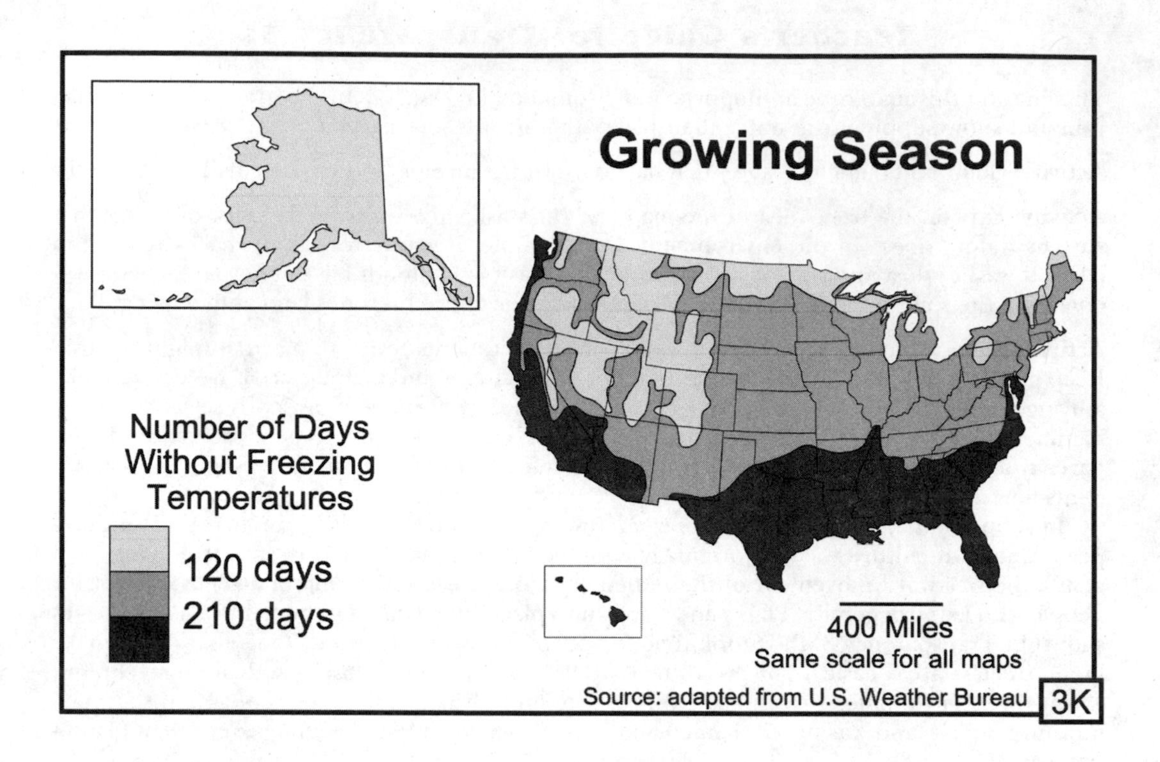
Growing Season
Number of Days
Without Freezing
Temperatures
120 days
210 days
400 Miles
Same scale for all maps
Source: adapted from U.S. Weather Bureau
3K

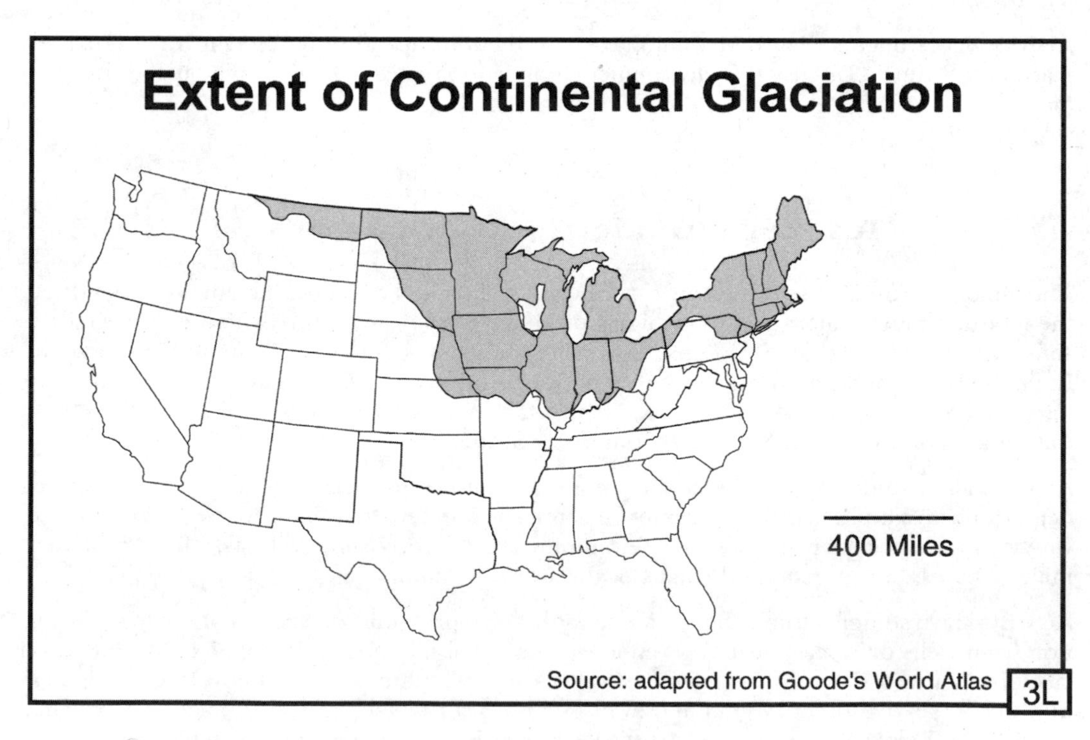
Extent of Continental Glaciation
400 Miles
Source: adapted from Goode's World Atlas
3L

Teacher's Guide for Transparency 3K

This map is like Transparency 3J but with more isolines (Transparency 4U shows how to make an isoline map). Each isoline represents a frost-free season of a specific length, and it separates places that have a longer growing season from those with a shorter one.

- The 4-month line separates the country into places that can grow grain from those that cannot (this is assuming that the places have all other necessary requirements, such as moisture and soil nutrients).
- The 7-month line separates places that can grow cotton and rice from those that cannot. It also is a dividing line between where people tend to need to spend more for heating they do for air conditioning and vice versa.

Activity: Project the Transparency and compare it with Transparency 4W (pre-Civil War plantations). Ask about whether there seems to be a northern limit to the plantation economy (the 7-month frost-free season line). Why? (because plantation-style farming seemed to work best on crops such as rice, in South Carolina; sugar cane, in Louisiana, with an almost 12-month growing season; or cotton, all across the South). Continue with Transparencies 3M and 3N if desired, or save them for another day.

Teacher's Guide for Transparency 3L

The glacier-limit line is another very influential divide in the United States.

Activity: Ask students which of these would expect if a mile-thick mass of ice pushed across the land:

- It would get cold.
- Trees and other plants would get crushed or pushed aside, animals would move.
- Hills might be worn down, loose soil and even some solid rock scraped away.
- Places with weak rocks might be gouged deeper than places with resistant rock.
- Some low places might be filled with soil and rocks that were pushed along by the ice.

And when the ice melts?

- Water would fill the low, gouged-out areas.
- Piles of rocks and dirt might be left where the ice stopped.
- Meltwater floods might carry sand and mud somewhere else.
- Dust might blow east when winds sweep across places where the ice just melted but trees and grass had not yet had time to cover the ground (this one is more sophisticated, but it is economically important and therefore worth noting for some advanced classes).

Now, let's put these abstract consequences on a map.

- What states have a lot of natural lakes and swamps? (Connecticut, New York, Michigan, Wisconsin, Minnesota, North Dakota – swamps and lakes are good for recreation, and, unfortunately, for mosquitoes, but bad for road building).
- What states have land that is more fertile than it would otherwise be because Ice-Age floods and wind brought extra soil into them? (upstate New York, Ohio, Indiana, Illinois, Iowa; most of the Corn Belt).
- What states have easier transportation because hills have been flattened and valleys filled? (Iowa, Wisconsin, and the northern parts of Illinois, Indiana, Ohio).

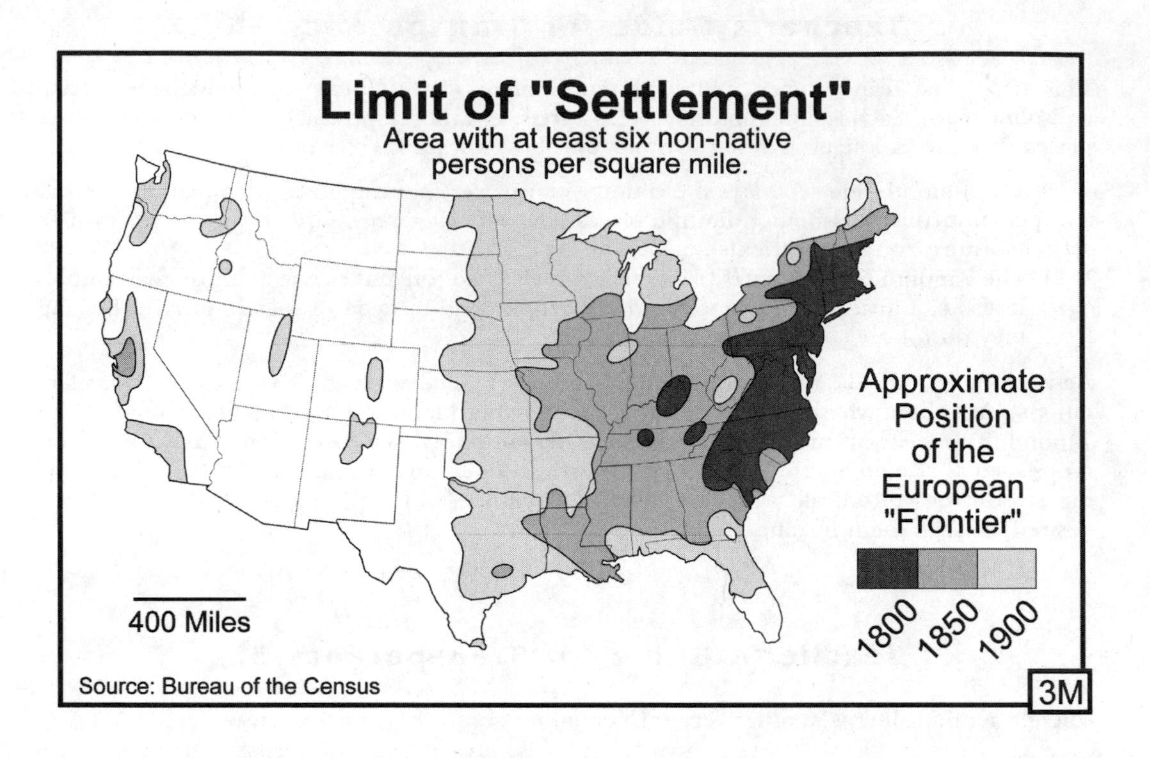
Limit of "Settlement"
Area with at least six non-native persons per square mile.
Approximate Position of the European "Frontier"
1800
1850
1900
400 Miles
Source: Bureau of the Census
3M

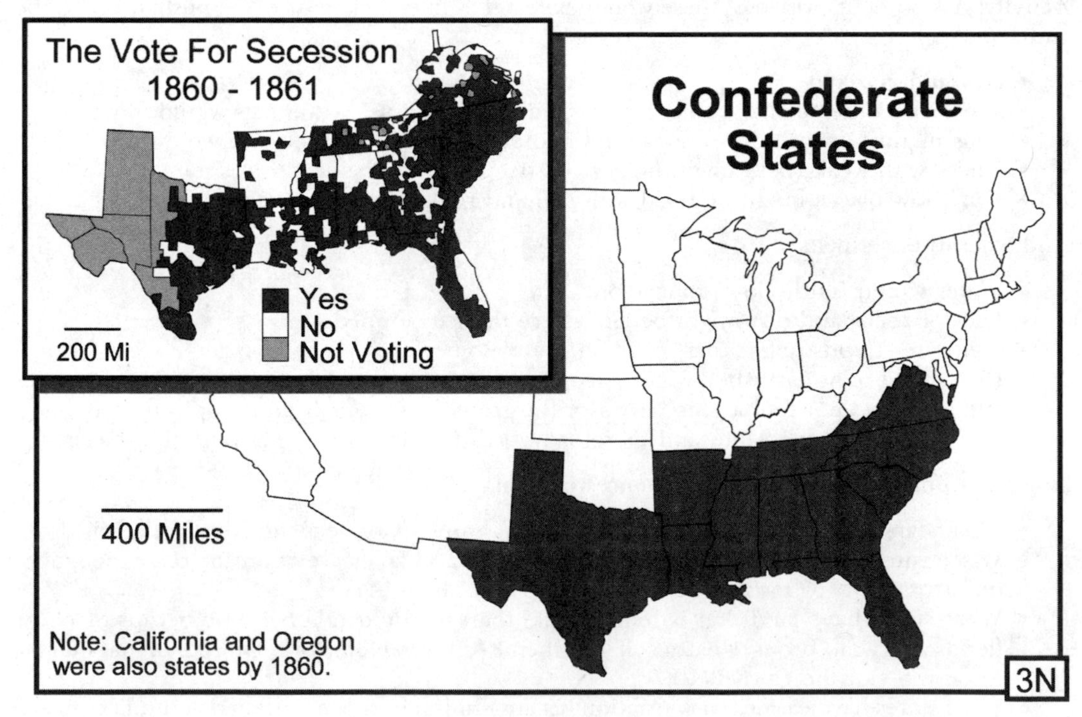
The Vote For Secession
1860 - 1861
Yes
No
Not Voting
200 Mi
Confederate States
400 Miles
Note: California and Oregon were also states by 1860.
3N

Teacher's Guide for Transparency 3M

The position of the settlement frontier at particular moments in history had a strong influence on who ended up where, the kind of economy they started, the kind of houses they built, and so forth.

Activity: Point to a specific place on the map, and ask students when the first immigrants from Europe, Africa, or Asia arrived in that area. Compare that data with a graph of immigration patterns (any good history text should have one; Activity CX of ARGUS provides data; the National Geographic Society's *Historical Atlas of the United States* has a good graph; and the Historical Statistics of the United States, from the Superintendent of Documents, Washington, DC 20402, has good data).

What immigrants moved to an area first? (Spanish to southern Texas, New Mexico, and California; British to New England; Scots-Irish to Appalachia; British and African to the plantation areas of the South; German and Scandinavian to the Great Lakes region; East European to many parts of the Great Plains and the mining/steelmaking regions of New England or Appalachia; Chinese and other Asian peoples to railroad and mining towns of the West).

What consequences of this can we still see? (styles of architecture, festivals, foods, clothing, religious denominations, etc.). This is the kind of topic that can be introduced in early grades and reviewed in later years as students slowly refine their ability to link historical events with present-day landscapes in specific places. This map is a key to making that link and organizing the resulting knowledge; it could be used a dozen times!

Teacher's Guide for Transparency 3N

This transparency shows a geographic pattern that is both effect and cause: it is the consequence of powerful forces, and at the same time it is an underlying cause of some persistent features of American geography, sociology, and politics.

Activity: Project the Transparency, and ask why the shaded states chose to secede (they were the ones that had a plantation economy and slavery).

- Add Transparency 4W to reinforce the conclusion; then add Transparency 3K and discuss the role of the growing season (see the Teacher's Guide for that Transparency).
- Add Transparency 3I (moisture regions). How does this help explain why plantations were where they were? (before deep-well pumps were invented, cotton could not grow west of the moisture-deficit line in middle Texas).
- Add Transparency 3M. How does this help explain why the Civil War began when it did? The Missouri Compromise of 1820 said that new states should enter the Union in pairs, one slave and one free, to maintain the balance of power. By the 1850s, however, the frontier reached the zero moisture-balance line. People who moved to the next states were not likely to grow cotton or use slaves. Therefore, the balance of power, especially in the Senate, would shift to the North. Many Southerners in 1860 must have thought: "it's now or never."

And the consequences? The sources cited in the Teacher's Guide to Transparency 3M above have modern maps of poverty, voting, education, labor unions, and many other features that have been affected by lingering consequences of the Civil War and Reconstruction. In short, some effects of that time are still felt a century and a half later.

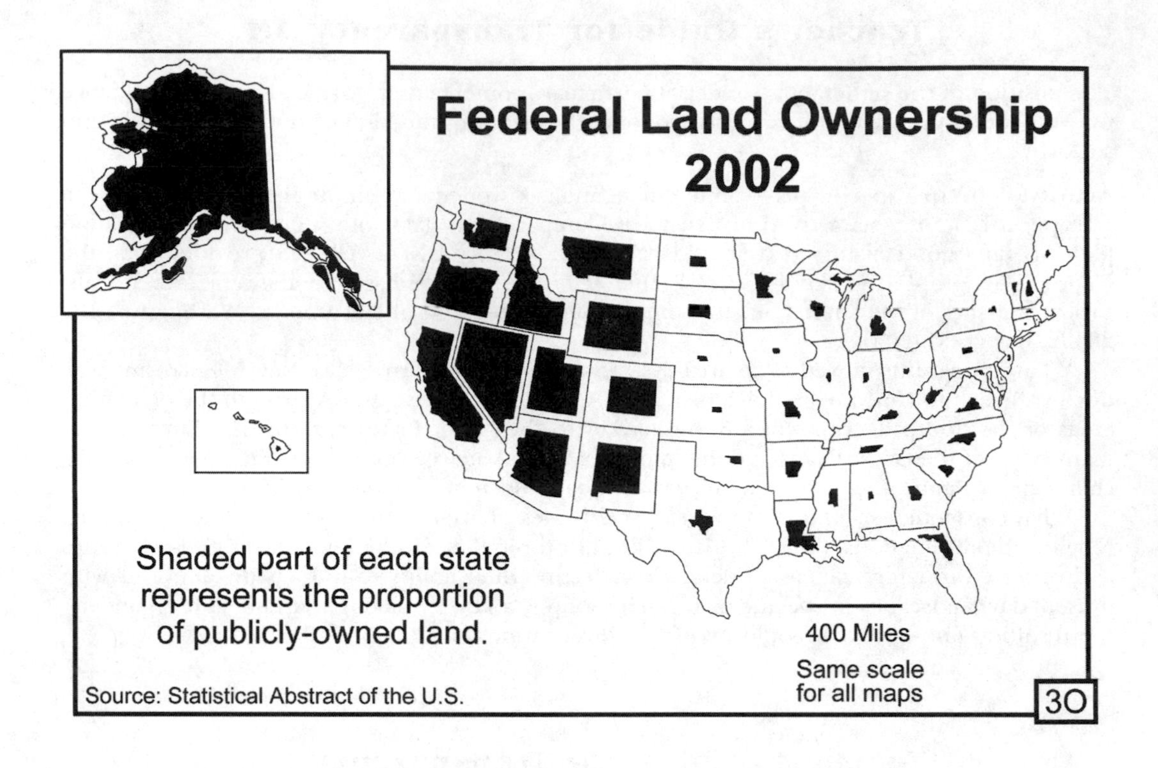
Federal Land Ownership
2002
Shaded part of each state
represents the proportion
of publicly-owned land.
400 Miles
Same scale
for all maps
Source: Statistical Abstract of the U.S.
3O

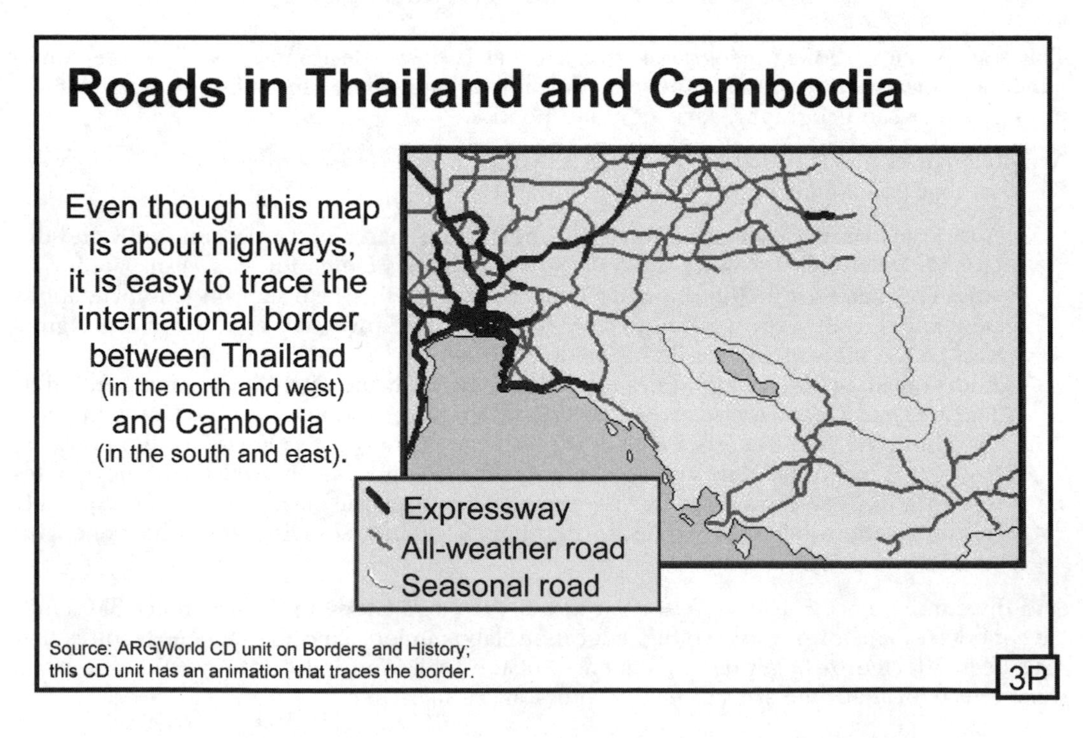
Roads in Thailand and Cambodia
Even though this map
is about highways,
it is easy to trace the
international border
between Thailand
(in the north and west)
and Cambodia
(in the south and east).
Expressway
All-weather road
Seasonal road
Source: ARGWorld CD unit on Borders and History;
this CD unit has an animation that traces the border.
3P

Teacher's Guide for Transparency 3O

The government "owns" more than half the land in many states. This land can be used in many ways – hiking, hunting, logging, grazing, even mining. The users often pay nothing, or only a small fraction of what it would cost if the land were privately owned. That is one side of the issue of government ownership.

The government can change the rules of use almost overnight. People who have come to depend on the use of government land might find their livelihood in jeopardy from one of those decisions. This is a negative side of using public land for private purposes, and it exerts a huge cost, both economic and psychological.

Activity: Have students imagine they do logging and are trying to decide how to invest the money they earn selling wood. They want to do things that would make their logging operation more profitable in the future. Have them think of options (buy a new saw, a new truck, train new workers, plant new trees, start a research station to find a cure for a tree disease, etc.).

"Now that you've made a list of things you might do, let's ask a really messy question. How many of those things make sense if you don't know if you'll be allowed to cut trees in this particular forest next year?"

As if these two sides of the public ownership issue weren't enough, here are some other sides: what other uses might the forest serve if it wasn't used for logging? Is logging on public land fair to people who must pay more to cut trees on private land? In the United States, this is a regional issue: most loggers in the Southeast use private land, and they rightly complain that they must compete on an unlevel playing field with loggers in the West or North. All this is a good topic for an extended role playing (see Table 3-4), with students doing research on various regions and then debating the proper use of government-owned land.

Teacher's Guide for Transparency 3P

If you think public ownership is a knotty issue, try to unravel the effects of colonialism or civil war. Some people blame European colonialism for every ill that affects Southeast Asia and Africa today. Others see is as an issue of the past with little effect on today. The truth, of course, is somewhere in between, and part of the solution to today's problems lies in using maps to find out what present features in a place are legacies of past colonialism.

The road patterns of many countries, for example, reflect past governments in that area. Roads are expensive to build and even harder to move. Once built, however, they affect what people can do and where they can do it. Obviously, it is hard to sell a product if you cannot get it to customers. In Africa and South Asia, roads and railroads built in the colonial era served very different purposes than might be desirable today.

Activity: Trace a pattern of major roads or railroads in an area. Then ask whether the road pattern seems to be better for:

- Getting products to the coast (important for a colonial power).
- Connecting people within the country (better for a country's modern needs).

Other purposes include bringing raw materials to factories, moving people into new areas for settlement, or allowing exploration for oil or minerals. This Transparency provides a context for addressing these questions. The difference in road patterns between these two countries is the result of centuries of history, with different colonial governments, postcolonial politics, and civil wars, and the resulting road patterns certainly have an effect on what people can accomplish there today.

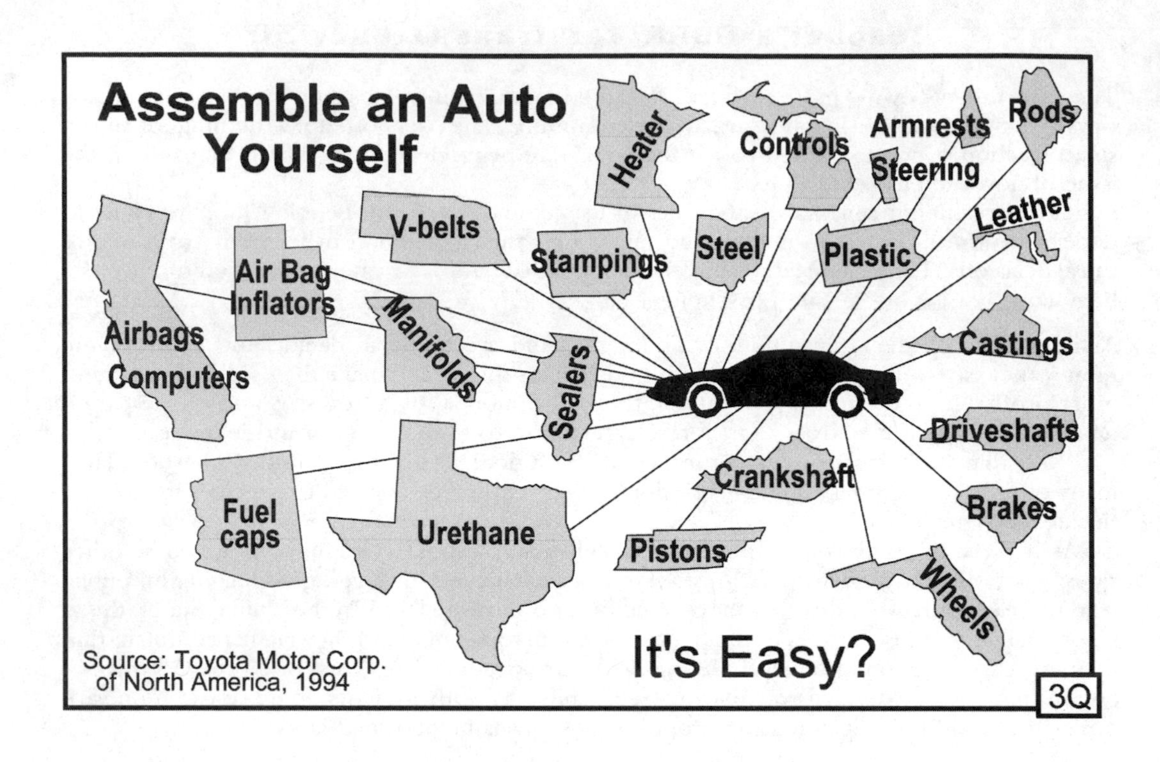
Assemble an Auto Yourself
Heater
Controls
Armrests
Rods
Steering
Leather
V-belts
Stampings
Steel
Plastic
Air Bag Inflators
Manifolds
Airbags
Computers
Sealers
Castings
Driveshafts
Crankshaft
Fuel caps
Urethane
Brakes
Pistons
Wheels
It's Easy?
Source: Toyota Motor Corp. of North America, 1994
3Q

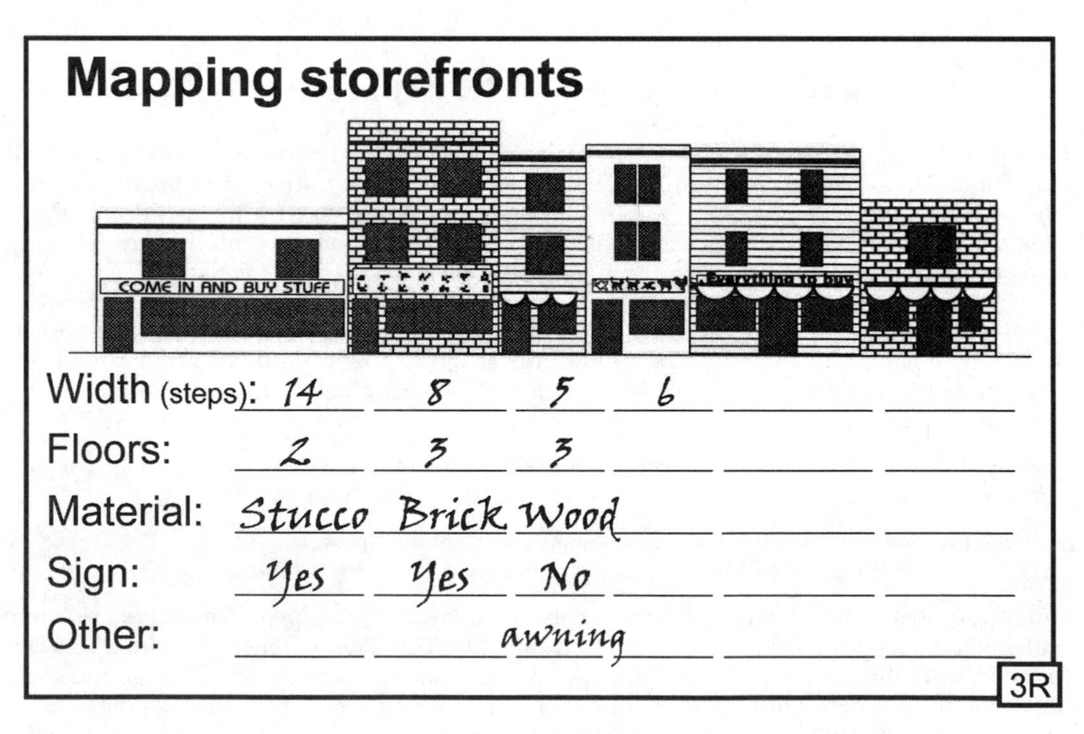
Mapping storefronts
COME IN AND BUY STUFF
Everything to buy
Width (steps): 14 8 5 6
Floors: 2 3 3
Material: Stucco Brick Wood
Sign: Yes Yes No
Other: awning
3R

Teacher's Guide for Transparency 3Q

In the past, industrial geography used to be fairly simple – people tended to find resources and make something with them. For example, the Shoshone people in Wyoming made arrowheads and tools from obsidian, a rare rock that occurs there. The Algonquin people made tools and jewelry from copper they found in what is now Upper Michigan. People of European descent made cloth in water-powered mills, many of which were in New England, because that region had the essential ingredients for natural waterfalls – plenty of surplus moisture and rugged terrain.

Nowadays, products and industrial processes are more complex. Factories often make only part of a product. Considerations such as skilled labor, venture capital, tax rates, school systems, even cultural amenities such as orchestras or art museums play a role in how people decide to locate factories. This makes the question for academic geographers, "Why is that factory located where it is?" harder to answer. It also adds complexity to the question for applied geography, "Where is the best place to put this kind of factory?" Finally, it raises a question for speculative geography, "Are factories in this country being put in appropriate locations?" — and it pushes that question into the limelight of talk show and Congressional hearings.

Activity: Ask students to list some things that factory owners might consider in deciding where to locate a factory. Don't settle for a safe "textbook" list of abstract terms such as raw materials, markets, and so on. Insist on their listing specific resources and the products that might be made from them. Have students describe how the age and income of the population can influence what products are wanted. They should include features such as crime, local politics, sewer charges, parks, air pollution, anything else students think might be important. Then have students rate their own community (or another place of interest) by those criteria. They should create maps and graphs to show how their community would compare with other alternatives.

Teacher's Guide for Transparency 3R

Map making is an exercise in gathering data systematically and using standard symbols to depict results. A way to practice map-making skills is to make a detailed survey of a small area in a community. This Transparency can help prepare students for that task. Have them fill in the blanks for a row of storefronts and buildings on a commercial block. The example comes from an old Pennsylvania mining town, but answers may not always be so clear. For example, how do you classify an old brick building with a new metal facade? Have students describe some buildings that they pass on their way to school, and then discuss how they might handle ambiguous categories in their final survey.

Activity: Ask students to walk a block and record what they see. For example, one entry might read: "52 feet north from corner, three-story brick building, sign says built in 1897, first floor used as hardware store, two mailboxes in a stairway imply that people live on the second floor, but a broken window on the third floor suggests that it is empty."

The next step is to impose some order on the mass of information. This is usually done by considering one variable at a time, choosing reasonable categories, and recording the information.

This is not an exercise in duplicating what a building inspector or other government official would record. There is no "right" answer. The goal is to record useful information in a systematic way and then to communicate it clearly. Planning the survey in class can get students thinking about categories. In short, mapping a block lets students wrestle with issues of observation and classification; those issues, in turn, underlie the development of better theory.

Latitude - what is it?

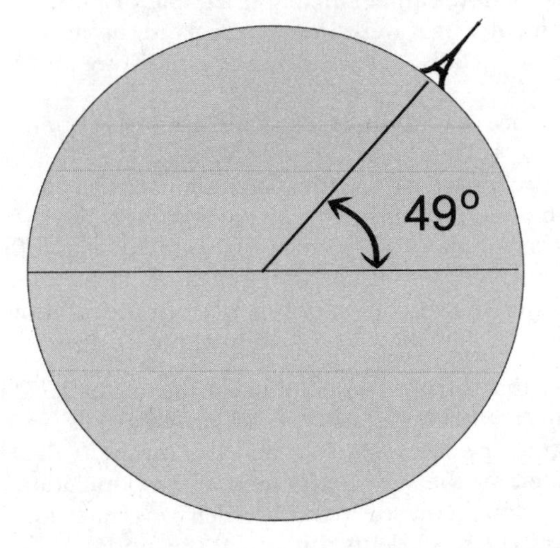

It's the angle north or south of the Equator

Paris, France, has a latitude of 49 degrees; it is that many degrees north of the Equator.

(What important border is at the same latitude?)

4A

Describing location

There are many ways to say *where* something is

Landmark - next to the old church

Topological - between the pond and the forest

Distance/Direction - three miles northeast of Lily

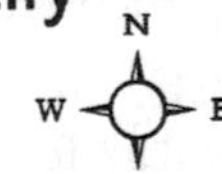

Address - at 1910 Maple Street

Global grid - at 34°S and 151°E

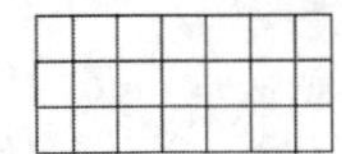

Map grid - in map sector 3B

4B

Teacher's Guide for Transparency 4A

The concept of latitude is straightforward, but a feature that's included on a map among a whole host of other features can always have the potential to be confusing. "Where do we start?" is a common question when students are confronted with complex maps or diagrams. This is not surprising – one definition of geography is "the science of what's happening in many different places at the same time."

Activity: Use washable markers to trace key features directly on the Transparency:

- Suppose we want to find the latitude of Paris (highlight the Eiffel Tower).
- What is the definition of latitude? (the angle – technically, the "angular distance" – of a place away from the equator).
- Here is the equator (highlight the horizontal line).
- Here's the line that goes up to Paris (highlight the diagonal line).
- And here is the angle we want (highlight the curving two-headed arrow).
- Optional: We could measure it with a compass (put a plastic compass on the diagram and show how to measure the angle).
- Refer to a globe (or atlas) and ask, What important border is at the same latitude? (the long border between the United States and Canada, from Minnesota to Washington).
- Optional: What does this tell us about Paris? (it has about the same amount of daylight in winter as North Dakota). Most Americans do not realize that almost all of Europe is well north of the latitude of New York City or Chicago.
- Now, let's check our understanding by trying another place. Where should I draw a diagonal line to show the latitude of Rio de Janeiro? It's at 23 degrees South latitude. Tell me when to stop (slide the point of the marker down from the equator along the outside of the earth until students tell you to stop; repeat with other cities until students can answer quickly).

Teacher's Guide for Transparency 4B

People have invented dozens of "languages" to describe the locations of things. This Transparency can be used to guide discussion of different ways to describe a location.

Activity: Use a 4″ × 6″ card to hide the sample descriptions on the right. Then, go down the list of terms and explain what each means, asking students for examples. Uncover the examples on the Transparency when appropriate.

An alternative is to cover the list and reveal each line as discussion proceeds.

Another way to use this Transparency is as a summary of a free-form discussion. Start by asking students how many different ways they can think of to describe the location of their school, a theater, a shopping mall, or some other well-known local place. Write their suggested terms on the board as the discussion proceeds; follow the order of the Transparency, which goes in a rough logical sequence from concrete (landmark) to abstract (latitude and map grid).

Discussion of this Transparency can also serve as background for the Activity that accompanies Transparency 4C.

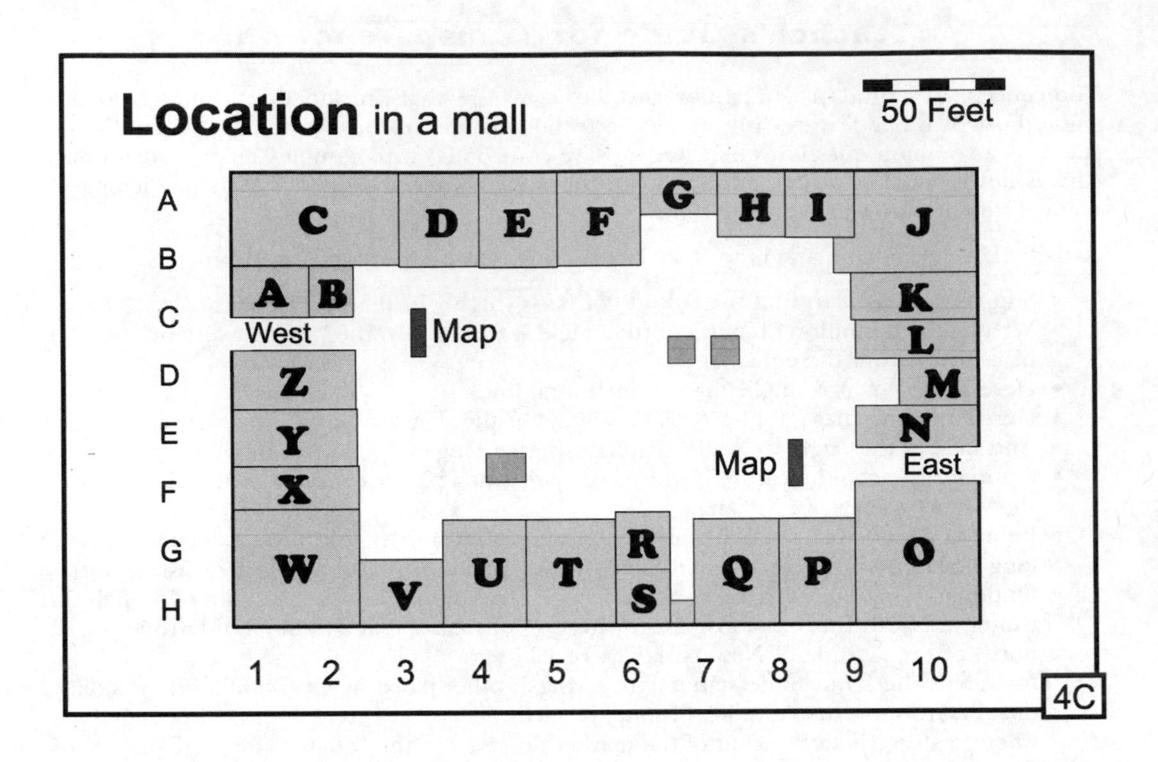
Location in a mall
50 Feet
A
B
C
D
E
F
G
H
C
D
E
F
G
H
I
J
A
B
K
L
M
N
West
Map
Map
East
Z
Y
X
W
V
U
T
R
S
Q
P
O
1
2
3
4
5
6
7
8
9
10
4C

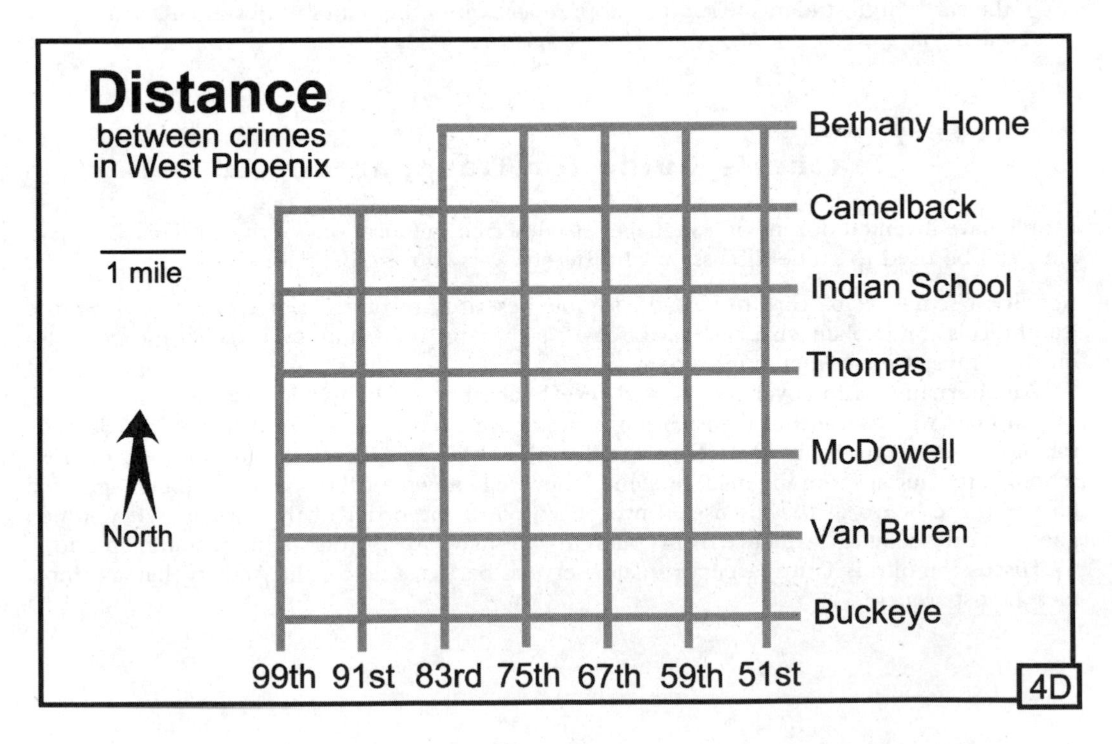
Distance
between crimes
in West Phoenix
1 mile
North
Bethany Home
Camelback
Indian School
Thomas
McDowell
Van Buren
Buckeye
99th
91st
83rd
75th
67th
59th
51st
4D

Teacher's Guide for Transparency 4C

This Transparency shows a typical shopping mall (maybe point out some features, like the entrances, maps, and information kiosks). In this mall, someone could describe the location of store U as:

- Between stores T and V.
- Directly across from stores D and E.
- The sixth store to the right from the west entrance.
- 100 feet west of the east "you are here" map and then south.
- At map coordinates H4.

These examples of location description use landmarks, topology, approximate distance, measured distance, or map coordinates to specify location (see Transparency 4B).

Activity: Have one student write down the location of a store and pass it on to the next student to their left. That student, in turn, "decodes the message," figures out what store is being described, writes its location in another "language" or mode of describing it, and passes their new description on. In this game of "telephone," see if the last person in line is still describing the same spot as the first.

Perhaps repeat this activity with a real map of a local shopping center, stadium, golf course, or other place of interest.

Teacher's Guide for Transparency 4D

After location, distance is typically the next basic spatial concept introduced. The ability to measure distance and translate map measurements into real-world terms is important for many purposes.

Activity: Write down a list of addresses and times on the board:

Camelback and 67th	9:15	9:20
McDowell and 59th	9:34	9:45
Van Buren and 91st	9:50	10:01
Indian School and 75th	10:17	10:25
Buckeye and 83rd	10:31	10:38

Imagine a store at each of these intersections; the time in the first column is when a clerk said that a robber entered the store, and the time in the second column is a phone company record of the time the robber cut the phone cord in that (imaginary) store.

Presuming the robber was trying not to attract attention by driving too fast, it would take 2 minutes to go 1 mile. The question is, could one robber have committed all of the crimes? In crime-novel jargon, are these robberies the work of a serial criminal or some copycats?

To answer this question, students have to figure out the shortest route between the locations, count the blocks, translate the blocks into time, and compare the results with the time elapsed between reported robberies. This Transparency uses the street names and pattern of West Phoenix, which has flat land and a very simple street pattern. The exercise could be made more complex (and more interesting) by using a location with one-way streets, diagonal roads, curving roads, parks, or other complications in the traffic pattern.

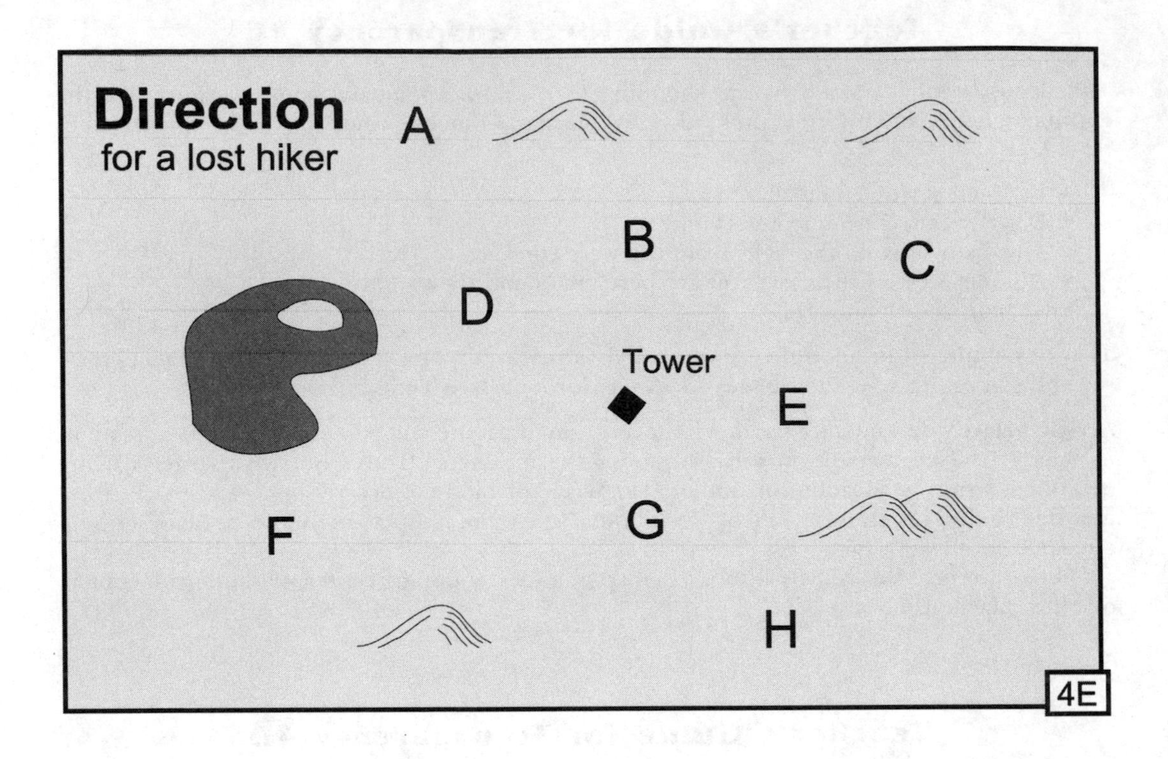
Direction
for a lost hiker
A
B
C
D
Tower
E
F
G
H
4E

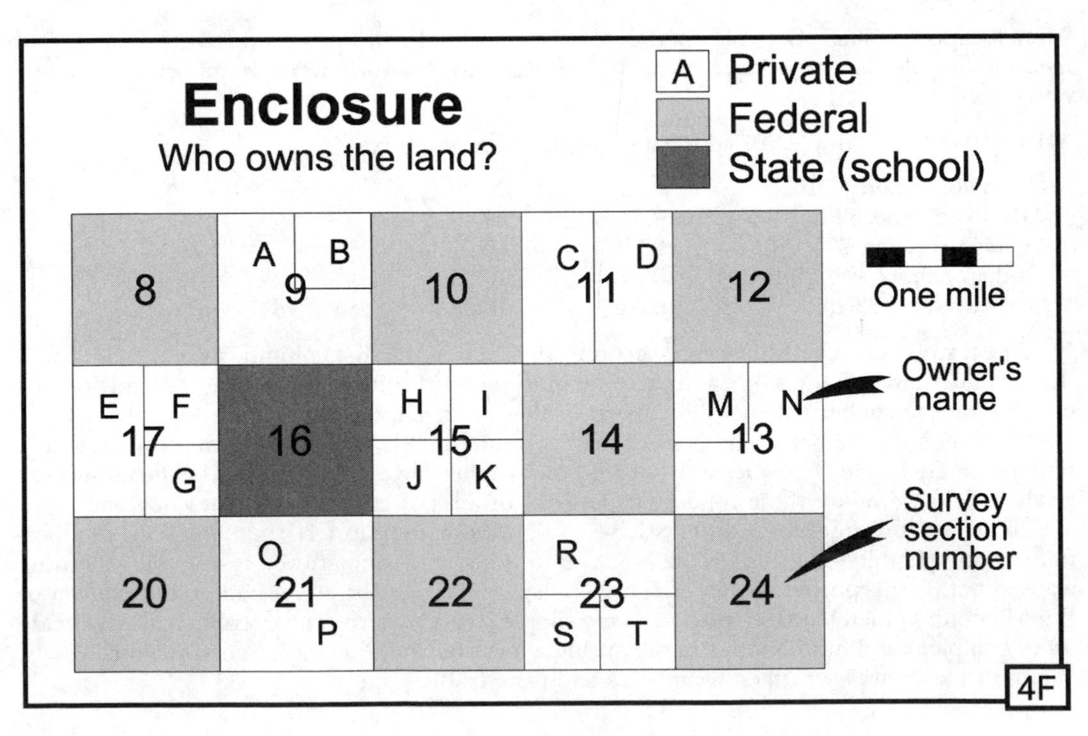
Enclosure
Who owns the land?
A Private
Federal
State (school)
8
A
B
9
10
C
D
11
12
One mile
E
F
17
G
16
H
I
15
J
K
14
M
N
13
Owner's name
O
20
21
P
22
R
23
S
T
24
Survey section number
4F

Teacher's Guide for Transparency 4E

Direction is the third basic spatial concept, after location and distance. This map shows a small area in the glacial hills of western Massachusetts.

Activity: Ask students to imagine that they are hikers in the terrain depicted by this map and can see landmarks in various directions. Then ask:

- Where are you if there is a tall tower directly south of you? (use a nonpermanent marker to draw lines southward from several of the lettered places; place B is the only one with a tower directly to the south).
- Where are you if there is a single hill to the north and a double one to the south? (using two clues helps reduce the chance of error).
- Where are you if there is a tower to the west and a double hill to the southeast? (intermediate directional names may be needed to communicate the location of a place).
- Where are you if you see a lake to the south-southwest and a hill to the east? (intermediate directions can become confusing, which is why many compasses have numbers, allowing more precise readings, usually in degrees clockwise from north).

Once students master the basic idea with this simplified map, use a topographic map of a local area (or some other area of interest). To make the exercise more challenging, omit the letters and have students try to mark where they are or use coordinates to describe their location. In upper grades, you could introduce the complication of magnetic versus true north. In short, like any fundamental skill, finding position by "direction backsighting" has several levels of difficulty and therefore can be treated at several grade levels.

Teacher's Guide for Transparency 4F

This map uses the "language" of the U.S. Public Land Survey to describe tracts of land. This survey was used in most areas west of the Appalachian Mountains (except in French settlements in Louisiana and along the Mississippi River, where a longlot system was used, and in Spanish land grants, in parts of Texas, New Mexico, and California).

This particular map shows part of the northern Rocky Mountains and has a distinctive "checkerboard" pattern of land ownership, because the government granted alternate sections of land to companies to build railroads. Hundreds of thousands of square miles of the American West have this kind of land ownership pattern, with a mix of private and public ownership.

Activity: Tell students a cousin has written them to say they have discovered gold in Section number 10; the cousin is asking the student to study the map to see who owns the area near the gold discovery (it is owned by the federal government and therefore may be available for lease). Repeat the request for other areas, such as the southwest quarter of Section 15 (owned by J); the NW 1/4 of Section 23 (owned by R); and so forth.

Activity: Ask students whether tract A is bigger or smaller than tract C. That's easy – now, which is bigger: a ranch that owns tracts A, F, and E and leases Section 8 from the government, or a ranch that owns tracts O, P, R, and S and has a lease on Section 22?

As with most skills-based activities, it might be better to use local topographic maps rather than this one to help teach the basic ideas of owner identification and area estimation. These can be bought from the U.S. Geological Survey (*www.usgs.gov*) or downloaded from *www.topozone.com*.

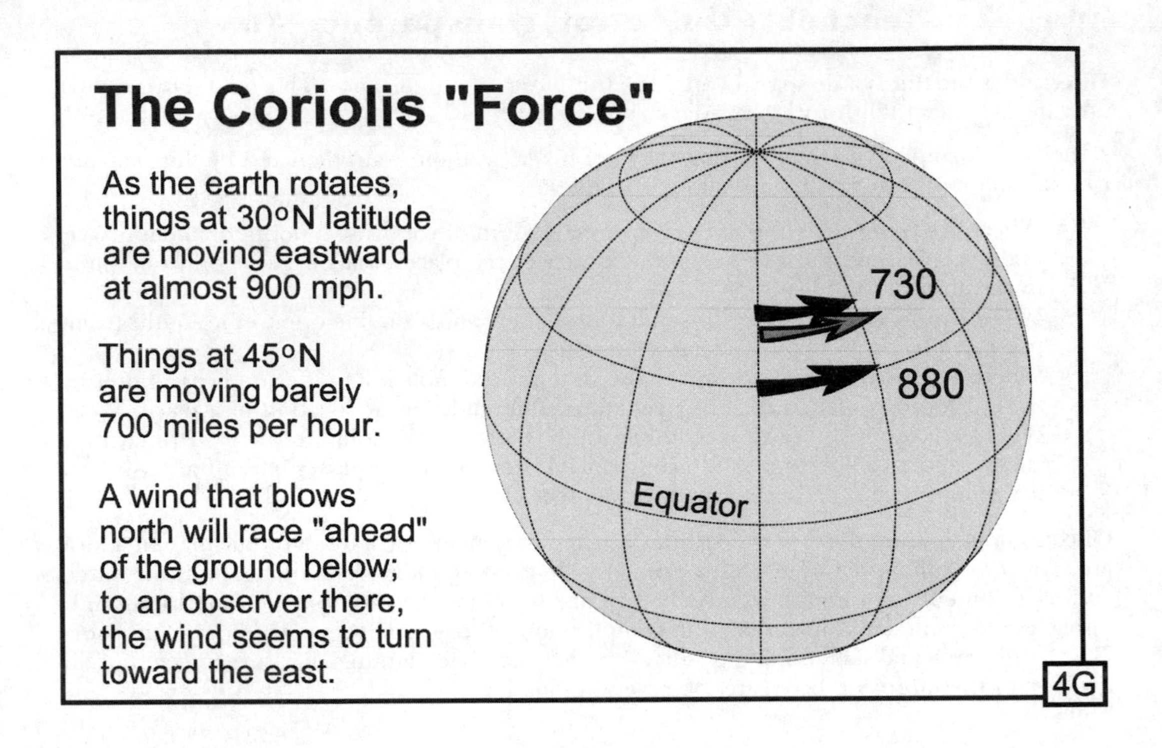

The Coriolis "Force"
As the earth rotates, things at 30°N latitude are moving eastward at almost 900 mph.
Things at 45°N are moving barely 700 miles per hour.
A wind that blows north will race "ahead" of the ground below; to an observer there, the wind seems to turn toward the east.
730
880
Equator
4G

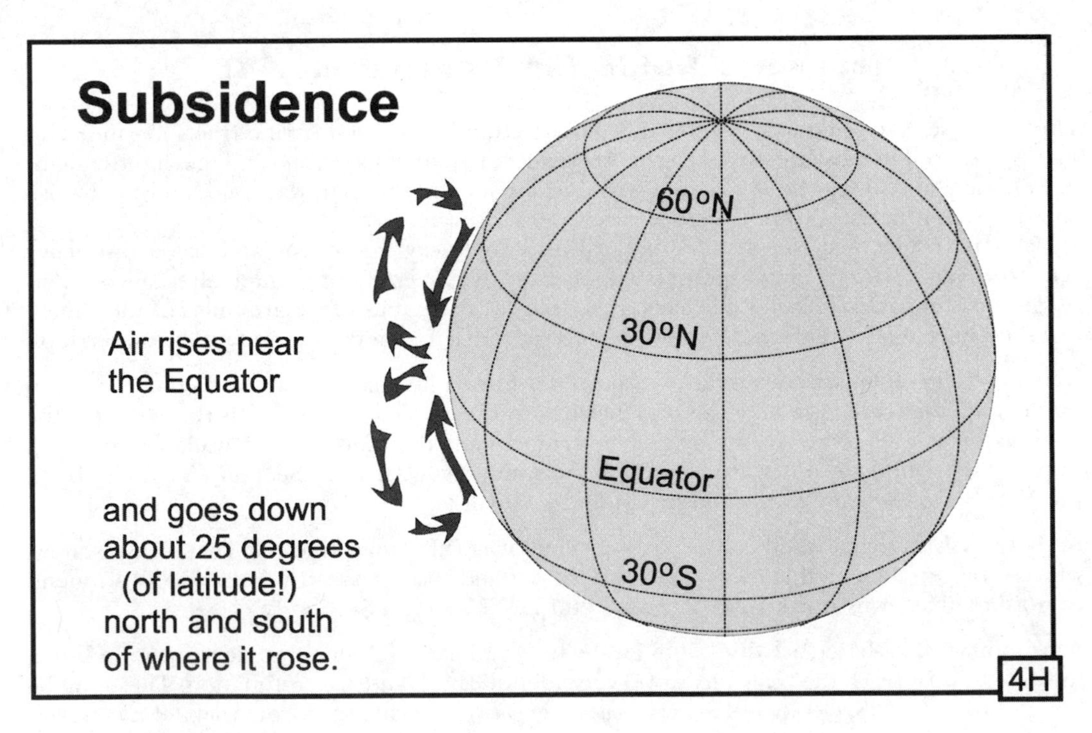

Subsidence
Air rises near the Equator
and goes down about 25 degrees (of latitude!) north and south of where it rose.
60°N
30°N
Equator
30°S
4H

Teacher's Guide for Transparency 4G

The Coriolis "Force" is an important but hard-to-teach "fact" about the earth. Part of the problem is that there are some erroneous popular "theories" about it (e.g., "notice how water swirls when you unplug a sink?") That swirling is just fluid dynamics, not Coriolis. The Coriolis effect operates on a scale of hundreds of miles, and it deflects any object in motion. The push is to the right of the direction of movement in the northern hemisphere and to the left in the southern. This is why large things like hurricanes always spin in a predictable way.

Activity: Use washable markers to highlight and draw arrows as you go through a logical sequence of steps such as the following:

The earth is spinning from west to east (sweep the pen from left to right across the diagram, but don't draw anything yet).

As a result, every place on earth is in motion. Here, at 30 degrees north latitude, the surface is moving about 880 miles per hour to the east (trace the arrow from left to right).

Here, at 45 degrees N, it is going only 730 mph (trace the arrow).

Remember, air at 30 degrees is moving nearly 900 miles an hour (point). If that air moves north, it is like merging into a slow lane on a highway. The air is going faster than the ground below (trace from the tip of the 30-degree arrow to the tip of the hollow arrow at 45 degrees; start north along the meridian and then make an exaggerated rightward curve, so that you end at the tip of the 45-degree arrow; a curving path is more accurate and also more evocative than a straight line).

To an observer on the ground, the wind seems to veer off to the right, toward the east (repeat the curve).

That is why Americans tend to look to the west to see what weather is coming; most of us live in the zone of eastward-moving air (continue the movement eastward at 45 degrees N).

The same thing happens at the equator; air is going nearly a thousand miles an hour there (draw a new arrow to the east).

As it rises and then moves north, this air also curves to the east (draw a curve that starts at the equator and is heading east by the time it gets to 30 degrees latitude).

These eastward flows are the jet streams, and they help make the weather that people feel below (shift to Transparency 4H).

Teacher's Guide for Transparency 4H

Trace and explain: Heat from the sun makes air rise at the equator. This makes thunder and rain (trace the upward arrow above the equator).

Then the air moves north and south (trace the northward arrow high above the surface).

As it goes, the air is turned to the east by the Coriolis effect (the inevitable result of motion on the spinning earth - remember Transparency 4G). That deflection in turn causes crowding, which pushes air down toward the surface (trace the downward flow).

Downward flow makes deserts here, about 25 to 30 degrees of latitude north (and south; draw clockwise ellipses over the arrows in the Northern Hemisphere to show the cycle; then shift to Transparency 4I).

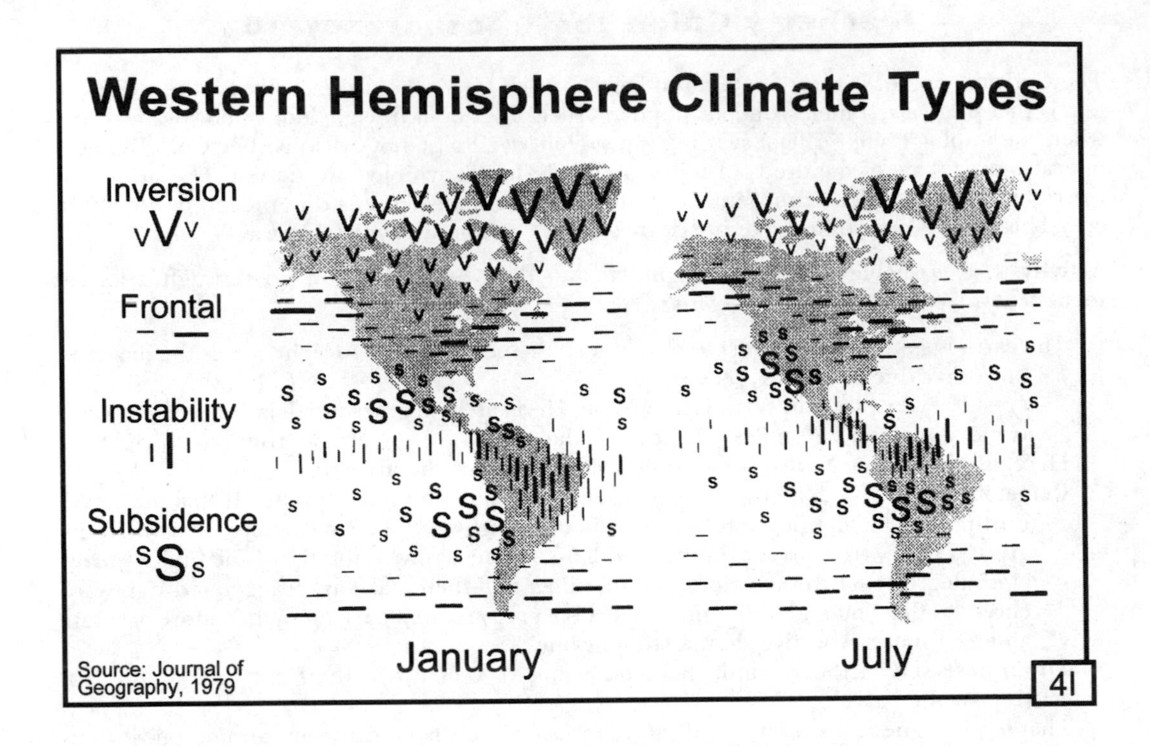
Western Hemisphere Climate Types
Inversion
Frontal
Instability
Subsidence
January
July
Source: Journal of Geography, 1979
4I

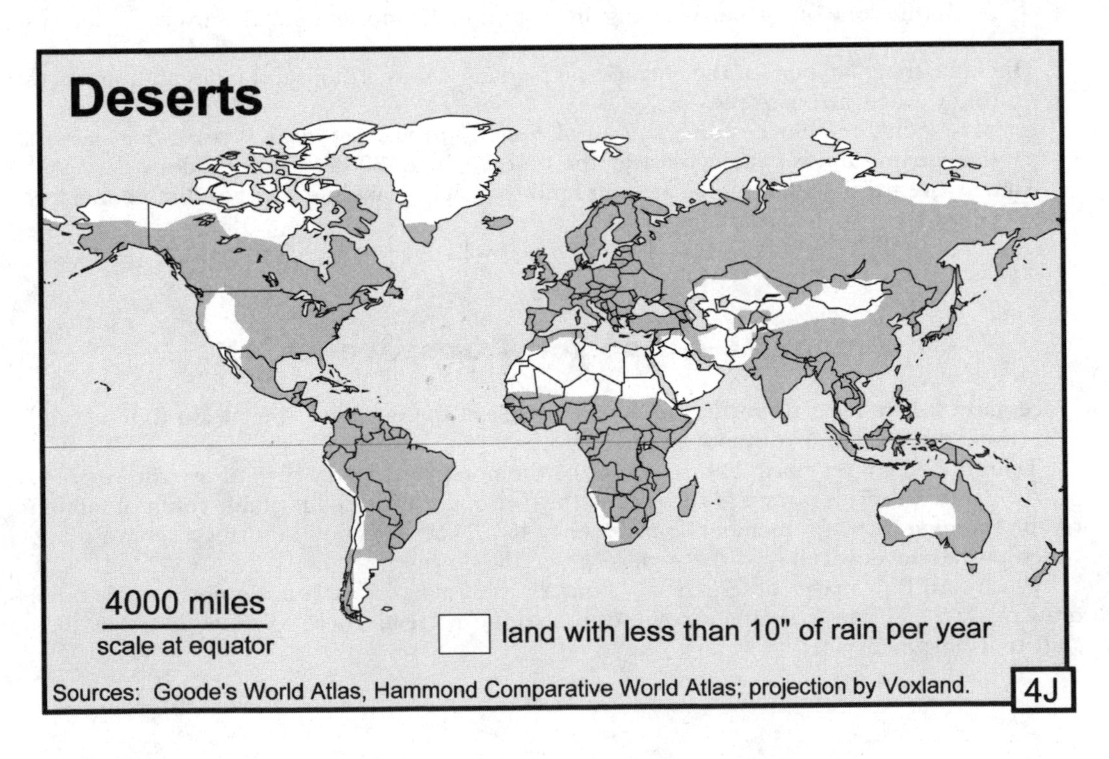
Deserts
4000 miles
scale at equator
land with less than 10" of rain per year
Sources: Goode's World Atlas, Hammond Comparative World Atlas; projection by Voxland.
4J

Teacher's Guide for Transparency 4I

To explain reasons for the process shown on this map, you could start with Transparencies 4G and 4H. Or, you could just begin by asserting that surplus solar energy near the equator sets in motion a long chain of events that have consequences all over the world. Either way, one especially important result of equatorial heating is a persistent downward movement of air at about 25 to 30 degrees north and south latitude (the actual position depends on a number of factors, including season, ocean currents, and arrangement of mountains).

Activity: Point to a specific place, such as southern California. Note that it is in a rainy, frontal climate in January; ask what kind of weather it has in July.

This kind of map summarizes information, so that we do not have to understand all of the forces that cause a particular result. The map shows the places in the world where downward movement of air occurs, which lets us trace the consequences of this downward movement without necessarily knowing the causes. That, in a nutshell, is one of the great strengths of maps, but only insofar as people are aware of one important thing: a map is just a communication device. It is not a fact! A map is a summary of what we understand about the world, not a record of what is "really" there.

Keeping map appreciation from becoming too unquestioning can be tricky. This is why I recommend working "backward" through some (not all!) maps to underlying assumptions and theories. The influence of the subsidence on human life is so great (and the resulting pattern is therefore so worth knowing) that Transparency 4I is a good map to use for such backward analysis - hence Transparencies 4G and 4H.

Merely knowing about sinking air, however, is not enough. It's the results that are important. Transparency 4J shows a global pattern that is one of those results: the world-class deserts that are created in the zones of subsiding air.

Teacher's Guide for Transparency 4J

Rain occurs when moist air rises. (Technically, condensation occurs when air cools; air cools when it expands; and air expands when it rises.)

Activity: Let's make a list of ways nature can make air rise (the sun can heat it until it rises all by itself; wind can push it up a mountainside; colder air can push it up).

Where does this happen? (surface heating? in Brazil, or in Texas, Alabama, etc. on a summer afternoon; wind pushing air up mountains? places like Oregon, Norway, or northern India; cold air? wherever the weather map says a cold front is coming).

Now make a list of ways nature could keep rain from happening (have the air be dry; have the air stay still or move downward).

Here is a map of land areas that get less than 10 inches of rain. Compare the map of downward moving air (Transparency 4I). How many of the world's deserts are "explained" by Subsidence? (Sahara, Arabian, Mojave, Atacama, Kalahari, Great Australian - all the deserts between 25 and 30 degrees latitude).

What explains the other dry areas? Well, where are they? (mostly way up north). So why are they dry? (third grade answer - the air is too cold to rain; junior high - cold air cannot hold much moisture; university - the partial vapor pressure at low temperature is too low).

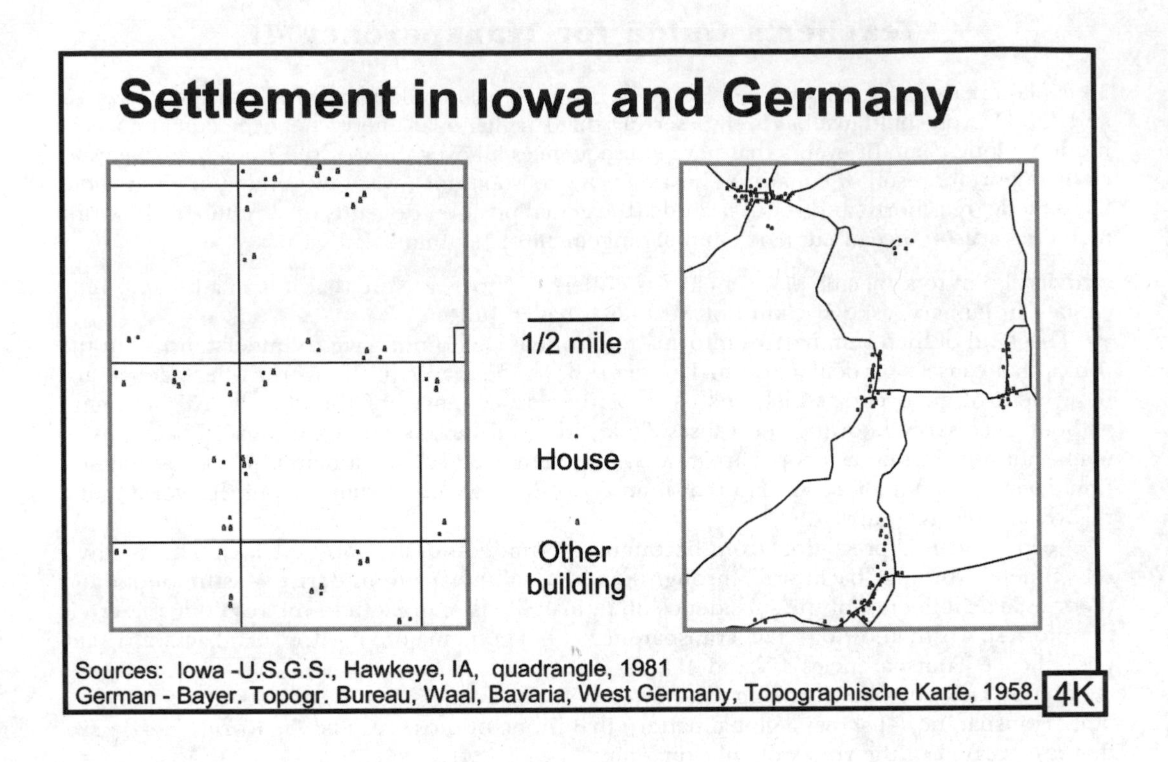
Settlement in Iowa and Germany
1/2 mile
House
Other building
Sources: Iowa -U.S.G.S., Hawkeye, IA, quadrangle, 1981
German - Bayer. Topogr. Bureau, Waal, Bavaria, West Germany, Topographische Karte, 1958.
4K

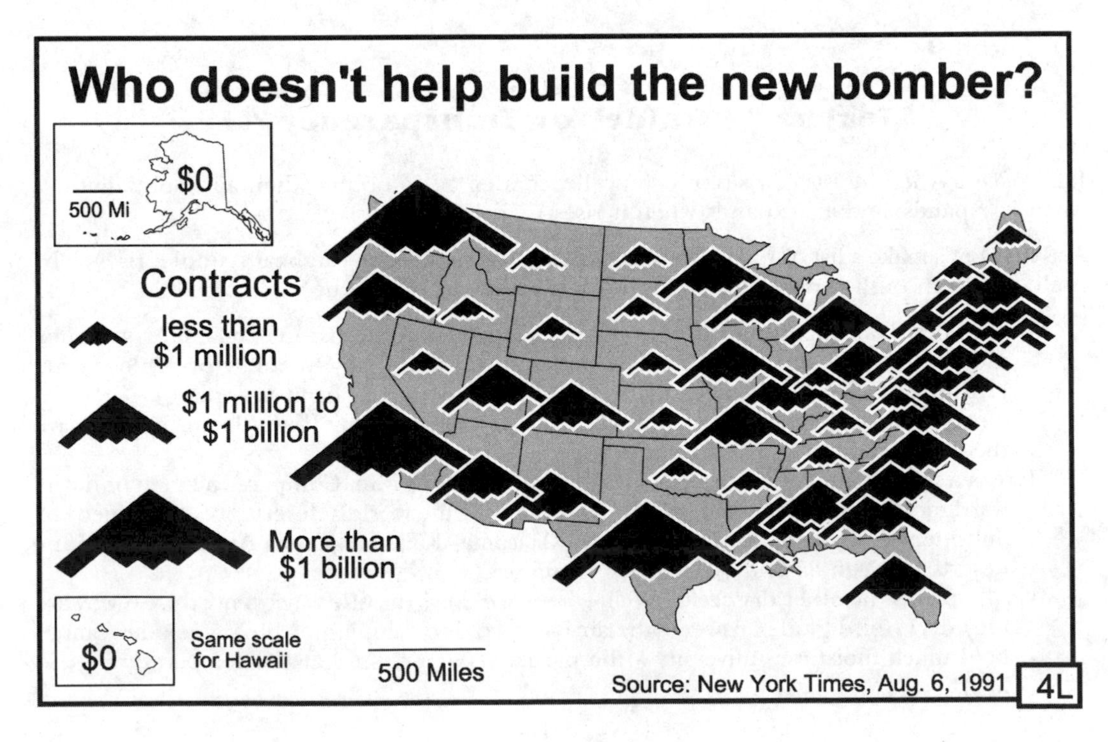
Who doesn't help build the new bomber?
$0
500 Mi
Contracts
less than $1 million
$1 million to $1 billion
More than $1 billion
$0
Same scale for Hawaii
500 Miles
Source: New York Times, Aug. 6, 1991
4L

Teacher's Guide for Transparency 4K

Architects have a saying: "We mold space into buildings, and then our buildings help mold us." Geographers extend this idea to the entire landscape. The first people to inhabit an area oftent have quite a bit of freedom to put property lines, fences, roads, buildings, and other features wherever they want. Once in place, however, those features have a powerful influence on what people in that area can do in the future.

This is especially true in the case of buildings. They are large and hard to move, and therefore they influence where other things are built around them. People in different places have different ways of arranging houses and other buildings.

Activity: Match maps of settlements with places on a map. Find pictures of different arrangements of houses - examples might include an Iowa farmstead, rowhouses in an East Coast city, adobe houses in Arizona, a cluster of concrete high-rise apartments in Moscow, a Bavarian chalet, a village of stilt houses in the Amazon rainforest, a Tibetan monastery, and so on. Post the pictures and a map with letters showing where the pictures were taken. Have students try to match houses with their locations (like Transparencies 3C and 3D, but with an emphasis on the arrangement of houses as well as their individual style).

Activity: Pass out maps of different arrangements of houses and have students measure the distance a mail carrier has to travel in order to deliver mail to 20 houses. What about emergency ambulance service: what is the average distance from the hospital to a typical house? This activity develops in students the concrete skill of measuring and causes them to consider the choices people make in how to arrange houses. Such arrangements have trade-offs - for example, cheap mail service might require people to live closer together than they would like.

This activity has a strong citizenship component, because the arrangement of houses has implications about social relations, energy use, and other costs. One cannot deal with these issues without examining the influence of the geographic arrangement of key features in our "built environment." The Measuring Distance unit on the CD deals with this idea.

Teacher's Guide for Transparency 4L

Tracing the sources of material for a complex product is another aspect of the same basic question about where to put things. With any complex product, there are trade-offs. Having all the parts factories close together lowers the cost of transporting parts. On the other hand, some parts might require specialized raw materials or labor skills that are available only in specific places. Moreover, there can be political considerations with a military product. The map shows that factories making subassemblies and parts for this bomber are located in nearly every state. This pattern makes it harder for Congress to reduce military spending for the bomber, because closing a weapons factory will have an impact on jobs in many Congressional districts around the country.

Activity: Examine this Transparency and try to figure out what airplane parts might be made at specific locations as a result of local features. For example, aluminum frames might be built in the Pacific Northwest because aluminum refining uses a lot of energy, and electricity from big river dams is cheap and plentiful there.

> (How do we know that? It is the scissors principle in action: in this case, a puzzling map about airplane parts makes it seem worthwhile to search an almanac, atlas, or computer data set for information about local places. Why haven't I provided all the "answers" for the activities in this book? Because a complete answer for this question would take at least half a dozen pages, which would be multipled by 80-odd activities, and this is supposed to be a short book about teaching geography, not a big fact book about geography!)

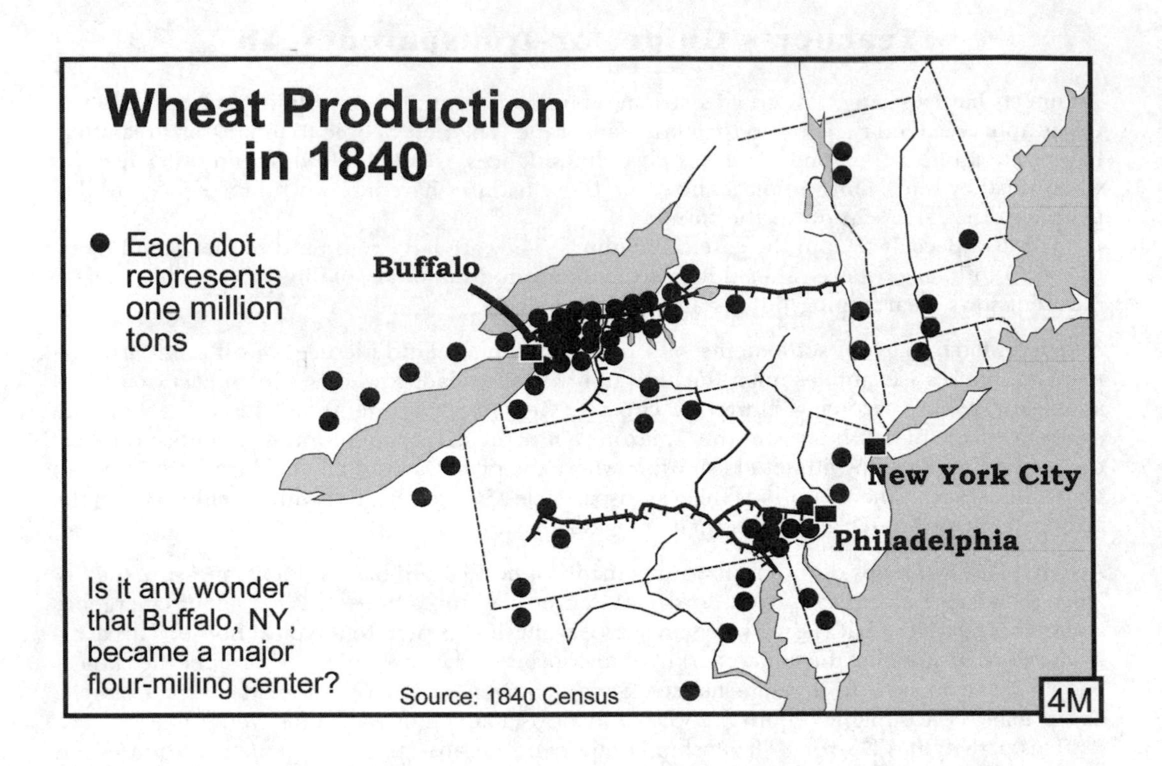

Wheat Production in 1840
Each dot represents one million tons
Buffalo
New York City
Philadelphia
Is it any wonder that Buffalo, NY, became a major flour-milling center?
Source: 1840 Census
4M

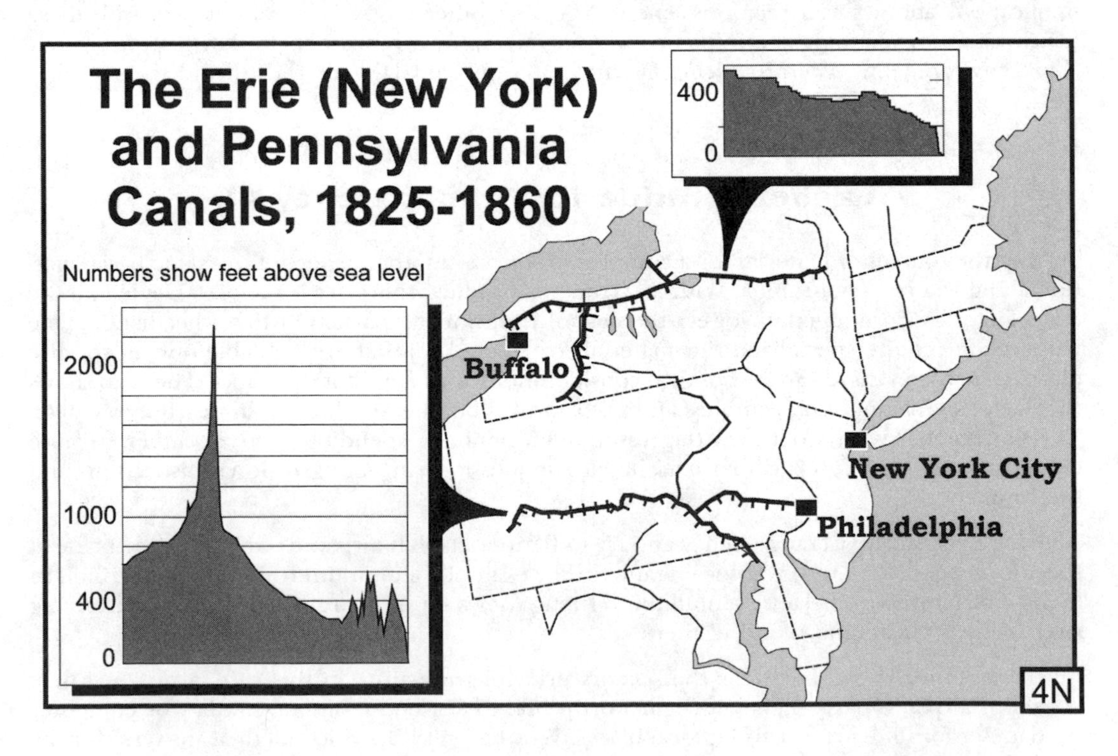

The Erie (New York) and Pennsylvania Canals, 1825-1860
Numbers show feet above sea level
400
0
2000
1000
400
0
Buffalo
New York City
Philadelphia
4N

Teacher's Guide for Transparency 4M

Subsistence farmers (those who grow their own food) usually raise a variety of crops in order to have a balanced diet. Wheat, by contrast, is usually a **commercial** crop: it is grown to be sold to others.

Activity: Ask students if they would describe wheat as heavy, like coal; or light, like diamonds or computers (this is a slightly tricky question, because the appropriate unit of measurement is not the weight of an individual grain of wheat or a diamond ring; it is the weight of comparable values of different products. For example, $100 worth of wheat weighs half a ton, whereas an equivalent value of computer chips weighs only a few grams). So, wheat is a heavy product.

To sell a heavy, perishable product such as wheat, people need either:

- A location close to the customers.
- A good transportation system to get the product to the customers.
- A factory that can transform the product into something lighter and less perishable (does that help explain why the legislature in North Dakota voted to subsidize a pasta factory?).

How does Buffalo, New York, illustrate several of these principles? (encourage speculation, then show Transparency 4N).

Teacher's Guide for Transparency 4N

This Transparency will be easier to read if you use a permanent marker before class to (lightly) color New York and Pennsylvania and their profile lines.

To change elevation, canal boats have to be lifted by locks or cable railroads. For that reason, 2,200-foot Allegheny Mountain was a major barrier in the path of the Pennsylvania Canal. The story of competition between New York and Pennsylvania is in the text; it is a justly famous illustration of three important geographic principles:

1. The role of infrastructure in economic growth (in this case, a canal made products easier to ship - see Transparency 4M).
2. The interaction of human activity and topography (which makes the Pennsylvania canal much more expensive than the Erie Canal through New York).
3. The "noble stupidity" of people with good intentions deciding to build something like a canal in an inappropriate environment.

Activity: Ask students to imagine life before cars or trucks. Canal boats were the cheapest way to move heavy things on level ground. A lock could raise or lower a canal boat about ten or fifteen feet in half an hour. Then ask students to look at the profiles and estimate:

- Elevation change on each canal (2,200 feet in Pennsylvania, 450 in New York).
- Number of locks it would take (five times as many in Pennsylvania?).
- Extra time needed to cross Pennsylvania (days, weeks?).

Ask students if they would rather have a farm in western Pennsylvania or western New York if they want to sell wheat to people in eastern cities or Europe. Then show Transparency 4M again. Kicker: Why is the Stock Exchange in New York City? (originally, to trade livestock and feedstock [wheat]). How does its location there affect the pattern of wealth today? (it's part of the reason why per capita income in Manhattan is nearly twice the national average).

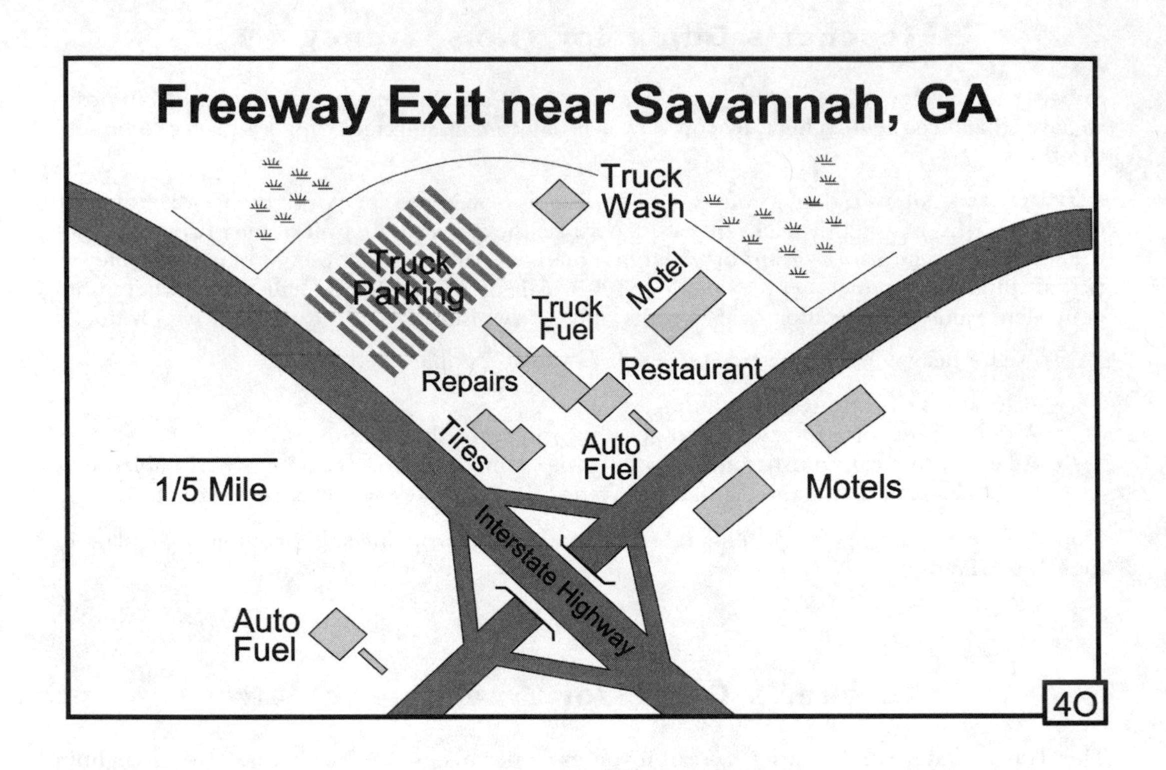
Freeway Exit near Savannah, GA
Truck Wash
Truck Parking
Truck Fuel
Motel
Repairs
Restaurant
Tires
Auto Fuel
1/5 Mile
Motels
Interstate Highway
Auto Fuel
4O

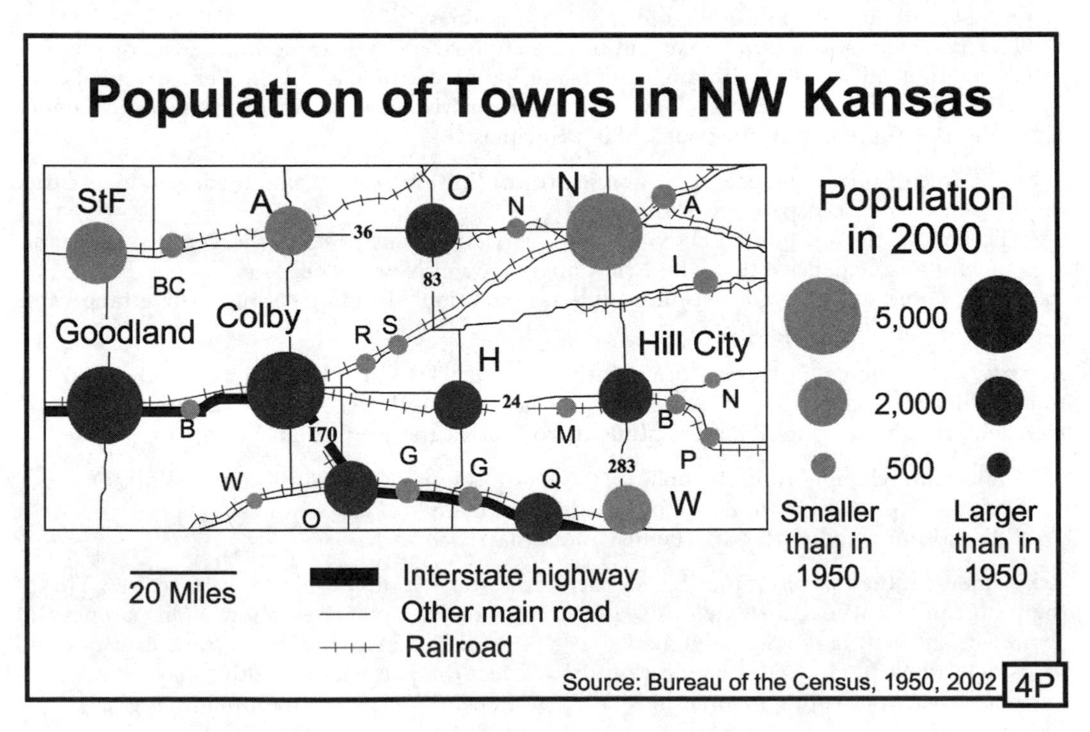
Population of Towns in NW Kansas
StF
A
O
N
N
A
36
83
L
BC
Goodland
Colby
R
S
H
Hill City
B
24
M
B
N
I70
G
G
Q
283
P
W
O
W
Population in 2000
5,000
2,000
500
Smaller than in 1950
Larger than in 1950
20 Miles
Interstate highway
Other main road
Railroad
Source: Bureau of the Census, 1950, 2002
4P

Teacher's Guide for Transparency 4O

Many eras in history have seen the development of unique, newly available technology and/or resources that then led to the dramatic growth of some distinctive landscape features:

- In the late Middle Ages, when ecclesiastical authority was still strong and construction technology was improving, people in practically every major city in Europe built impressive Gothic cathedrals.
- In the 1400s and 1500s, when sailing ships became substantially more seaworthy, powerful port cities emerged in many places with suitable natural harbors.
- In the Industrial Revolution of the 1700s and 1800s, people built towns and textile mills near waterfalls that could be used for power.
- In the 1800s and early 1900s, railroads and autos enabled people to live in "dormitory suburbs" outside of the cities where they worked.
- In the late 1900s, limited-access highways were built between major cities, land around highway exits became a resource, and clusters of service buildings appeared near most exits.

Activity: Ask students to make a sketch map of the areas around local highway exits. The goal is to locate motels, gas stations, convenience stores, restaurants, souvenir shops, tire-repair shops, and other features with sufficient accuracy that the class would be able to compare maps and make generalizations (e.g., what percentage of exits have a gas station? what is the average distance from the highway to these features? does the number of features depend on whether there is a town nearby?).

Variant activity: Make maps from aerial photographs rather than in the field. Such photos might be available from the county highway department, various planning offices, or farm service agencies (look under U.S. Government, Dept. of Agriculture in the phone book).

Teacher's Guide for Transparency 4P

Hill City, Kansas, illustrates the flip side of the new-resources coin. This town had a great location on a major railroad and on a highway that went from Kansas City to Colorado. Then Congress authorized construction of the Interstate Highway system. I-70 parallels Highway 24 through eastern Colorado, enters Kansas, follows the old highway past Goodland and Colby, and then turns southeast before it reaches to Hill City.

The impact of this new development was almost immediate. Hill City became a much less desirable location for factories, warehouses, truck-repair shops, restaurants, hotels, and a host of other functions. The result is clearly visible on a map of population in recent years. Goodland and Colby continue to grow steadily, while Hill City and many other "bypassed" towns have grown only slowly or actually lost population.

The picture is complicated by the movement of service jobs at a local scale. Some towns got so small that they could no longer support a school, grocery store, doctor's office, and so on. The remaining residents are obliged to drive to nearby towns where the services are still available. As a result, one town might gain customers and grow even as towns around it lose population. The school in Morland (M) closed in 1994; its students now go to Hill City. Other "decliner" towns include Bogue, Palco, and Nicodemus (which was an "exoduster" town, originally settled by ex-slaves after the Civil War).

Activity: Have students use census data to make population maps similar to this one for other areas of interest (and try to explain the pattern). A map like this has practical value, especially to those who make decisions about where to locate stores and offices.

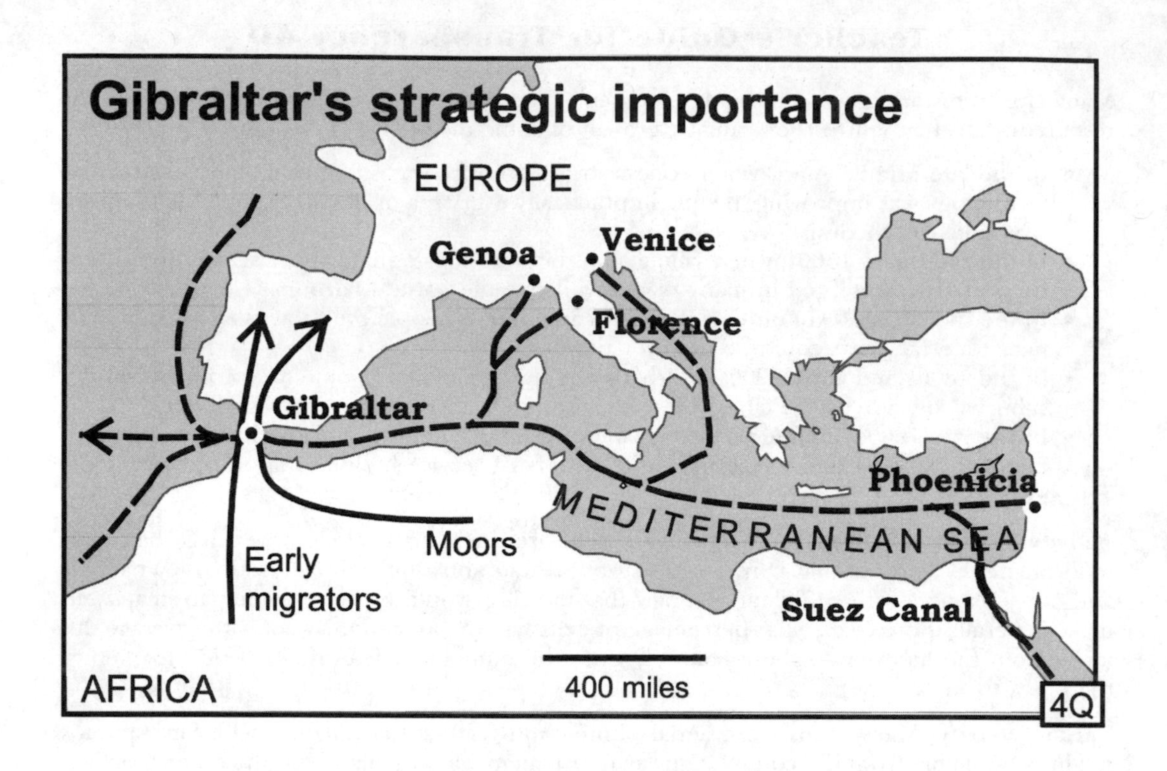
Gibraltar's strategic importance
EUROPE
Genoa
Venice
Florence
Gibraltar
Phoenicia
MEDITERRANEAN SEA
Moors
Early migrators
Suez Canal
AFRICA
400 miles
4Q

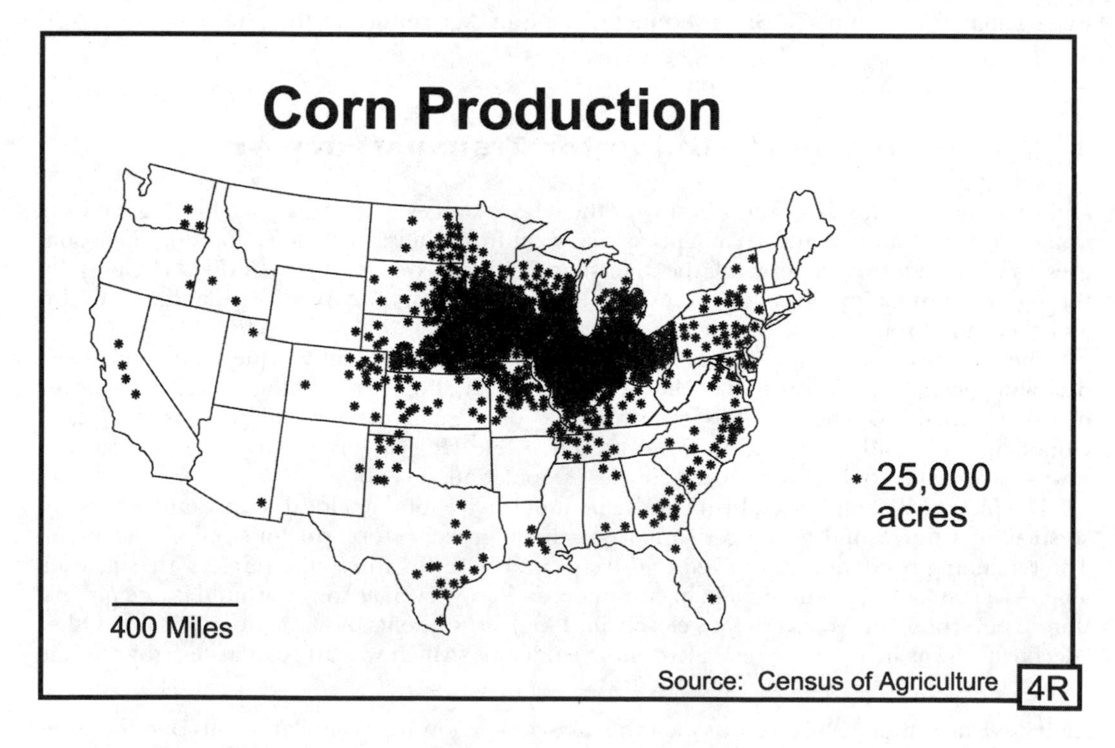
Corn Production
25,000 acres
400 Miles
Source: Census of Agriculture
4R

Teacher's Guide for Transparency 4Q

For several thousand years, Gibraltar was a site of great strategic importance. Interestingly, its importance seemed to change with each major shift in transportation technology:

- In very early times, the narrow Strait was the best way to get from Africa to Europe.
- For the ancient Phoenicians, who built some of the world's first sailing ships, the Strait of Gibraltar was the only way to venture out of the Mediterranean Sea and reach the coasts of Europe and North Africa.
- For the Moors, with their powerful armies, the Strait was a bridge to invade Europe and help bring about the end of the Roman Empire.
- For explorers from Florence, Genoa, and Venice, the Strait was the gateway to the Atlantic and the riches that lay across the ocean.
- For the British, Gibraltar was a potential tollgate for their tankers carrying oil from the Middle East through the new Suez Canal.
- And now, in an age of intercontinental ballistic missiles, internets, and supertankers that are too big to fit through the Suez, the Strait of Gibraltar has practically no strategic significance at all (it is like the Erie Canal, Transparency 4N).

Activity: Look at a world map and try to identify other "chokepoints," places where an army or navy would be able to control movement between two large bodies of water or masses of land (Harpers Ferry is one; see Transparencies 1C and 1D). The strategic significance of a chokepoint depends on what resources and populations are on either side. Once you have identified some possibilities, have students gather information from an encyclopedia, almanac, or online source. The goal is to see how important the chokepoint was at particular times in history.

Teacher's Guide for Transparency 4R

The Corn Belt is a classic example of what geographers call a **formal region**; it is a sizeable area where people responded to similar environmental conditions in similar ways and created a landscape that is more or less homogeneous - a vista in western Ohio looks a lot like one in eastern Nebraska!

Activity: Ask students what environmental conditions might contribute to reducing the profitability of corn production and thus limiting the extent of the Corn Belt. Discussion (aided by a map overlay) should uncover at least three significant environmental factors:

- Growing season. Most corn needs at least 90 days to mature. Add a few weeks to allow for climatic variability, and note that the Corn Belt seems to stop in the north where the growing season drops below 4 frost-free months (compare Transparency 3J).
- Moisture. Most corn needs a slight moisture surplus during the growing season (compare Transparency 3I).
- Good land. Fairly flat land is needed in corn production, because some of the machinery cannot be used when the land is too hilly, and soil erosion is more of a problem in corn production than with some other crops. Glaciers from Canada helped level land in what became the Corn Belt and improved the soil depth and quality (compare Transparency 3L).

Activity: Use map comparison to try to identify key environmental conditions that limit the geographic extent of other important agricultural regions, such as the wheat-growing areas of Canada or Siberia, the rice lands of China, the grazing regions of Argentina or Australia, and so on.

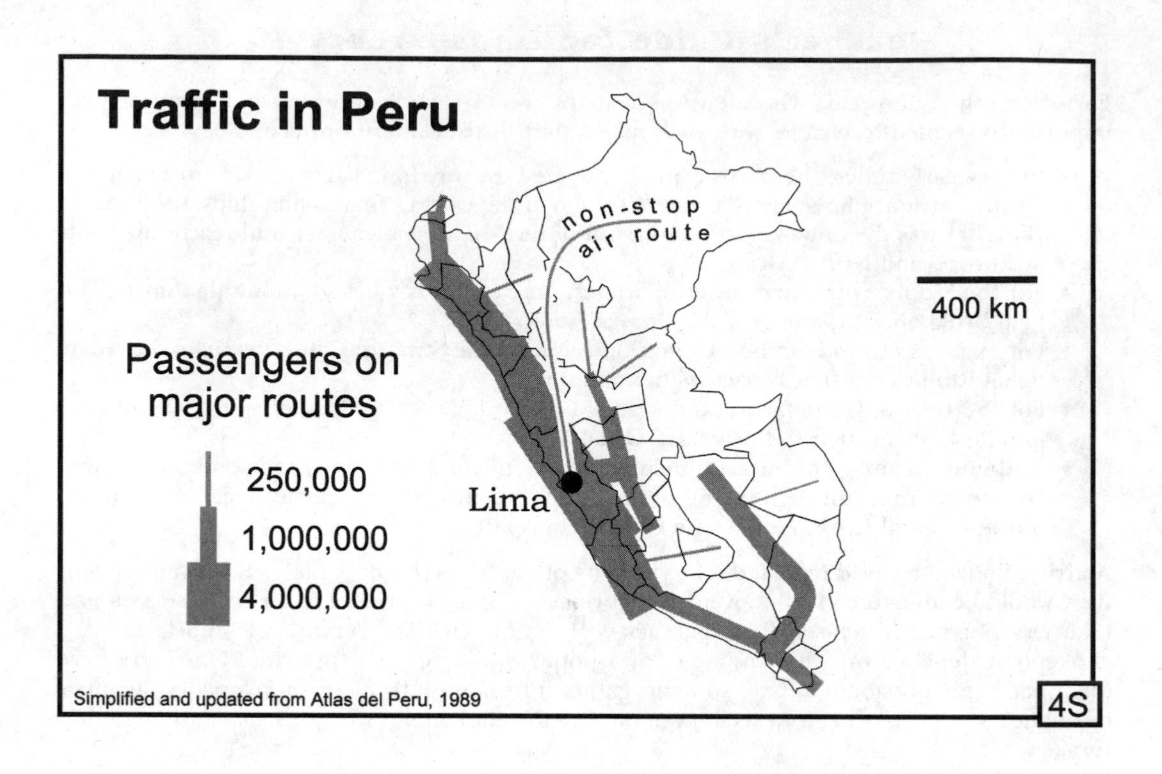
Traffic in Peru
non-stop
air route
400 km
Passengers on
major routes
250,000
1,000,000
4,000,000
Lima
Simplified and updated from Atlas del Peru, 1989
4S

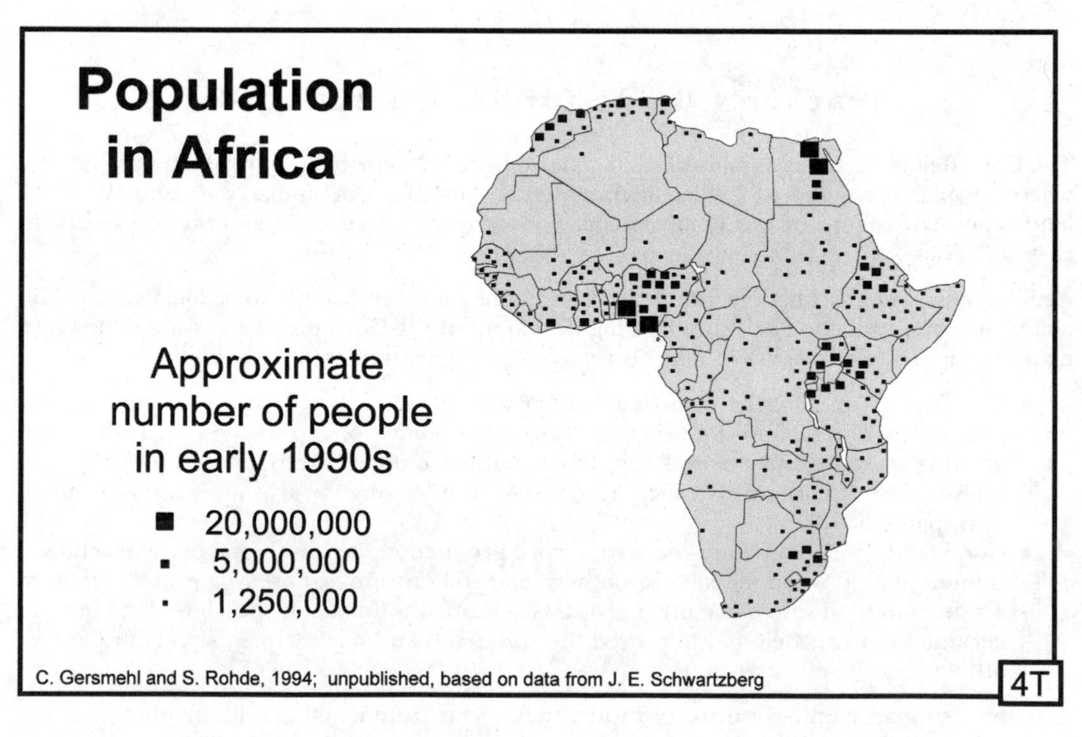
Population
in Africa
Approximate
number of people
in early 1990s
20,000,000
5,000,000
1,250,000
C. Gersmehl and S. Rohde, 1994; unpublished, based on data from J. E. Schwartzberg
4T

Teacher's Guide for Transparency 4S

Peru is a country with a transportation and communication problem: the large cities are near the Pacific Coast, but many resources are in the eastern part of the country. In between are three great mountain ranges: the Cordillera Occidental (CORE-dee-yay-rah means mountain range, ox-ee-den-TAHL means western), Cordillera Central (cen-TRAHL), and Cordillera Oriental (oh-ree-en-TAHL means eastern; to get the pronunciation right, roll your r's a little).

The influence of the mountain ranges is apparent on this map of the functional region of Lima (LEE-mah, the largest city). The width of the lines indicates the traffic on major roads as they follow valleys between the mountain ranges.

Activity: Divide an area into regions based on road patterns. Find highway maps and have students trace big roads, so that they stand out from the clutter. Study the maps to see what might influence the road pattern. Arkansas, for example, has (at least) four distinct road regions:

- An eastern region of flat land and straight roads
- A southwestern region of hilly land and curving roads
- A west-central region of east-west ridges and east-west roads in the valleys
- A northwestern region of irregular high hills and twisty roads

These are **formal regions** (areas of "homogeneous" terrain and road patterns). Arkansas also has a number of **functional regions** (areas that are linked by traffic flow). Try to figure out where people shopping in Fayetteville, Fort Smith, Texarkana, Little Rock, Blytheville, or Memphis would likely be coming from. Take account of city size, distance, and the pattern of roads in dividing the state into the Little Rock shopping region, the Fayetteville one, and so forth (see also Transparency 10A).

Teacher's Guide for Transparency 4T

This map shows clustering of the population of Africa. The continent has about 700 million people, living on about 12 million square miles. For perspective, both the area and the population are about three times the size of the United States. Unlike in the United States, however, a significant proportion of the people in Africa still raise much of their own food; thus, patterns of soil fertility and rainfall influence where people live.

Activity: Ask students to compare this map of population with a map of precipitation (Transparency 4U) and have them make generalizations about relationships between rainfall and population density. (As with many questions about map patterns, this can have several levels of complexity. At a large scale, students should note that the dry Sahara and Kalahari deserts are virtually uninhabited. At the same time, very rainy areas near the equator also have lower populations. This is partly because rain tends to wash nutrients out of the soil, leaving it less able to produce food for humans. The issue of sparse population near the equator also touches on the subject of Transparency 4V; many human and crop diseases thrive in hot and wet places.)

A logical extension of this Activity would be to look at rates of population growth. The Population Reference Bureau (PRB; *www.prb.org*) publishes annual figures for population and population growth, country by country; the CIA World Factbook (*www.cia.gov*) has country profiles and data – because the CIA info is public domain it also available on a great many other websites. As with the map in this Transparency, the PRB and CIA data are "best guess" figures, because different countries take their censuses in different years. That is the reason for using the word "approximate" in the legend for this map: the exact population at a given time of every country would be impossible – and in many ways pointless – to determine. That we have to learn to "live with" such inexactitude is another important lesson of geography.

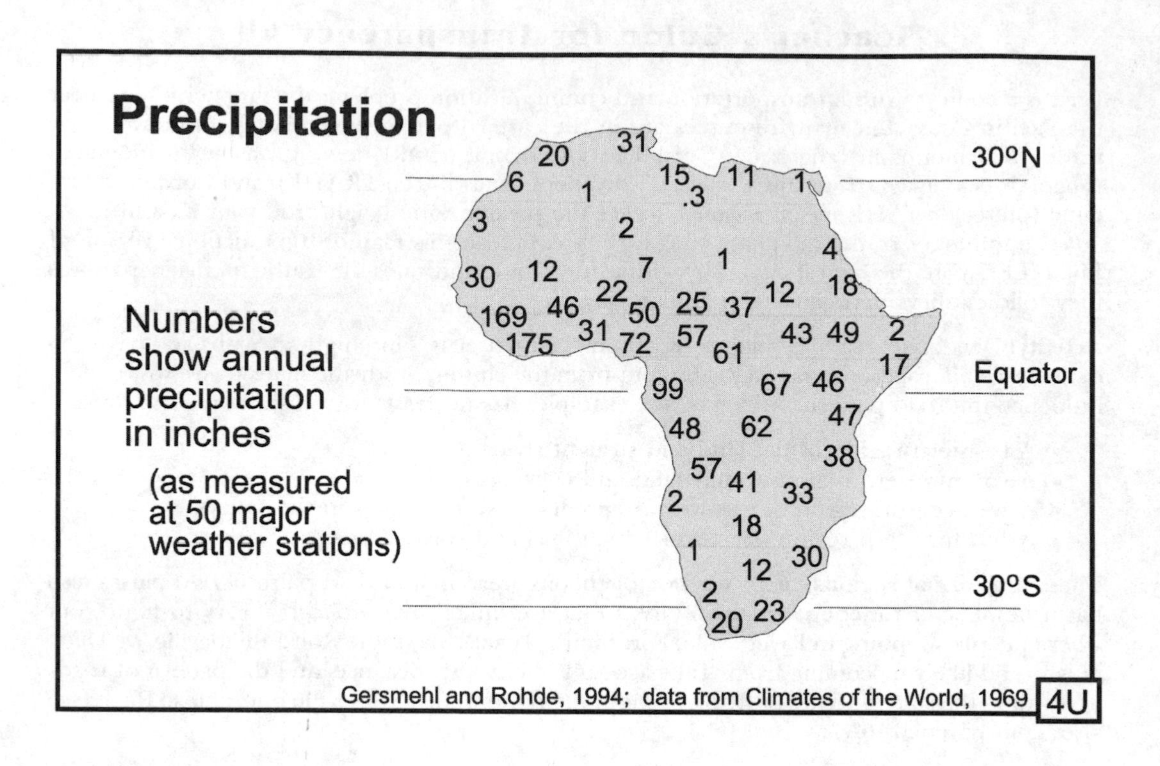
Precipitation
Numbers show annual precipitation in inches
(as measured at 50 major weather stations)
30°N
Equator
30°S
20
31
6
1
15
.3
11
1
3
2
1
4
30
12
7
22
12
18
169
46
50
25
37
175
31
72
57
61
43
49
2
17
99
67
46
48
62
47
57
41
38
2
33
18
1
30
12
2
23
20
Gersmehl and Rohde, 1994; data from Climates of the World, 1969
4U

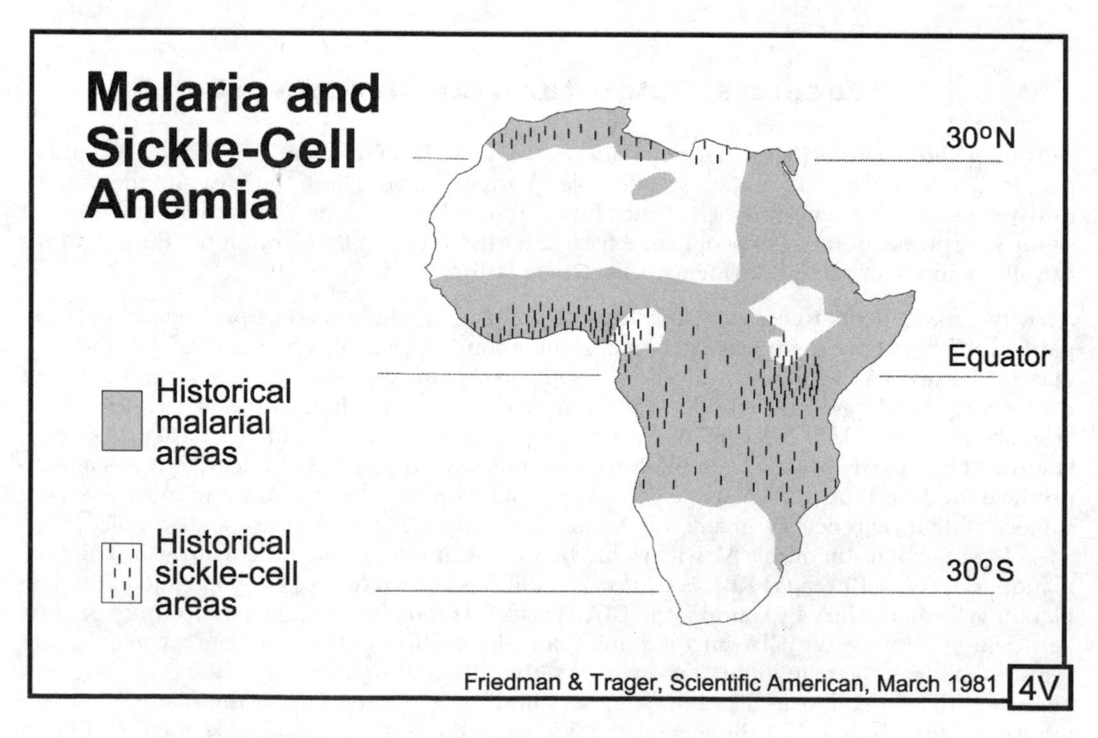
Malaria and Sickle-Cell Anemia
30°N
Equator
30°S
Historical malarial areas
Historical sickle-cell areas
Friedman & Trager, Scientific American, March 1981
4V

Teacher's Guide for Transparency 4U

This map in this Transparency is only half finished, in terms of how it conveys its message. The numbers are accurate for each location, but the overall pattern is difficult to discern easily. The map symbolism known as isolines (a specific map "language") was invented to handle this problem – it translates a mass of numbers into a visually coherent pattern (but hopefully only if such a pattern actually exists in the original data!).

Activity: Identify places with high and low values, and draw isolines to separate them. To teach students how to draw isolines, it sometimes helps to make a quick regional map out of the data. Make copies of the map for each student (or pair or small group). Ask students to put a box around the number at each place that gets more than five feet (60 inches) of rain in a year. Then have them circle the number at each place that gets less than 12 inches. The result is a crude regional map that identifies several wet and dry regions.

The next step would be to draw isolines. An isoline is a visual separator – it separates places with substantially higher and lower numerical values. Demonstrate this by tracing on the Transparency with an erasable marker: "an isoline at 12 inches of rain would enter Africa between the 3 and the 30 on the west coast, go directly through the 12, and continue east between the 7 and the 22, through the next 12, and between the 4 and the 18 near the Red Sea. Another (shorter) one enters from the east between the 2 and the 18, curves around this 2, and exits between it and the 17."

Have students put two more "12-inch isolines" in the appropriate places on the map. Then, have them draw 60-inch lines to separate the boxed places from the places with less rainfall.

Finally, have students color the 60-inch-plus areas with a dark color, the 12-to-60-inch areas with a medium color, and leave the less-than-12-inch areas unshaded (this follows the convention of isoline maps: higher values darker, lower values lighter). The result should be a map much like the inset on Transparency 2I.

The resulting isoline map is easy then to compare with Transparencies 4T and 4V.

Teacher's Guide for Transparency 4V

Malaria is carried by mosquitos. How might a comparison of maps 4U and 4V help medical researchers identify the cause of the disease? (mosquitos need water and they thrive in places with a lot of rain or melting snow, especially if the terrain allows swamps to form).

Another disease, sickle cell anemia, is a genetic disorder. People get seriously ill if they inherit the sickle cell gene from both parents. People who get the gene from only one parent are "carriers" of the trait but have few symptoms — they can pass the disease on to their children.

Ordinarily, this kind of disease would slowly disappear from a population, but the sickle cell trait has an important side effect. A carrier of the gene is partially immune to malaria. This gives people with the sickle cell gene an advantage in tropical regions, where they might increase in number even though some of their children have the sickle cell disorder.

When these people move outside of malarial regions, their "carrier status" is no longer an advantage. This background information helps explain why sickle cell anemia in the United States is a disease that affects mostly people whose ancestors came from Africa. It therefore is a problem primarily in the rural South and some large urban areas of the North.

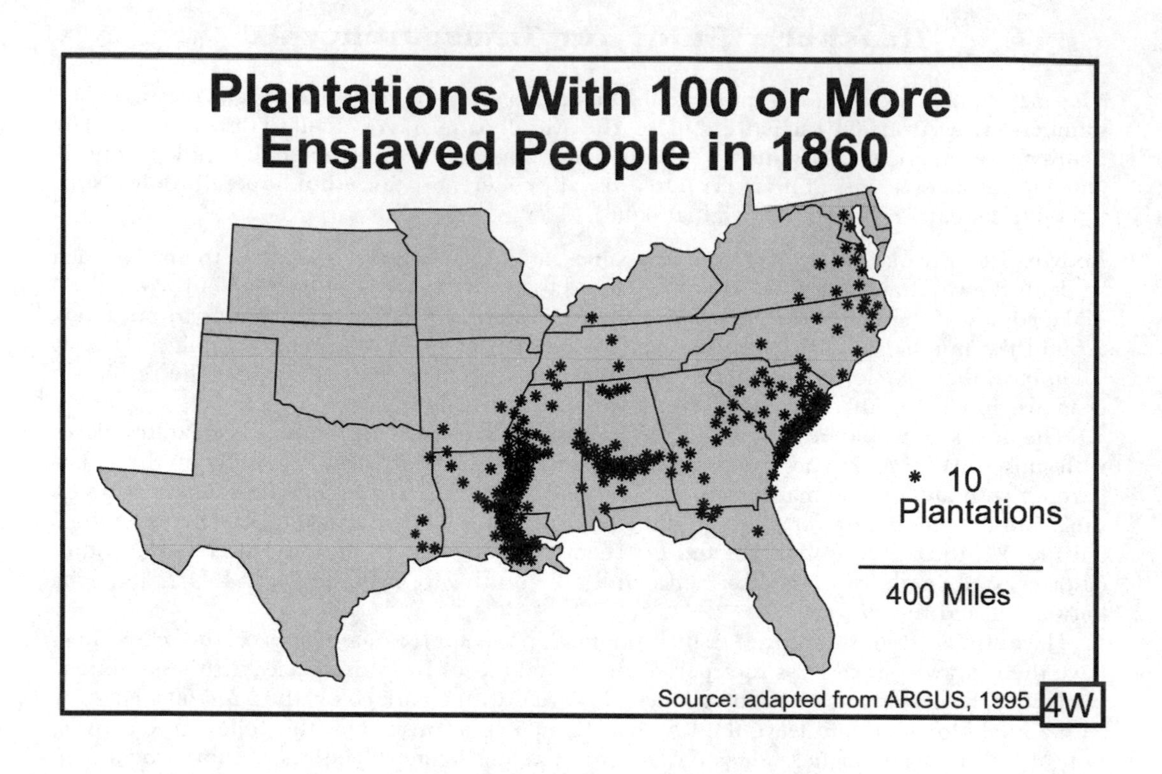
Plantations With 100 or More
Enslaved People in 1860
* 10
Plantations
400 Miles
Source: adapted from ARGUS, 1995
4W

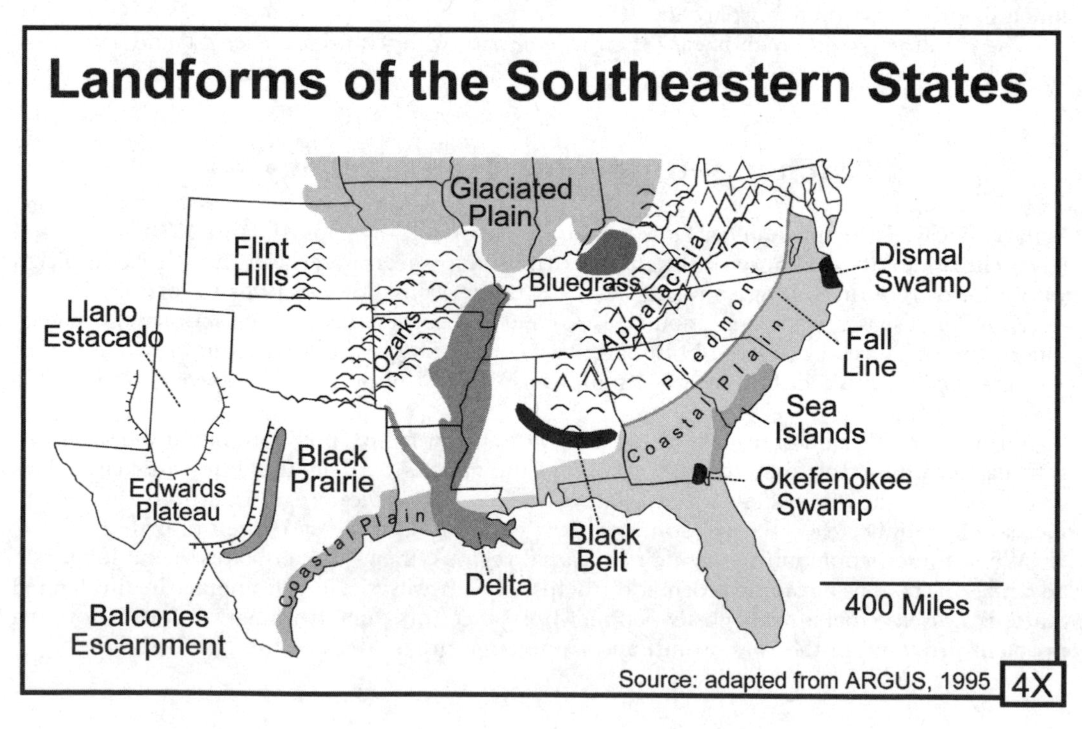
Landforms of the Southeastern States
Glaciated
Plain
Flint
Hills
Llano
Estacado
Ozarks
Bluegrass
Appalachia
Piedmont
Dismal
Swamp
Fall
Line
Coastal Plain
Sea
Islands
Black
Prairie
Edwards
Plateau
Coastal Plain
Okefenokee
Swamp
Black
Belt
Delta
Balcones
Escarpment
400 Miles
Source: adapted from ARGUS, 1995
4X

Teacher's Guide for Transparency 4W

This simple dot map broadcasts one fact to the reader: the South was not a land of wall-to-wall giant plantations before the Civil War. The lifestyle depicted in *Gone with the Wind* was the exception, not the rule.

In fact, more than a third of the counties in the South voted against secession, because they had little economic or political stake in the institution of slavery. People in many parts of the South joined the fight only when armies were entering their territory and threatening their families. Some of them actually fought on the Union side.

Why is this history important for us today? Because some economic and demographic consequences of the plantation and Reconstruction eras are still evident on maps of income, education, voting, and other variables. Therefore, an understanding of where the big plantations were located can help us put a number of present-day issues into context.

Activity: Compare this map of plantations with a map of landforms (Transparency 4X). What landform regions were most likely to have large plantations? (the Low Country of South Carolina, the Black Belt of Alabama, and the Delta of Mississippi and Arkansas; see below).

Teacher's Guide for Transparency 4X

Activity: Compare this map of landforms with a map of plantations (Transparency 4W). What traits do the plantation regions have in common? (flat and fertile land). Much of the South, however, had red sandy or clayey soil on hills. People who tried to grow cotton in these relatively infertile soils were "rewarded" with low yields and severe erosion. Many cotton "oldfields" were already abandoned by the time of the Civil War. Some were still used for cotton after the war, but by the mid-1900s most of the land in the states of the former Confederacy was used for pine trees or pasture.

Activity: To compare maps in a more rigorous way, throw 20 grains of rice or small stones on this map. Note their locations, and then find exactly the same locations on Transparency 4W. The next step is to tabulate the results; count the grains in each landform region, and then count the percentage of those locations that had large plantations.

The reason for using stones or rice this way is similar to throwing dice; it takes the selection of sample points out of the control of the analyst. Another way to do this is to make a sampling grid: find (or make) graph paper with lines about half an inch (or one centimeter) apart. Put the grid behind the map, and put dots on every major line intersection. Then put dots on the other map in exactly the same locations.

If the maps are the same scale, you can "cut out the middleman" and lay the maps directly on top of each other. This is the core principle of a Geographic Information System (GIS). A GIS is a way of storing information in a computer according to a common frame of reference, so that maps can be overlaid and compared with absolute precision. The overlaying process can be accurate to many decimal places. The results, however, are only as accurate as the original maps, which lead back to a main point of this book: that maps are a means of communicating what we understand about the world. This is definitely not the same thing as saying that maps are an accurate depiction of the world.

In short, you must know geography to avoid misinterpreting output from a GIS!

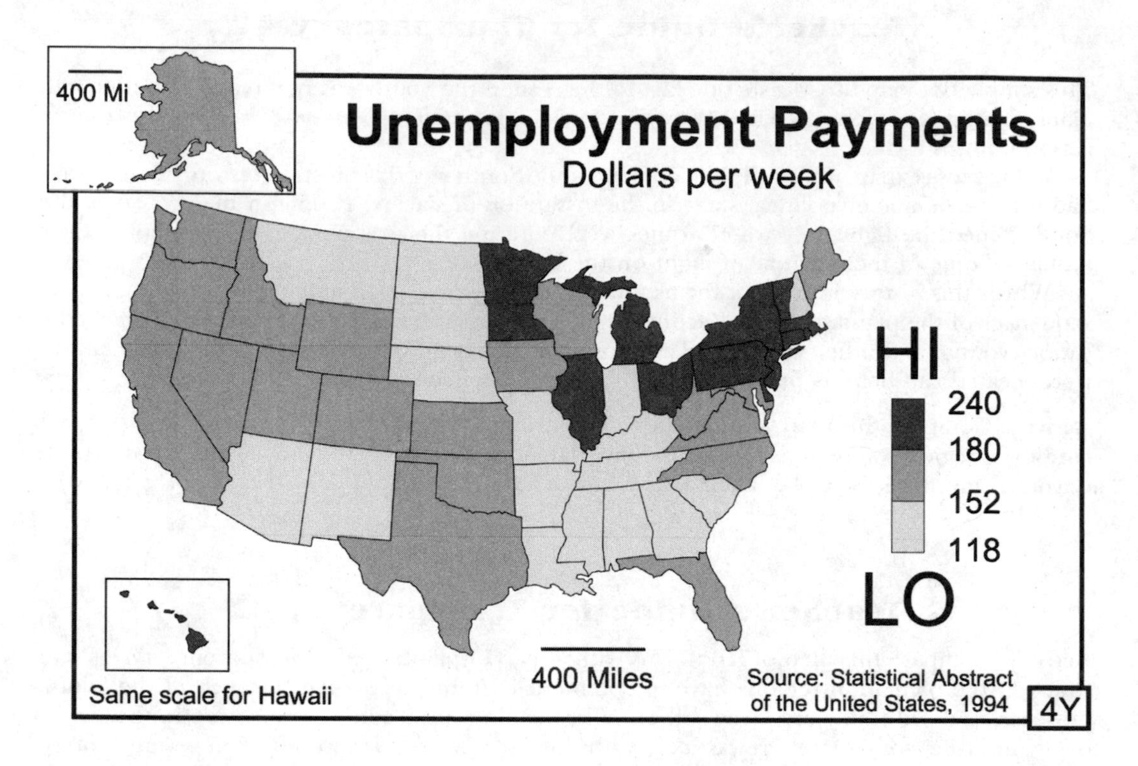

400 Mi
Unemployment Payments
Dollars per week
HI
240
180
152
118
LO
Same scale for Hawaii
400 Miles
Source: Statistical Abstract of the United States, 1994
4Y

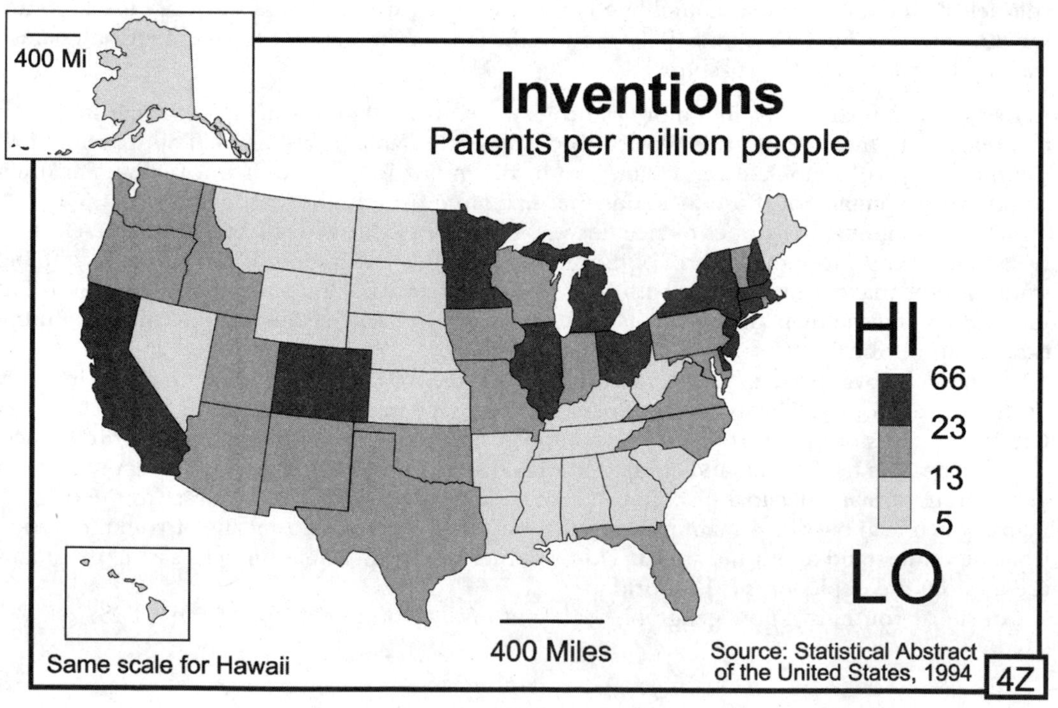

400 Mi
Inventions
Patents per million people
HI
66
23
13
5
LO
Same scale for Hawaii
400 Miles
Source: Statistical Abstract of the United States, 1994
4Z

Teacher's Guide for Transparency 4Y

This Transparency raises some intriguing questions about perspective. One of the most subtle and yet most important questions about mapping welfare is: from whose perspective should we make a map?

One approach is to examine the data from the point of view of state government. To do this, the map maker divides the dollar figure of total state welfare payments by the population of the state. The result is an estimate of the size of "the welfare burden" per person in each state. States with high payments per person may have higher taxes or may compensate by spending less on other things, such as parks or police.

Another approach is to examine the data from the recipients' perspective. To do this, record the unemployment payment per recipient. These are the numbers we should compare in order to see how reassuring the "safety net" is to people contemplating what would happen if they lost their jobs.

There are other perspectives: unemployment payments per welfare-office worker (an estimate of the size of their caseload), or per taxpayer (an estimate of what it costs those who pay). In short, there are many ways to combine information in order to put it in perspective.

Activity: Find some thematic maps in magazines or newspapers. Ask students to try to figure out what data were combined to make the map. For example, a map of corn yield per acre likely starts with a measure of the total amount of corn and a measure of field size in acres. A map of death rates may start as a count of deaths and a total population. Don't be surprised if students make mistakes in trying to reconstruct original data. In truth, many maps have ambiguous legends, which makes it even more important for students to be able to reconstruct the steps cartographers make to prepare information for a map.

Why? Because it is also a fact that citizens (or politicians!) who cannot identify the original data that underlie a map cannot intelligently use the map to guide decisions.

Teacher's Guide for Transparency 4Z

An invention happens when a person who is adequately prepared for it has the right kind of idea. Inventors register their ideas or creations with the U.S. Patent Office. This office investigates each invention and decides whether it is indeed innovative. If so, the patent granted allows the inventor to sue others who might try to profit from or copy that invention without permission. The goal is simple: to encourage inventiveness by guaranteeing that inventors will have a chance to gain the financial rewards for their new ideas.

Important inventions have altered the geography of places, often quite dramatically.

Activity: Ask students to list ways in which the invention of the automobile made some places more important and others less so. For example, it generated jobs in Detroit. It caused petroleum to become a valuable commodity, which in turn helped places like Texas and Kuwait. Rubber tires became necessary, which helped Akron and Malaysia. The automobile made it possible for people to create and live in suburbs; it also made cities smoggier. It made canal towns less important. And so forth. Put plus and minus signs on a map to show the changes. Then do the same for other important inventions, such as air conditioners, fax machines, computers, and the Internet.

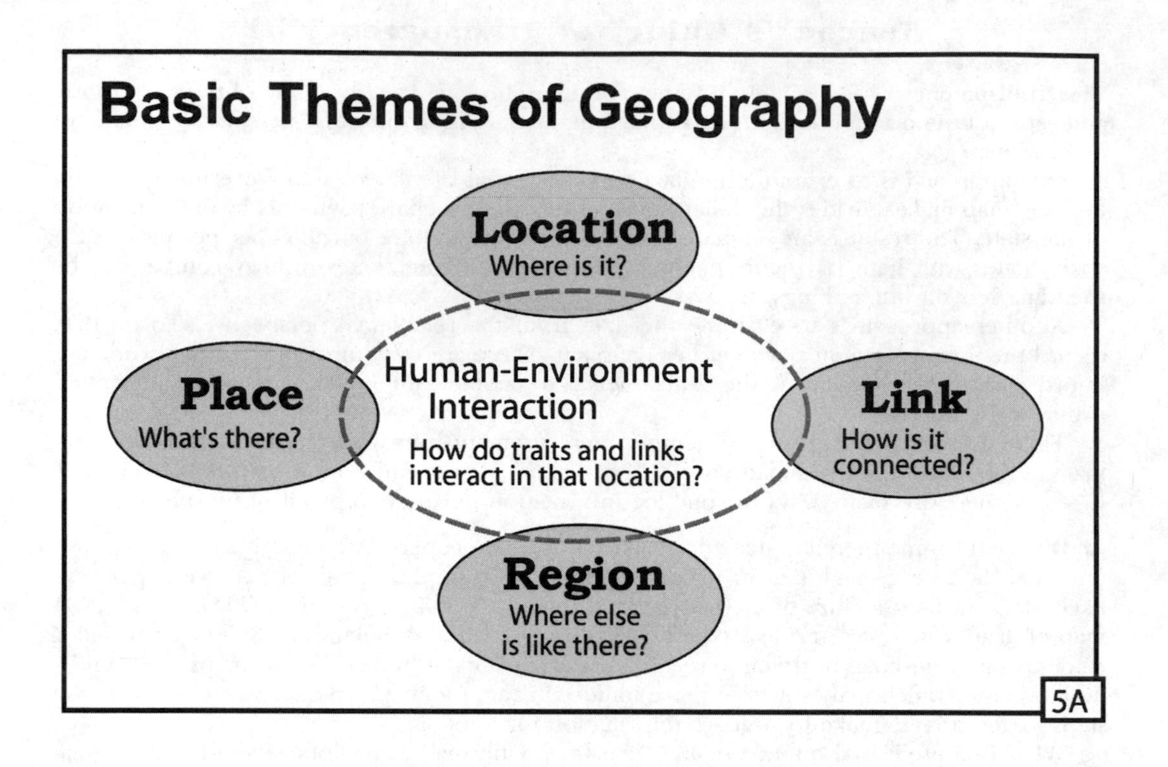
Basic Themes of Geography
Location
Where is it?
Place
What's there?
Human-Environment
Interaction
How do traits and links
interact in that location?
Link
How is it
connected?
Region
Where else
is like there?
5A

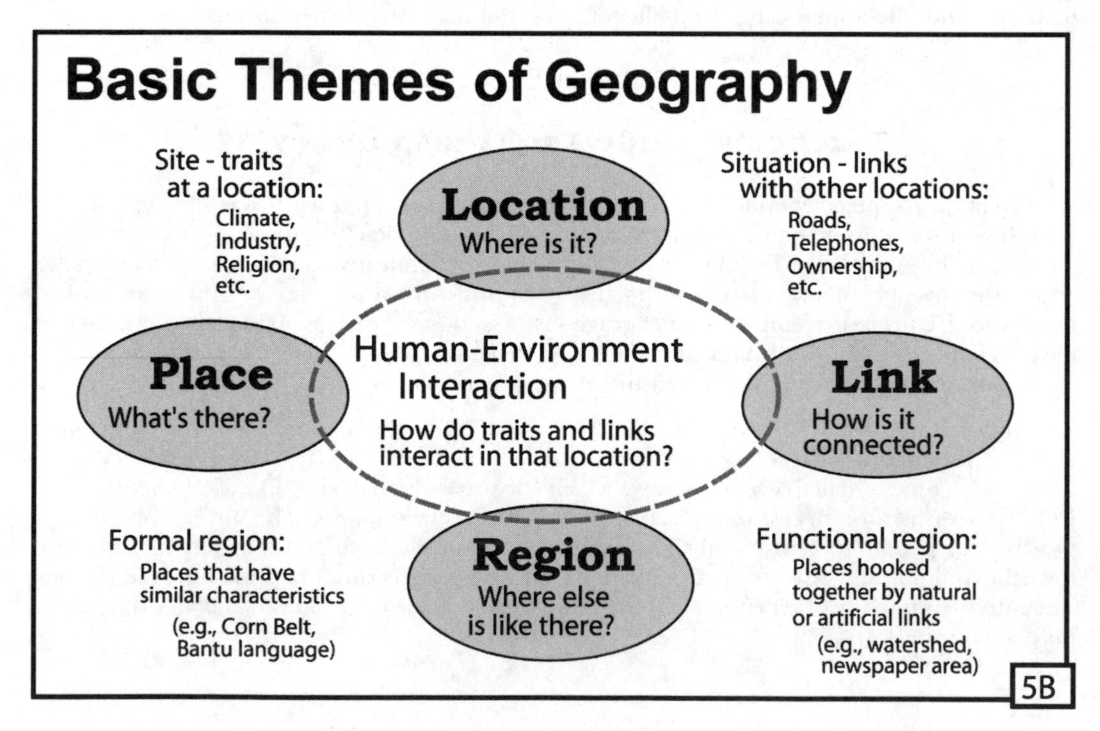
Basic Themes of Geography
Site - traits
at a location:
Climate,
Industry,
Religion,
etc.
Location
Where is it?
Situation - links
with other locations:
Roads,
Telephones,
Ownership,
etc.
Place
What's there?
Human-Environment
Interaction
How do traits and links
interact in that location?
Link
How is it
connected?
Formal region:
Places that have
similar characteristics
(e.g., Corn Belt,
Bantu language)
Region
Where else
is like there?
Functional region:
Places hooked
together by natural
or artificial links
(e.g., watershed,
newspaper area)
5B

Teacher's Guide, Transparencies 5A and 5B

Five Eastern haiku
(to be read individually):

These pictures can do
what bee does for cherry tree:
start a seed growing.

Letterbrush and hand
bend to inner discipline,
mind now free to soar.

Small mushroom can kill;
look closely, pick when ready,
cook just long enough.

On a foggy day,
shared bouquet of plum blossom
helps teach what bees mean

Young bluebirds can watch
an eagle soar, but must move
their own wings to fly.

Point a finger; when
all can name the trees they see,
then talk about them[es].

and

A Western free-verse poem
(to be read downwards):

These diagrams? they can be
a visual aid to help me
speak with a colleague,
a board, a principal.

They can also keep me mindful
of the mental discipline
I've chosen willfully
to practice in this class,

but they're likely to be
poison in the class room,
unless we'd "done geography"
for at least half a year

We need to build a stock
of shared experience
for these transparencies
to tie together.

Better yet, start sketching it,
ad lib, one cheerful day,
and ask the students,
to make it better,

using themes to help teach them
how to observe, and only then
naming the themes that we use
to see more clearly.

There are no suggested Activities based on these two Transparencies, because I really do not think they have much place in a typical class. See the text for reasons.

Now, what do we do with the rest of this space? (It is bad stewardship not to use it all, according to nearly every teacher's grandmother!) How about writing an anecdote, example, or a joke you think might be useful to reintroducing a theme that has been absent from a class discussion for too long?

Location:

Conditions:

Connections:

Region:

(OK, if you insist) Human-Environment Interaction:

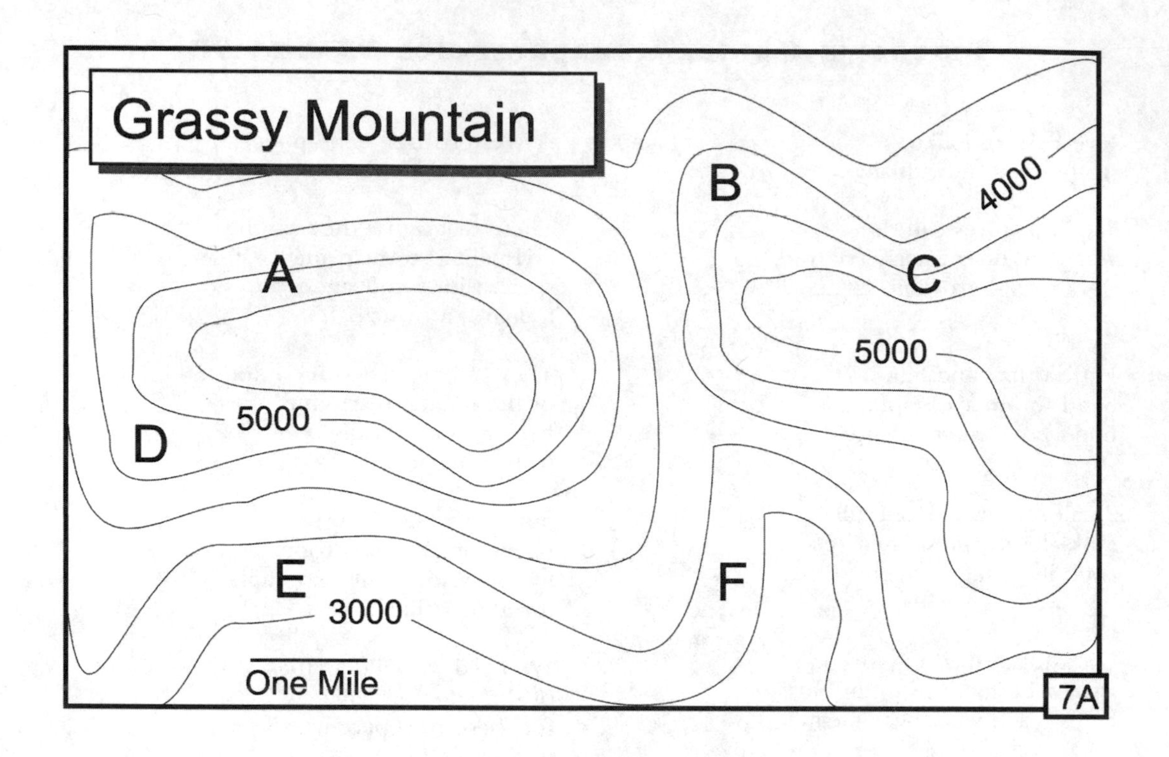

Map Interpretation Skill Quiz

This map shows an area in the Appalachian Mountains. The area has the narrow valleys and rounded summits that are typical of that region. On the day of the test, you will be given part of a map like the one above. The italicized words and letters in the questions might may be different on the actual test, but the questions will be the same. Practice your skills with this map and find topographic maps of other places and practice the same kind of questions with them.

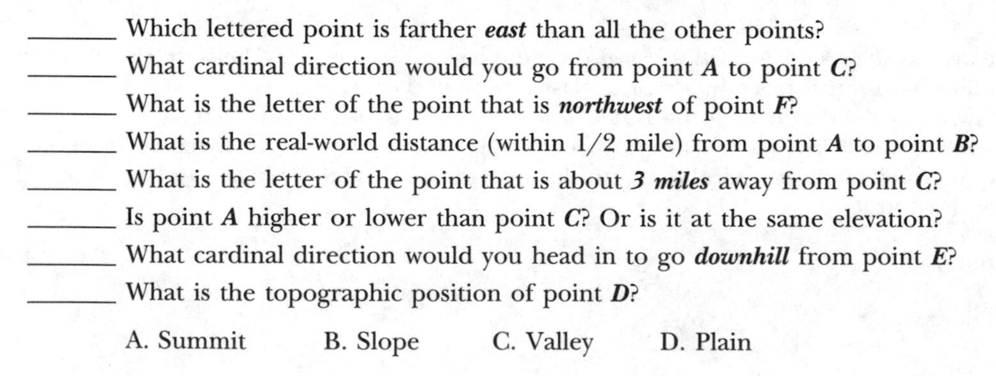

_______ Which lettered point is farther ***east*** than all the other points?
_______ What cardinal direction would you go from point ***A*** to point ***C***?
_______ What is the letter of the point that is ***northwest*** of point ***F***?
_______ What is the real-world distance (within 1/2 mile) from point ***A*** to point ***B***?
_______ What is the letter of the point that is about ***3 miles*** away from point ***C***?
_______ Is point ***A*** higher or lower than point ***C***? Or is it at the same elevation?
_______ What cardinal direction would you head in to go ***downhill*** from point ***E***?
_______ What is the topographic position of point ***D***?

A. Summit B. Slope C. Valley D. Plain

Teacher's Guide for Transparency 7A

This is a quiz that can be handed out ahead of time and modified on test day in two ways: by replacing the underlined letters, words, or phrases, or by using a different map of the same general kind. These sample questions show how the same ideas can be adjusted for different grade levels.

Elementary: You will be given part of a map; look at it and write your answer on each line:

________ What is the name of the town in map sector D4 on the map?

________ In what letter-number map sector is point D located?

________ What cardinal direction would you go from point A to point C?

________ What is the letter of the point that is north of point F?

________ How far is it in the real world (within one mile) from point A to point B?

________ Is point A higher or lower than point C? Or is it the same elevation?

________ What cardinal direction would you head to go downhill from point C?

________ Is point D on a hilltop, on a slope, or in a valley?

Intermediate: You will be given part of a map with some lettered points on it:

_____ Print the letter-number designation for the sector that contains point B.

_____ Print the letter that is closest to latitude 34° 11′ N and longitude 104° 23′ W.

_____ Print the true azimuth (within 10 degrees) from point B to point C.

_____ Print the letter where you would be if you saw point F to the southeast.

_____ Print the shortest road distance (nearest mile) from point A to point C.

_____ Print the RF scale of a map with a verbal scale of one inch equals 500 feet.

_____ Print the approximate elevation of point G.

_____ Print the letter that best describes the topographic position of point D.

A. Summit B. Ridge C. Spur D. Slope E. Draw
F. Bench G. Valley H. Hole I. Plain J. Pass

Advanced: You will be given a portion of a map with some lettered points on it.

_____ Print the magnetic compass reading (within 5 degrees) from point B to A.

_____ Print the letter of the map point from which you would see point A at a magnetic compass reading of 263 degrees.

_____ Print the amount of time (within 10 minutes) it would take to go from point A to point B at a speed of 40 miles per hour.

_____ Print the real-world area (nearest square km) represented by 3 sq cm on this map.

_____ Print the elevation difference between point C and I.

_____ Print the average gradient (nearest 5 percent) of a direct path from point A to point B on the map.

_____ Print the elevation of the highest point along a straight line from point F to point J on the map.

_____ Sketch a side profile and print YES if you would (NO if you would not) be able to see point C while standing on the ground at point A.

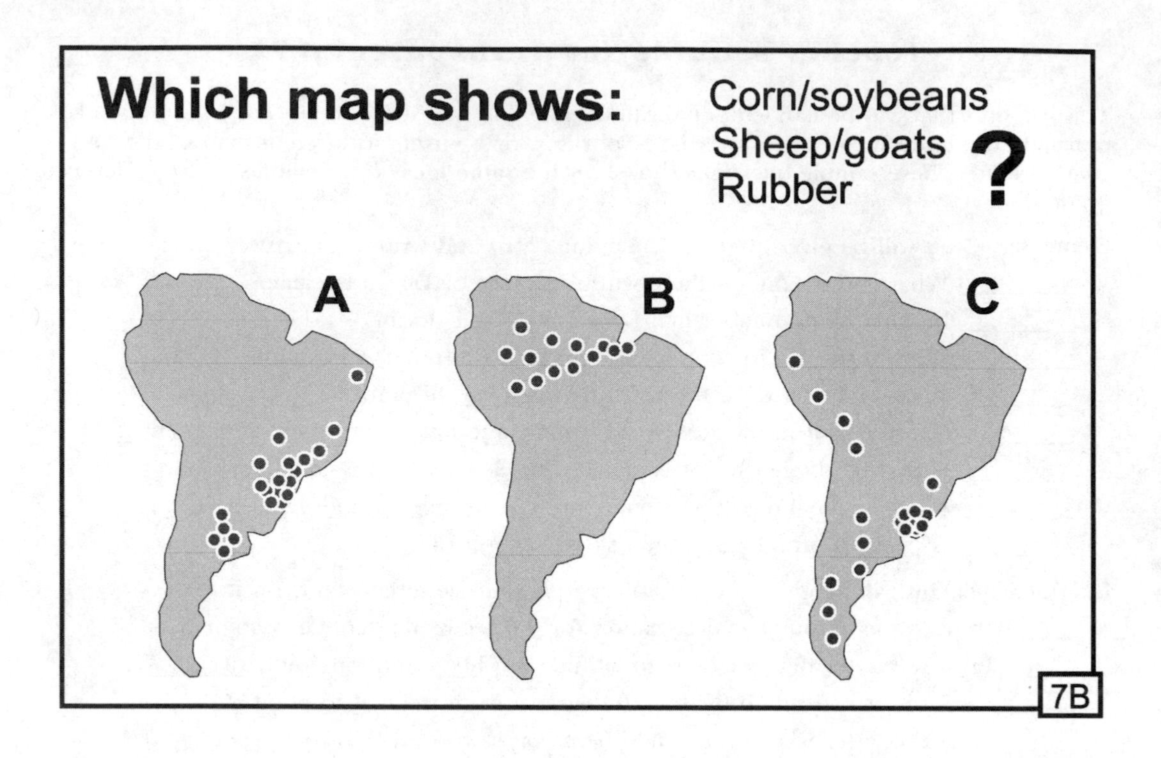

Economic Geography Matching Quiz

Match the appropriate lettered map with the following crops and briefly state your reasons:

Corn and soybeans require at least adequate rainfall, fertile soil, and a frost-free season that is at least 3 months long. They also require an economy that can use them for animal feed or export them to other countries. Which map shows where these crops are grown in South America, and why do you think so?

____ (letter)

Sheep and goats need grass or shrubs that grow throughout the year, or land that can be used to grow hay for winter feed. Because these animals are not very profitable, they are seldom raised on land that could get better use, unless there is a long tradition of raising these animals. Which map shows where these animals are raised, and why do you think so?

____ (letter)

Rubber comes from a tree that grows in the equatorial rainforest. Which map shows where this crop is harvested, and why do you think so?

____ (letter)

Teacher's Guide for Transparency 7B

A matching question can be phrased in many different ways, depending on what kind of knowledge a teacher wants to evaluate. The primary focus in this example is on the link between agriculture and climate. That is why the question provides a short explanation about each crop (so that the question does not just test familiarity with the crops).

If, on the other hand, a teacher wants to emphasize recall of plant and animal traits as well as of the climate pattern, the test might simply ask:

Which map shows where corn and soybeans grow? Why do you think so?

Or just: write the letter of the map that shows where rubber is produced.

Or, to emphasize a full set of spatial relationships, ask students to draw a line for the equator, draw in dots to indicate desert areas, draw in upside-down V's to show the Andes Mountains, and then have them cite those geographical features as they match the maps with the choices for what's produced, justifying their choices. South American agriculture clearly shows the influence of environment, but that is not the only variable. The mass of sheep and goats in Uruguay reflects a long cultural tradition of that as the primary land use even while neighboring countries with similar land have grown other crops.

Other geographic patterns that can be tested with matching questions include:

- Climate - *World Weather Guide, USA Today Weather Almanac, www.noaa.gov*
- Mineral deposits, rocks - *Goode's World Atlas*, a good geology text, or *www.usgs.gov*
- Plants and animals - maps in *Reader's Digest North American Wildlife*
- Soil types - maps in various physical geography textbooks, or nrcs.gov
- Environmental hazards - *State of the Earth Atlas, Atlas of Environment, www.epa.gov*
- Houses - *This Remarkable Continent, Atlas of Cultural Features*
- Settlement - *Atlas of American History* (at least five companies publish one)
- Ethnic groups - *We the People*, Facts on File *Atlas of Contemporary America*, or census reports for specific states or metropolitan areas, *www.census.gov*
- Languages - same as above, and *Atlas of the World Today*; spoken samples in some CD-ROMs like *EnCarta* or *101 Languages* (it wouldn't hurt anyone to learn how to say or recognize "Hello," "Excuse me," and "Thank You" in several languages)
- Industry - *Oxford Economic Atlas, Goode's World Atlas*
- Crimes - *Atlas of Contemporary America*
- Sports - Rooney, *Atlas of American Sport*; Shortridge, *Atlas of American Women*; Rand McNally *Baseball Atlas* (also has maps of stadiums, use the same way as Transparency 4C)
- Major department stores - Laulajainen, *Spatial Strategies in Retailing*
- Politics - *Latitudes and Attitudes*, Martis *Historical Atlas of Political Parties*, or U.S. Geological Service, *Electing the President: 1789-1992*; call 1-800-USAMAPS.
- Musical styles - some new CD-ROMs, or make a tape
- Vacation spots - travel guides by AAA, Mobil, Fodors, Smithsonian

Even more topics might include migration streams, population pyramids, places named in current news stories, environmental issues, maps from state atlases and history books, excerpts from novels, diaries, poems, and so on. Though full, this list is not exhaustive; ARGUS and other major curriculum projects usually have quite a few matching style activities. BUT, please, choose well; do we really want to focus on flags, tourist attractions, minor politicians, funny names, or other trivia?

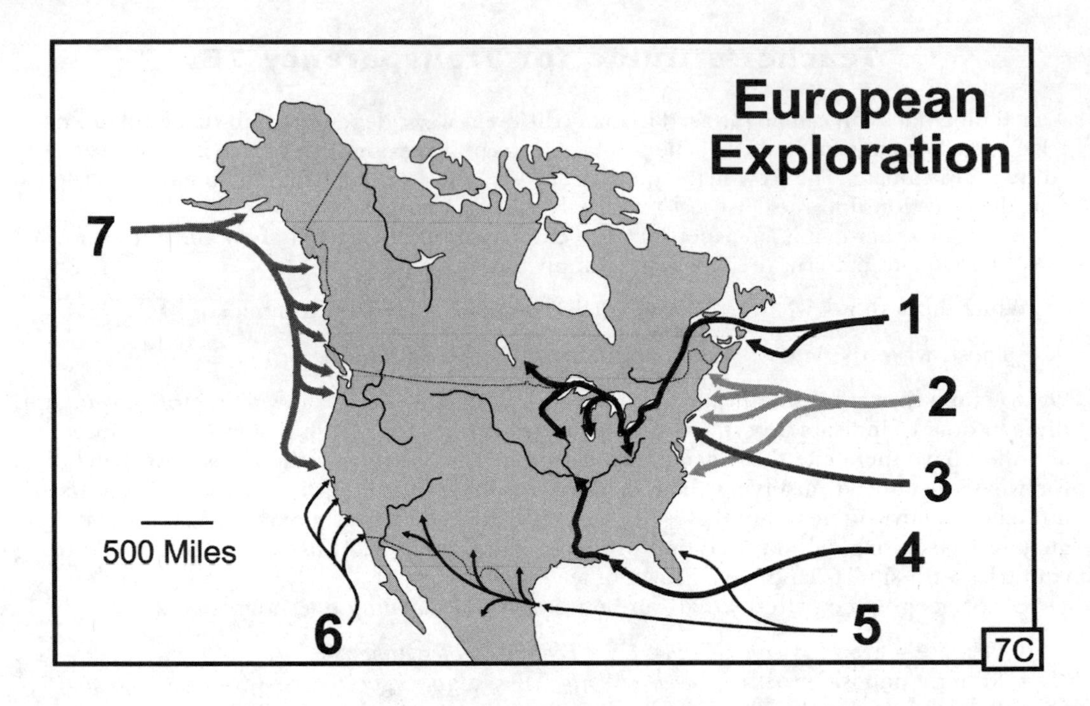

The map above shows seven general pathways used by explorers and settlers from five European nations. Describe the process and consequences of the settlement depicted by arrow number 5.

Stated this way, the instruction for what the students must do is too vague. It could, however, be followed by a clarifying question, which would narrow the range of possible answers and provide some guidance for students. Note the wide range of possible directions that the follow-up question could take:

What country does this arrow represent? Why did those pioneers choose the kind of land they tended to move to?

What kind of people came along this route? Was it a large or small number? Why would they have chosen to leave their home country at that time in history?

What were their major goals in North America? How did that influence their choice of land?

What were the major environmental conditions in the region they entered?

What Native American people were already lived in the areas they entered? What did those Native American people do for a living?

How did these immigrant people divide land? Has that land division had consequences for later users of the land, and what were the consequences?

What kind of houses and other structures did these immigrants build? How well did their structures fit with the local environment?

What did these immigrants do for a living? What resources in the area made that possible?

What kind of political organization did they adopt? How well did that organization fit with the local environment?

How would you describe the success of this migration? What aspects of the local environment helped or hindered the people moving in?

What aspects of the modern landscape in that region are results of the time when these particular people entered North America?

Teacher's Guide for Transparency 7C

Here are some sample essay assignments, aimed at different levels and outcomes:

Provide a map with several numbered locations; ask which location would be the best choice for siting a new facility – a store, highway, gas station, sewage treatment plant – and why. "I would put the restaurant at 3 because there's a big highway and more people driving there."

Provide a map and ask at which numbered location a particular hazard occurs – hurricane, earthquake, frost, drought – and why there. Or reverse the question and ask what hazards are likely to be in a specified location. What might people do to reduce the risk? "______ is prone to earthquakes, so residents should look for wooden frame houses on a solid foundation rather than brick houses with big windows. Or at least they should ask if a house is reinforced, because brick houses have problems in earthquakes."

Provide a map of an urban area and ask where a given feature – truck stop, high-rise office, baseball stadium, oil refinery – might be located and why that location is suitable. "The oil refinery should probably be at place C. It's near a river, and a railroad, and it is east of the city, where prevailing winds would blow smoke away from the city."

Provide a map of a continent with a location marked by an X. Ask what kind of people live there, what they do for a living, and what traits of the local area (or what influences from other areas) help give that place the traits it has. "People there might raise cattle for a living, because it is too dry for most crops and I don't think there is any mineral deposit there. Probably there are not very many people, because dry areas without big rivers usually do not have big cities."

Provide a map with a marked location. Ask students what they would take with them if they were traveling (or moving) to that location. Ask them to justify their list by citing features in the local environment that would fit the items they were taking. "I'd take a raincoat, because that place has a lot of rain." "I probably want to learn some French before I move there, because that is what government officials speak, but after I got there I'd start to ask my neighbors to teach me some words in the local language – people around there speak many different languages."

Provide a map with a marked location. Ask what kind of problem is happening in that area and what people can do to reduce the problem. "This area lost jobs when clothing factories closed. The department of education should send someone to help train people for new jobs." "That area has some problems because it got a lot of Vietnamese refugees; there weren't many teachers who spoke Vietnamese, and some of the children had a hard time at first. There should be night classes in English and the city should try to find some land to turn into a park with a Vietnamese theme – the city could hire some of the immigrants to help design and run the park. That would give them jobs while they learned more English, and it would also give English-speaking people a chance to find out about the new immigrants."

Note that these questions have four feature in common.

1. Each question deals with a real place.
2. Students must understand at least one broad-scale geographic pattern.
3. Students must think about how environmental, economic, and social factors are linked together in that place.
4. The question tests knowledge and skills that would be of use to a vacationer, student, worker, employer, or citizen making everyday decisions.

Community Profile - Components

Reference map - Where is the place?
What is it like? What is connected to it?
What resources are there?

Population - What kind of people are there?
How many? Where from? How old?
How is the population changing?

Economy - What bigjobs bring money in?
What services are available?
How is the economy changing?

Features - What else is interesting?

7D

Community Profile - Rubrics for evaluating

Research
Relevance of information? 3 very, 2 sorta, 1 not very
Importance of information? 3 very, 2 sorta, 1 not very

Graphics
Proper data for mapping? 2 definitely, 1 questionable
Proper map symbols? 4 definitely, 2 questionable

Presentation
Organization? 3 well organized, 2 OK, 1 disorganized
Neatness? 3 very attractive, 2 clear, 1 somewhat messy
Good grammar and spelling? 2 yes, 1 some mistakes

Bibliography
Sources cited properly? 3 yes, 2 somewhat, 1 errors

7E

Teacher's Guide for Transparencies 7D and 7E

An individual or group project is a good way to start independent geographic exploration. By its very nature, a project is a "synthesis" activity – it requires students to bring a variety of knowledge and skills to bear on a specific issue. The key is to make the expectations very clear at the outset of the activity. Expectations should span the range of learner outcomes, from application of factual knowledge to skills in organizing and presenting results. It helps to make an explicit list of evaluation criteria, such as those listed on the Transparencies or in this expanded list:

Stating a problem and writing some plausible hypotheses

A. How well did you state your problem (the landscape feature or map pattern you are trying to explain, or the issue you want to explore)?
B. How well did you state your hypothesis (your idea of what might be a valid explanation of the feature you are studying)?
C. How well did you choose an appropriate study area? Is it typical? Are data available?
D. How well did you find useful factual information about the subject? Did you focus on major features; were you able to avoid the "tourist trap" of noting only the exotic or unusual?

Mapping your data

E. How well did you choose your map scale and projection, so that the base map would not introduce unnecessary distortion in your message?
F. How well did you choose background information, so that it would help to put the thematic message into perspective, but wouldn't make the main theme hard to see or interpret?
G. How well did you manipulate your data (e.g., by dividing total amounts of production by population in order to get amount per capita)?
H. How well did you choose symbols for your thematic message? Do the symbols fit the innate nature of what you are mapping?

Analyzing your map

I. How well did you analyze the geographic patterns on your maps? Did you compare your map pattern with other appropriate maps if available?
J. How well does your analysis fit with generally accepted theories of geography? Did you cite others' observations or analyses if appropriate?
K. How reasonable are the conclusions you drew from your analysis?

Presenting your results

L. How well did you express your conclusions? Did you express them in ways that allow for reasonable disagreement on points where the data are inconclusive?
M. How well did you identify questions that warrant further investigation?
N. How well did you organize your presentation? Is your report logical and interesting?
O. Did you write clearly, concisely, with good grammar, and no unnecessary jargon?
P. Did you cite your sources completely and correctly?

The list of outcomes can vary; the Transparencies assume a rather straightforward community profile. A different list of outcomes would be appropriate for a geographic investigation of a community issue (e.g., proposed locations for a park, a solid-waste disposal area, or a transit system). Here, the focus would be on finding and evaluating relevant information about land use, population density, the economy, and other traits of communities that might be affected by the plan.

Evaluating Buildings along a Street

STREET ______________________________

BLOCK ______________________________

Compared with what we saw at the first stop, the buildings that I see on this block are:

SMALLER __ __ __ __ __ LARGER

CLOSER TOGETHER __ __ __ __ __ FARTHER APART

BETTER CONDITION __ __ __ __ __ WORSE

________________ __ __ __ __ __ ____________

________________ __ __ __ __ __ ____________

7F

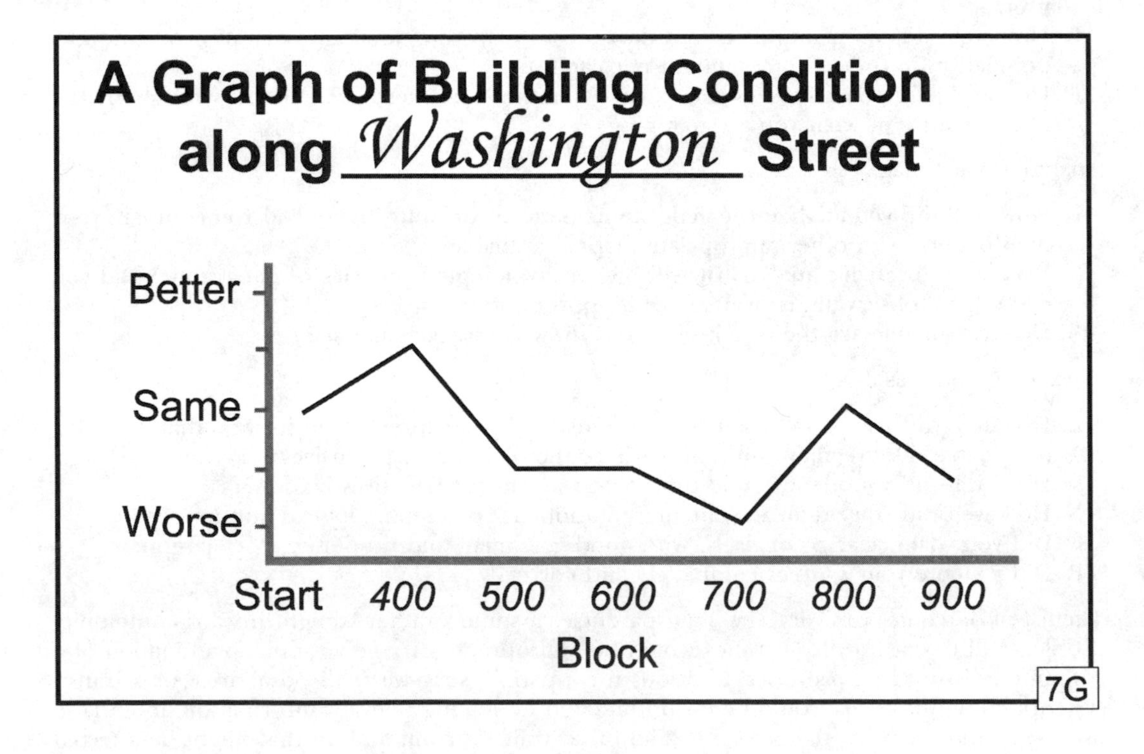

Teacher's Guide for Transparencies 7F and 7G

In preparing to lead a class on a field trip, discuss with the class the goals of what will be surveyed, specific procedures, criteria to be used, and the product. Here is an example: a field trip along a city street.

Goal: Observe the condition of buildings (or height of trees, wildlife diversity, ethnic origin of residents, land productivity – in short, practically any of the kinds of impressions that people might notice and use to compare places). These observations will be made in such a way that we can put the data on a graph or a sketch map of the route.

Procedure: Describe what we see at the first stop, and then use that description as a yardstick for evaluating the next stops.

Criteria. Criteria are the guidelines of what will be included and why. The real purpose of this discussion is to get students to realize that the criteria used in making a map or graph are human inventions. As such, they can range from useful to worthless, from fair to biased, and so on.

Product: A map or graph of some measurement of the real world, gathered in a systematic fashion and subject to evaluation by a teacher or peer group. On your way to the field-trip destination (e.g., a new Wildlife Research Center or a stadium), stop the bus and announce the starting point of the graph.

Activity: At the first stop, the goal is to write and/or sketch what you see, with enough accuracy that your description can be used as a yardstick for comparing the next sites. Note the height, density, and condition of buildings in the middle of the block. Clearly record any signs or other specific clues that you think are especially helpful. This set of buildings will be a reference for other observations. From here on, you are to record how the buildings (or the trees, the soil, the churches or restaurants, or whatever is being observed) at each stop compare with what you saw at the first stop.

When you get back to the classroom, you will make a set of three graphs to show the size, spacing, and condition of the buildings along the route.

This kind of disciplined field observation can be even more effective if students try to devise their own classification ahead of time rather than just following a predetermined set of categories. A teacher could have a list like the above in mind while conducting a rather open-ended discussion before the field trip. One could even have the students design an observation form like the example in Transparency 7F. All of these steps help tie the field experience into the classroom.

A disciplined form of observation does not automatically preclude serendipitous events while in the field. Offer a prize for the funniest sign seen, the most attractive sketches that students make in trying to record what they see, or the most interesting haiku a student writes about the trip. The students can even design quizzes for each other; immediately after the trip, ask them to write questions to elicit memories of what they saw.

By all means evaluate the product of a field trip; this is the most important step in setting the tone for future field trips.

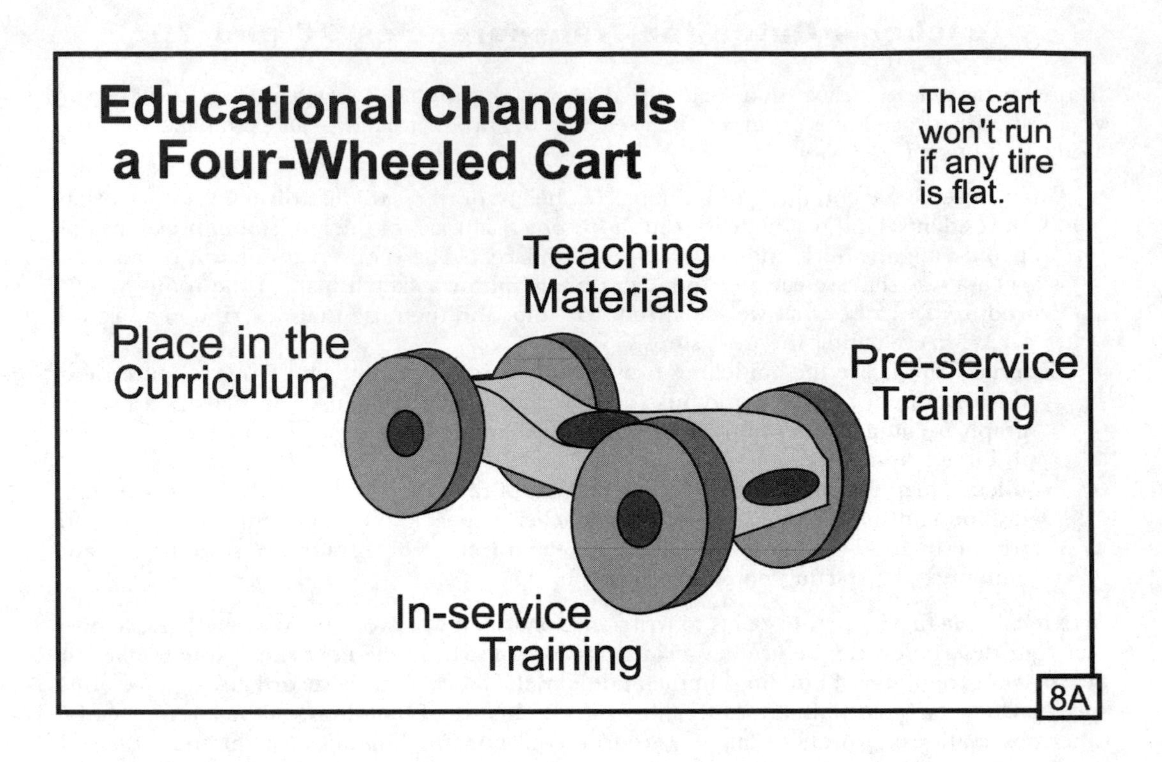

What Can Yellow Pages Tell Us?

(Figure the number of some things per thousand people; two examples are already done)

	Deming		Palm Springs	
Thousands of People	11		40	
Motels	11	____	134	3.4
Real-estate Offices	7	0.6	168	____
Video-rental Stores	4	____	5	____

8B

Teacher's Guide for Transparency 8A

This Transparency is meant primarily to be used in administrative meetings, teachers' conferences, and on other occasions where there might be a reason to put the process of materials development and teacher training into perspective.

Teacher's Guide for Transparency 8B

To thrive economically, every place needs at least one bigjob - an occupation that brings money into a community (BIGJOB is an acronym that stands for Basic Income-Generating JOB - other textbooks may call this a basic industry, economic base, or core activity, but all of these terms can be misleading in other contexts - hence the new term).

Most bigjobs produce something for sale to people outside the community. "Produce" and "sale," however, can also have different meanings in different places. People in some places, for example, might produce corn, pulpwood, or iron ore. In other places, people might make athletic shoes, movies, or missile parts. Still others might sell vacations, radio talk shows, heart transplants, or oil leases. And so forth, through a bewildering variety of products, both goods and services.

The question is: how do we teach students to identify what the bigjobs in a community are? (An alternative is having students memorize lists of places and their products, which would be tedious for them.)

We could start by trying to find a book that classifies communities. For example, *Places Rated Retirement Guide* (by Boyer and Savigeau) identifies Palm Springs, California, as one of the top places in the country to retire. Deming, New Mexico, is on the same list of retirement destinations, but it is near the bottom.

Both Deming and Palm Springs have some distinctive landscape features that contribute to their role as places for people to retire in and spend their pensions, Social Security payments, and stock dividends. With all of that money coming in, many residents of both towns earn their living by providing services to retirees and vacationers. Those workers have that community's bigjobs, and their places of work are part of the landscape that people have "built" in order to do those jobs. The Transparency has data about some occupations that serve various kinds of retirees. Comparison clearly shows that the two communities are different even though they both depend on retirees for their primary income.

Activity: A telephone book (printed or computerized) is a good resource for information about a community. Many occupations and structures in a community are related to its bigjobs - the goods and services that the people produce to sell or exchange with people in other places in order to get things they cannot produce. Have students pick a community and try to identify its bigjobs by looking at the yellow pages of the phone book.

For example, if you suspect that a South Carolina community gets some income from forestry (a good guess for any rural area in the Southeast), you might count the yellow-page listings under headings such as "forestry consultants," "logging equipment and supplies," or "timber sales." It also pays to look in the pages devoted to goverment listings to see which offices and agencies have a lot of phone numbers and divisions. Look especially at occupation-related departments such as Mine Safety, Natural Resources, or the Bureau of Alcohol, Tobacco, and Firearms.

To put the results in perspective, calculate the number of various things per thousand people in the community being studied. Then, compare those ratios with national or world averages (Transparency 8D). The product of this Activity might be a short report (e.g., to a group of investors) or a poster showing the bigjobs in a community.

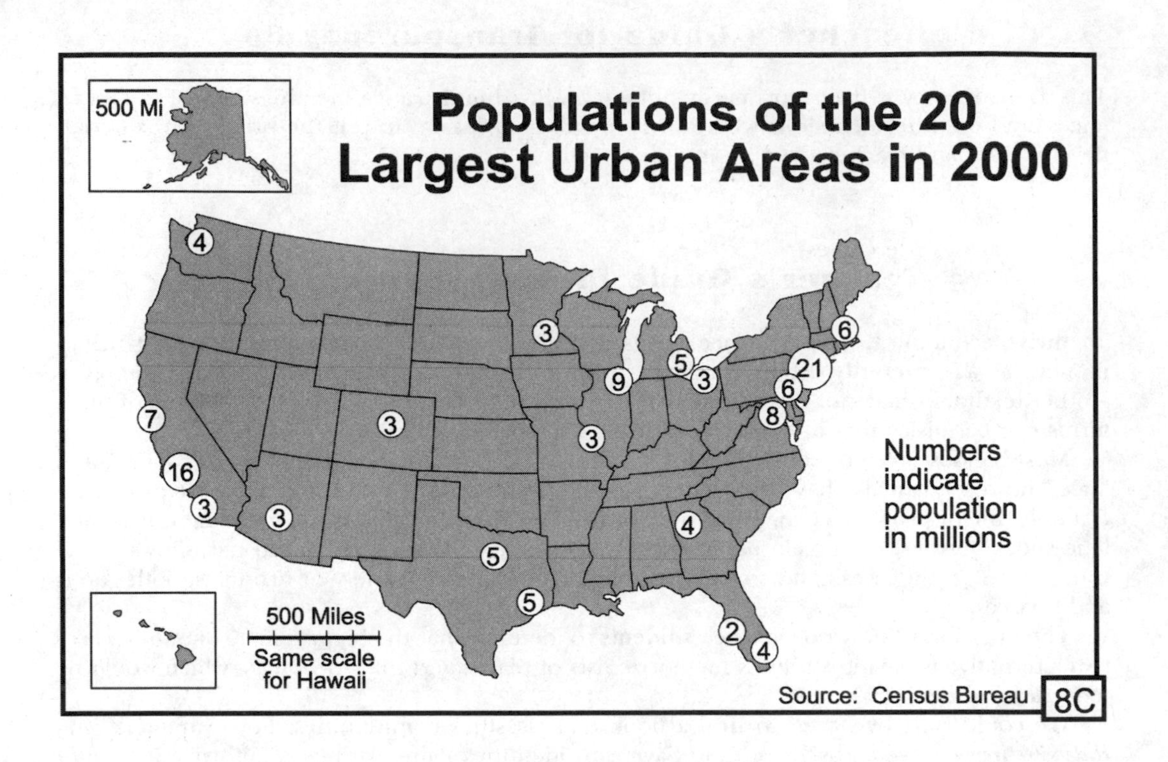

500 Mi
Populations of the 20
Largest Urban Areas in 2000
4
3
6
21
9
5
3
6
8
7
3
3
16
3
3
4
5
5
2
4
Numbers
indicate
population
in millions
500 Miles
Same scale
for Hawaii
Source: Census Bureau
8C

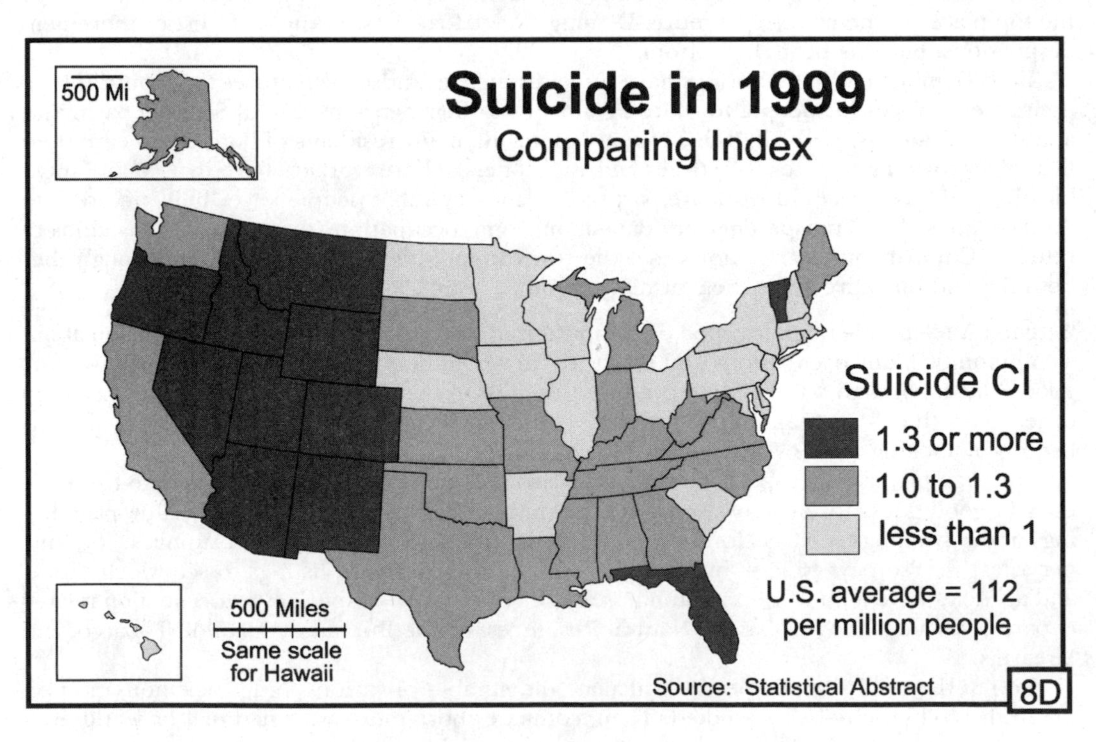

500 Mi
Suicide in 1999
Comparing Index
Suicide CI
1.3 or more
1.0 to 1.3
less than 1
U.S. average = 112
per million people
500 Miles
Same scale
for Hawaii
Source: Statistical Abstract
8D

Teacher's Guide for Transparency 8C

Since we can observe or measure places in so many different ways, it is not surprising that there are many different kinds of thematic maps. Each major "map vocabulary" (choropleth, isoline, dot, cartogram, etc.) is designed to show particular kinds of information. For example, a valid way to show *amounts* of something such as population or steel production is with a scaled-symbol map (sometimes called a proportional symbol map or a graduated circle map, though it does not always use circles). In this kind of map, symbols of different sizes show greater and lesser amounts in different places.

Activity: Find a large outline map of the United States (or project this Transparency or another country or continent on a large piece of paper and trace the outline). Lay the outline on a table and stack pennies or poker chips on each city to indicate its population. Use one penny or chip for each million people. The resulting stacks are a dramatic illustration of population geography. The activity also teaches some basic place location. You can provide city names and have students look them up in an atlas, or provide a table of names and numbers but have the students create the base map. Proportional symbol maps can be used for any topic that involves counts or other absolute numbers; other kinds of maps are better for percentages or other ratios (Transparency 1B).

Teacher's Guide for Transparency 8D

One way for advanced students to compare places is to calculate a Comparing Index (other books call this an Index of Local Importance or a Localization Quotient - we prefer CI, not just because it is a more accurate description and a simpler term, but also because its abbreviation, CI, gives teachers a wonderful nickname to use for the index, the "C-ing I" which sounds like "seeing eye").

This index is a simple way to show how a local area compares to others. A census report, for example, can tell how many people in an area do such things as work in clothing factories, speak Korean, or are between 15 and 19 years old. These "raw" numbers, however, cannot say whether that local figure is typical or unusual. To make raw numbers easier to compare, we figure some ratios (it's a great math-across-the-curriculum Activity!).

For example, in 2004, Bangladesh had about 140 million people and 56,000 square miles of land. Most of the land was usable for farming, although some was subject to flooding. Iowa had about the same amount of land (also mostly usable for crops), but fewer than 3 million people. To make those numbers easier to compare, calculate the amount of land per thousand people: Iowa had about 20 square miles per thousand people (in 2004), whereas Bangladesh less than half of one square mile.

To make the comparison even more obvious, divide the ratio in each area by the world average. That gives the number called the Comparing Index. In our example, Iowa had a land-availability CI of about 2 (20 square miles per thousand people, divided by the world average of 10). This means Iowa had twice the world average land per thousand people. Meanwhile, the figure for Bangladesh was 0.4 divided by 10, for a CI of about 0.04 - it had one 25th of the average land per thousand people.

Activity: Give students a table of data (e.g., population density, per capita income, life expectancy) for 10-15 states and national figures (or provide country figures and the world averages, for an international focus). Have them compute CIs by dividing state values by the national values. Warning: To be valid, CIs must use ratios, not raw data.

Activity: Make a choropleth map of CIs, to allow comparison with other maps (compare Transparencies 8D and 3O; it makes you wonder, no?).

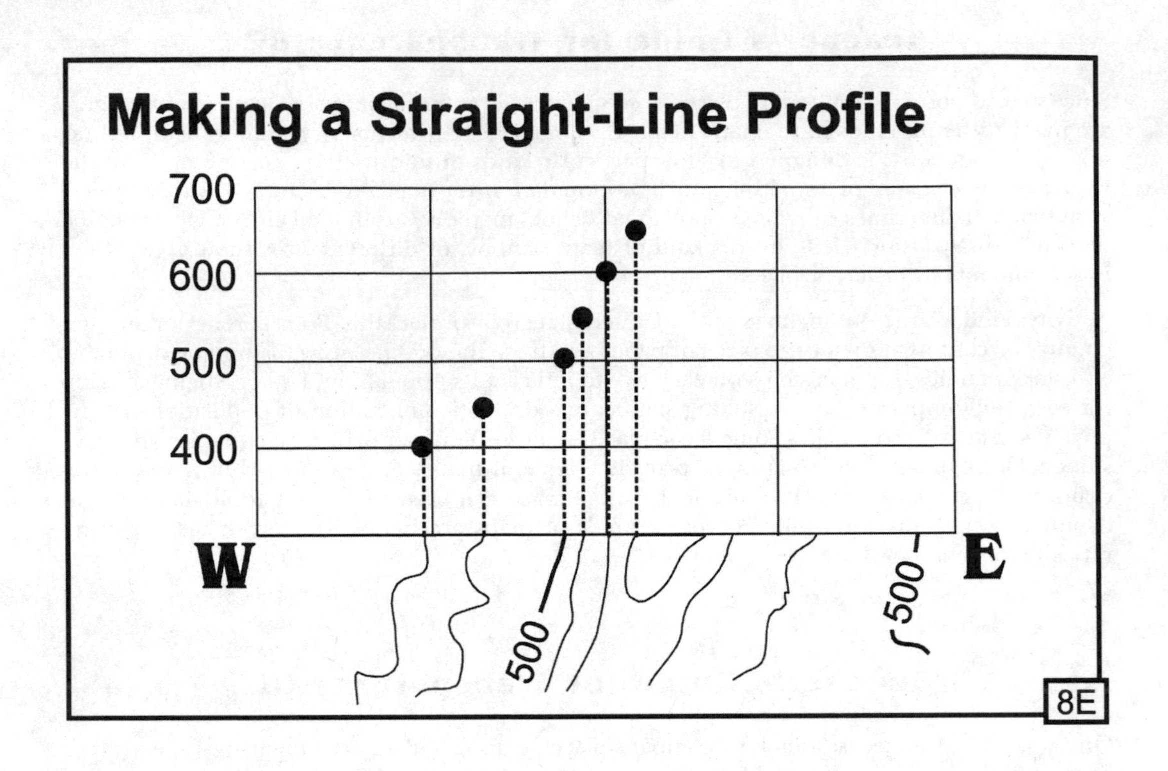
Making a Straight-Line Profile
700
600
500
400
W
E
500
500
8E

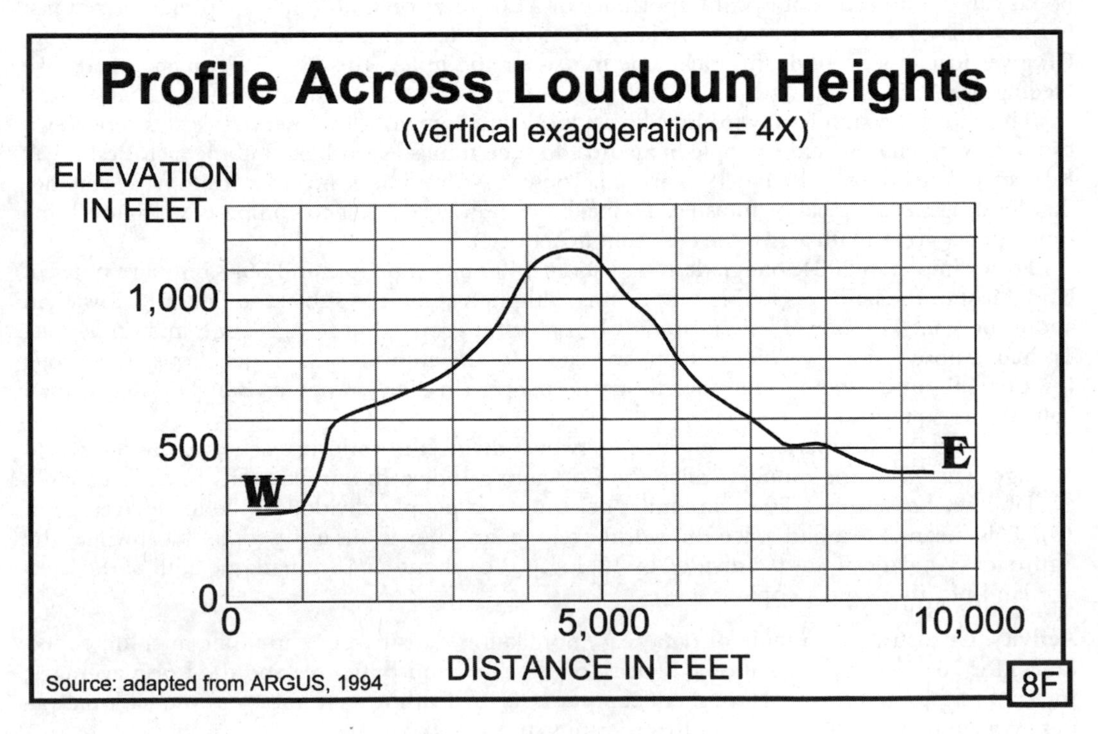
Profile Across Loudoun Heights
(vertical exaggeration = 4X)
ELEVATION IN FEET
1,000
500
0
W
E
0
5,000
10,000
DISTANCE IN FEET
Source: adapted from ARGUS, 1994
8F

Teacher's Guide for Transparency 8E

Here is a quick recipe for drawing a side profile of a topographic map (for more detail, see the ARGUS or ARGWorld CDs, or NCGE Pathways Publication Number 1, The Language of Maps):

1. Find (or make) some graph paper with grid intervals that match the scale on your map: latitute-longitude (or if it has a grid, you might try to match it).
2. Position the edge of the graph paper along your chosen profile line. At every place where a contour line crosses or touches the horizontal axis of the graph, go "up" from the bottom of your graph and place a dot at the proper elevation.
3. Note the positions of hill tops and valley bottoms along the profile line. Place dots to show the horizontal positions and elevations for these key features. (Someone who is fairly good at profile drawing can sometimes skip the other steps and concentrate on these information-rich locations).
4. Draw a smooth line connecting the dots. Other contour lines on the map can help shape the details of the profile - it should slope steeply where contours are close together and gently where they are far apart. This result is a line that shows the shape of the land as seen from one side.
5. Optional: Label key features of your profile with names. Add other information (e.g., forest cover, density of houses, underlying geology, etc.) with appropriate symbols if you wish to show the relationship between surface topography and other kinds of landscape features.

You could use this same grid with other maps, but it is usually better to design one to fit the specific scale and contour interval of the map you are using (the contour interval is the vertical distance shown by two adjacent contour lines).

Activity: Draw a side profile of a region of interest - a ski slope in Colorado, for example, or a local hill that might be used as a golf course or a minibike trail. You will need a topographic map; they can be downloaded for any place in the country from *www.topozone.com* or you could write to the Map Distribution Center, U.S. Geological Survey, Federal Center, Denver, CO 80225, for a (free!) index map of your state.

Teacher's Guide for Transparency 8F

This is a side profile of the ridge at Harpers Ferry (also shown in Transparency 1C). Some of the slopes are as steep as the stairs in an office building, and the ridge is a thousand feet high. Before the invention of internal combustion engines (for trucks and tractors) or dynamite (for clearing land and building roads), getting to the top of a hill this size and shape was a real challenge. You could walk up, or perhaps ride a horse, but moving something like a 1,200-pound cannon to the top was very difficult.

The town of Harpers Ferry is located at the only easy gap through this ridge for more than a hundred miles. The gap was created by the Potomac River. Washington, DC, the national capital, is only a short distance downstream. Is it any surprise that several of the bloodiest battles in United States history were fought in the vicinity of this town?

Activity: Find topographic maps of the area around Harpers Ferry, Chattanooga, Vicksburg, or other major Civil War battles. Have students draw side profiles of the terrain. Then discuss ways in which the terrain influenced the course of the battles (and why that would not be so important today). Looking at the interaction between terrain and history helps move the focus away from mere body counts.

Or, if you prefer a nonmilitary focus, do profiles of ski slopes or hiking trails.

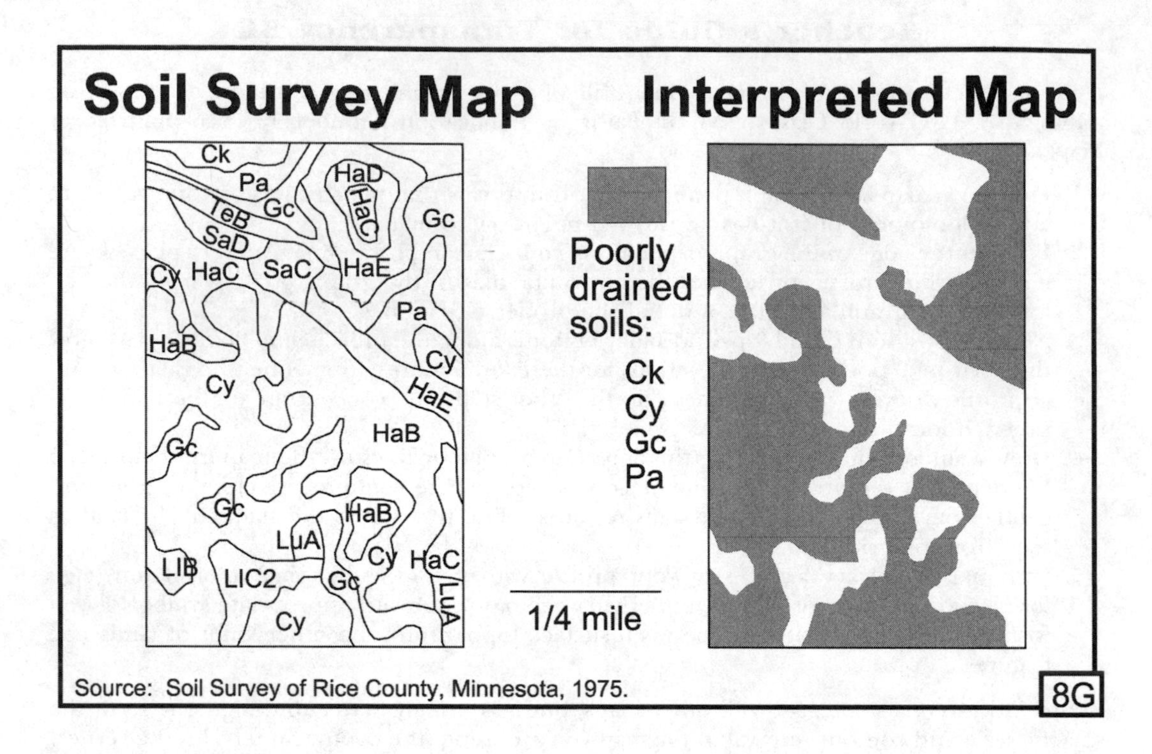

Describing a Newspaper Article

about an environmental issue

Prominent	___	___	___	___	___	Hidden ("buried")
Illustrated	___	___	___	___	___	Verbal
Factual	___	___	___	___	___	Opinionated
Logical	___	___	___	___	___	Disorganized
________	___	___	___	___	___	________
________	___	___	___	___	___	________

8H

Teacher's Guide for Transparency 8G

Names can be misleading. For example, a soil survey is not just about soil.

Every person, family, and society has to make decisions about where to store things. Ask students to list things their parents store in "weird" places – you know, like a flashlight in the underwear drawer, or a can of tuna under one leg of a table. In most cases, the "weird" place turns out to make sense in the context of available space and family needs.

The U.S. government chooses to store environmental information of all kinds in its soil surveys. Want to know what areas are suitable for campgrounds? The rainfall of the biggest storm that comes only once in 20 years? Where good habitats for deer are? What places would be hazardous for buildings with basements? What date is safely past the last frost 9 years out of 10? A soil survey can tell you. About 2,000 counties have them, and (write this down) they are usually provided *free* to teachers (look for the Natural Resource Conservation Service under Department of Agriculture in the phone book listings for the federal government).

The soil surveys have colorful maps of general conditions in an entire county as well as dozens of maps of smaller areas. The detailed maps have soil codes, which make them complicated, but they are really an aerial photograph that shows every road, tree, house, and pond. Post the map for a local area – students can ignore the soil codes as they enjoy trying to identify the features they are familiar with.

Activity: Give students a copy of a soil survey map and a list of soils that are suitable or unsuitable for a specific purpose, such as building houses or growing corn. The survey will give you that information, in easy-to-read tables that are tied to the map via short letter-number codes (it will make sense once you open the survey – just ignore the technical stuff about Munsell colors and plasticity indices and so forth; one goal of using a survey is to learn how to focus on the information you need). Have students color in the suitable areas. Then (here's the kicker), go on a field trip to see if people are using the land in appropriate ways. Or take pictures and bring them in for the class to study.

The Transparency shows a piece of land that was used for a high-income housing subdivision in Minnesota; some of the houses were hard to sell, because water got into their basements and caused wall damage. If the builders had consulted the soil map, they would have seen that construction on some of these soils was unwise. (What teacher of environmental science could resist: "Isn't it fun to see rich people do stupid things?")

Teacher's Guide for Transparency 8H

Activity: Describe a local issue – an environmental question, for example, or a proposal to build a mall. For several weeks, have students cut articles, editorials, and letters from the newspaper and try to classify the clippings according to several criteria: pro or con, logical or emotional, fact-based or just opinion, and so on. **Variation:** Use number scales rather than either/or categories – for example, give each article a number, say from 1 for strongly in favor to 5 for strongly against, and so on.

Activity: Cut maps out of the newspaper, post them, and have students rate them according to several criteria: clear or confusing, easy or hard to read, correct or questionable in use of conventional cartographic "grammar," fair or unfair in presenting issues. If an important story is not sufficiently accompanied by maps, write a letter to the editor asking why the paper did not put an issue of such obvious importance into its geographic context.

How to Give Directions

1 Name a compass direction, the route to go on, and a landmark ahead.

Go south on 14th, past the Library.

2 Give a distance and a warning feature.

go 3 miles cross Elm St.

3 Name a landmark and a "too far" feature.

at the Post Office;

(you see Oak St. you are too far.)

4 Describe the turn and the new direction.

turn left, to go east on Green St. . . .

••• Continue to destination.

8I

Teacher's Guide for Transparency 8I

Using a road map is a fundamental geographic skill. Satellites, global positioning devices, and electronic maps are starting to become the tools we use in personal navigation; still we need be able to translate a map view of the world into "egocentric" directional words such as left, right, and straight.

For 20 years, my research has involved interviewing government officials in small towns. For a typical interview, I drove toward a town, stopped near the edge, called the official, reminded him or her that we had an interview scheduled, and then asked for directions to the office. In my sample of 300-plus counties, the range of idiosyncratic approaches to giving directions was astonishing. For example, they might start by asking me:

Are you familiar with [town-name]?
No, I've never been here before.
Do you know where the supermarket is?
No, I've never been here before.
Well, go out on the highway and head toward the high school.
I haven't been here before. Which way is that?
Toward town.
OK, I can manage that.
Go about three, maybe four stoplights or signs,
OK, what then?
Then turn left where the old Amoco station burned down in '97.
What's there now?
You know, next to where the guy married Dr. Jones's daughter
How far is that from here?
About four or five stoplights. Then go left; we're behind some stores. You can't miss it.
Wanna bet?

Activity: Hand out copies of a map showing a destination of interest. Then ask students to plan a route from a different location to the destination. Then have them describe that route as if they were giving directions to a traveler. In general, a good set of directions should have:

- The number and/or name of the road the driver should take.
- The compass direction and a clearly visible landmark to go toward.
- The distance in miles (which can be estimated visually or read on an odometer) rather than stoplights or blocks, although that type of thing could be supplemental.
- A "warning sign," preferably a cross street, that comes just before a turn.
- A prominent landmark at key intersections (e.g., Library).
- Street name or route number of the street to turn on ("get in the left lane; you turn left on highway 10 when you get to the Library").
- The direction of the turn, in both egocentric and geocentric terms ("turn left, so you'll be going northwest").
- A prominent street or landmark that says "you've gone too far" (e.g., "if you see the Citizen's Bank, you've gone past the turn").
- And so forth until the arrival at the destination.

Teacher's Guide for Transparency 8J

This one is fairly self-explanatory; I saw some figures in a newspaper and was curious about where the countries were. (Selling high-tech weapons to dictators bothers me. Maps like this will help me judge candidates in the next election. That's geography for life!)

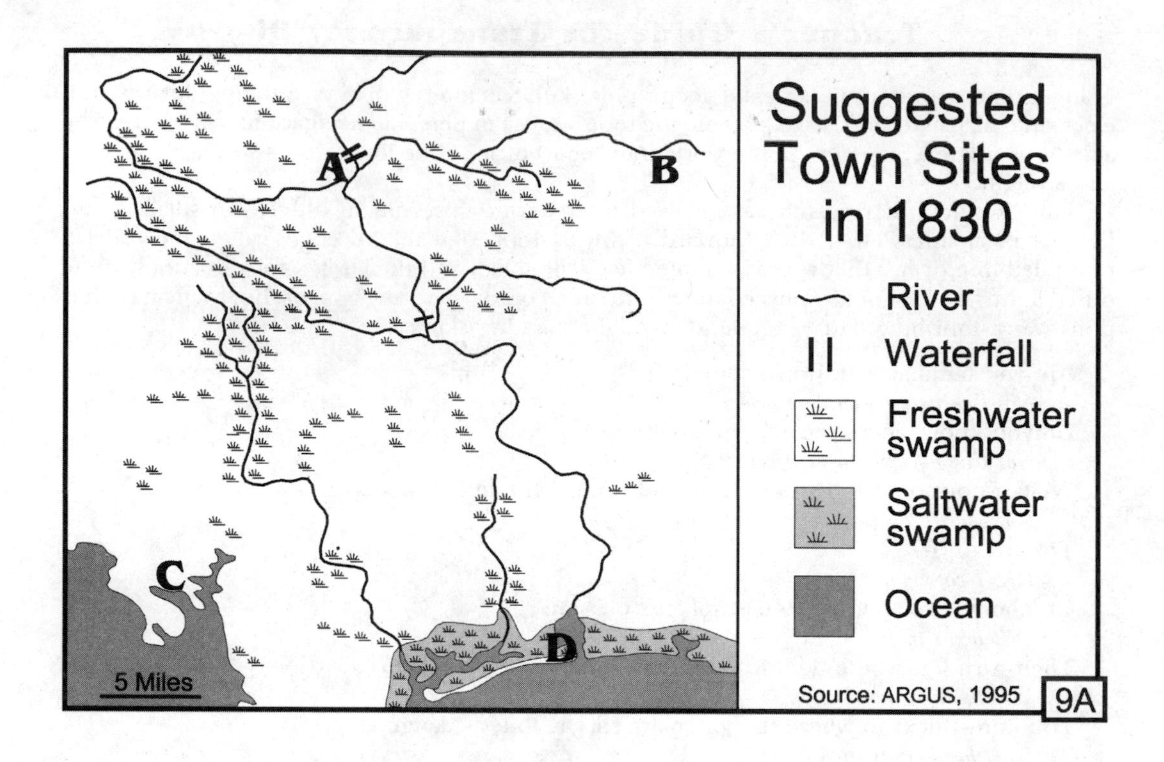
Suggested Town Sites in 1830
River
Waterfall
Freshwater swamp
Saltwater swamp
Ocean
A
B
C
D
5 Miles
Source: ARGUS, 1995
9A

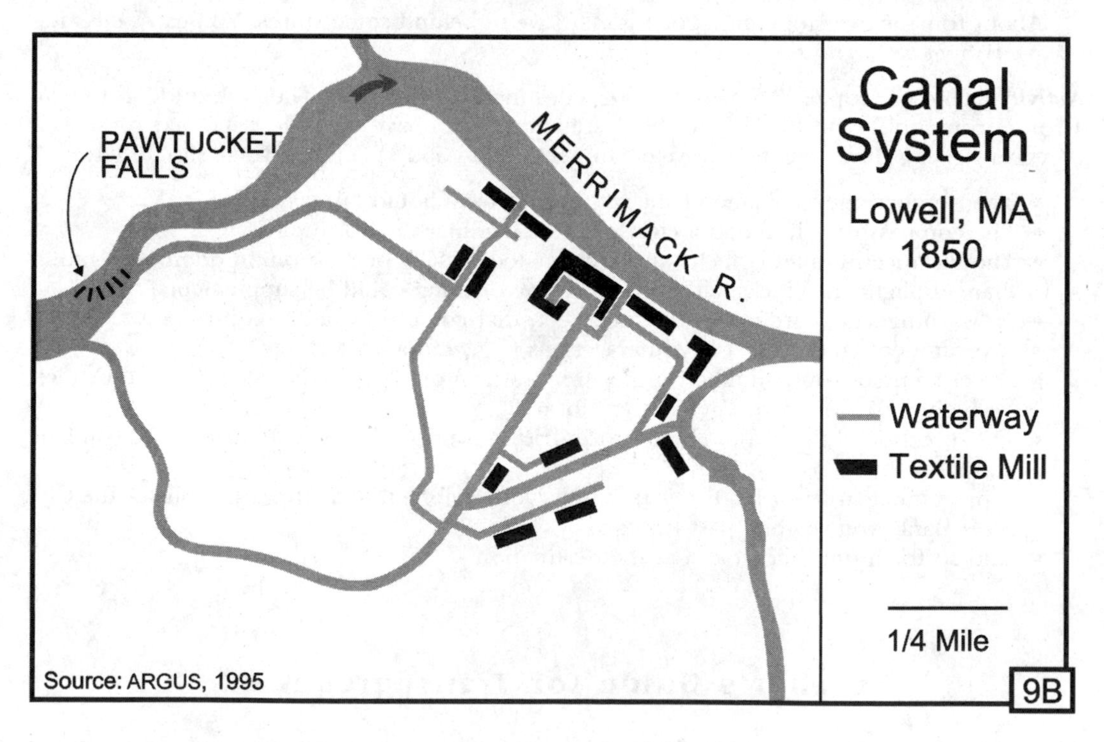
Canal System
Lowell, MA 1850
PAWTUCKET FALLS
MERRIMACK R.
Waterway
Textile Mill
1/4 Mile
Source: ARGUS, 1995
9B

Teacher's Guide for Transparencies 9A and 9B

The location of a town is a historical event and is a geographical fact. At a specific point in time, a specific group of people made a decision. That historical fact is often immortalized by a plaque or statue in a park or other prominent place.

Activity: Describe the origins of a town that is familiar to your students. Then ask whether there would be a town there if the originating event (General Dingbat building a fort, Dewey Cheatham building a trading post, or whatever actually happened) had *not* occurred in that place at that time.

In some cases, the answer is no. But in other cases, a town would have developed at that location anyway, because it had clear advantages: a good site for a fort, near valuable mineral deposits, and so on. The advantages of particular locations fall into two broad categories: site advantages (favorable conditions in the local area) and situation advantages (good connections with other areas). How advantageous they are, or whether they can be maintained, changes over time, for many reasons.

Activity: Tell students to imagine they are on a boat heading for America in 1830. Then project Transparency 9A and ask students to discuss which lettered location they would choose for a settlement (this well-known Activity is from the 1968 High School Geography Project, as modified for the 2003 ARGUS materials).

Site A? It's not very promising from an agricultural point of view, because it is swampy and hard to reach. A waterfall, however, would be an important asset in the early 1800s. People at Site A could take water out of the river that comes from the north. They could build canals to carry it across the site, use it to drive waterwheels in a mill, and then release it back in the other river.

Site B? It has dry land and fresh water. It may be the best of the four places for farming, but it is not very accessible from the ocean. Flooding seems to be no threat, and being away from the swamp would have been a health advantage in a time before antibiotics.

Site C? It has good ocean access, reasonable protection from storms, and perhaps some cropland. It would probably be a preferred site for a port or fishing village, although there is no obvious source of fresh water. The jaggedness of the coastline implies a rocky and rugged site (though this is not certain).

Site D? It is probably the most dangerous from a construction point of view. It is exposed to ocean storms and has little dry land; it may not even have a source of fresh water. It might be a good fishing area, however, and it would probably be the preferred choice if military security were still a concern.

Aha! This is not a hypothetical situation – sites A through D are all real places. Rotate Transparency 9A 90 degrees counterclockwise, so that the top becomes the west edge. The map now shows the area around Salem and Lowell, Massachusetts. By 1830 the government of the young United States was well established. The frontier was beyond the Mississippi River, and farmers on rich land in Ohio and Illinois were sending grain and meat eastward. New England farmers were leaving their rocky land and moving to cities or to the frontier. Immigrants from Europe included people who knew how to build factories and work in them. Site A is Lowell, which soon became one of the most famous milltowns in New England. Thousands of people worked in mills driven by water that was brought through an elaborate system of canals (as shown in Transparency 9B). Site B is Derry, a small farming town that has become a minor industrial center since Interstate 93 was built north from Boston. Site C is Salem, a fishing town with a distinguished history and a good harbor but little chance for growth. Finally, site D is now a resort service area for people coming to the beach (during favorable weather!).

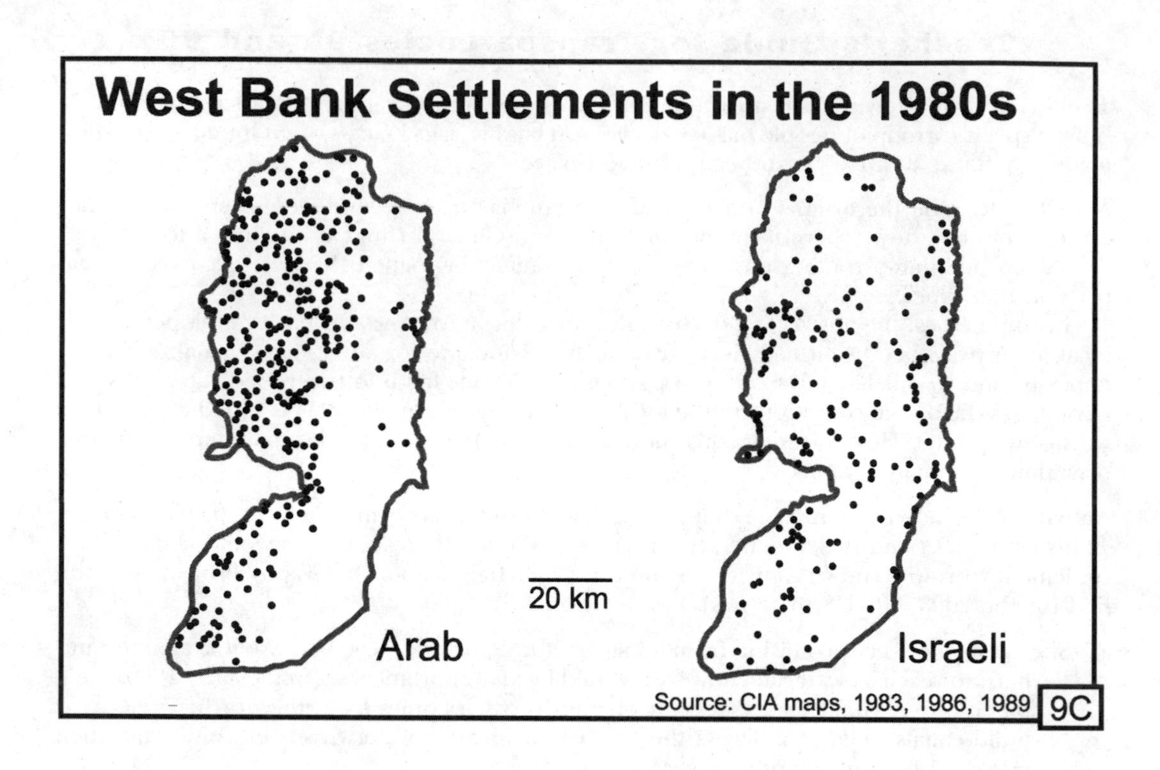
West Bank Settlements in the 1980s
20 km
Arab
Israeli
Source: CIA maps, 1983, 1986, 1989
9C

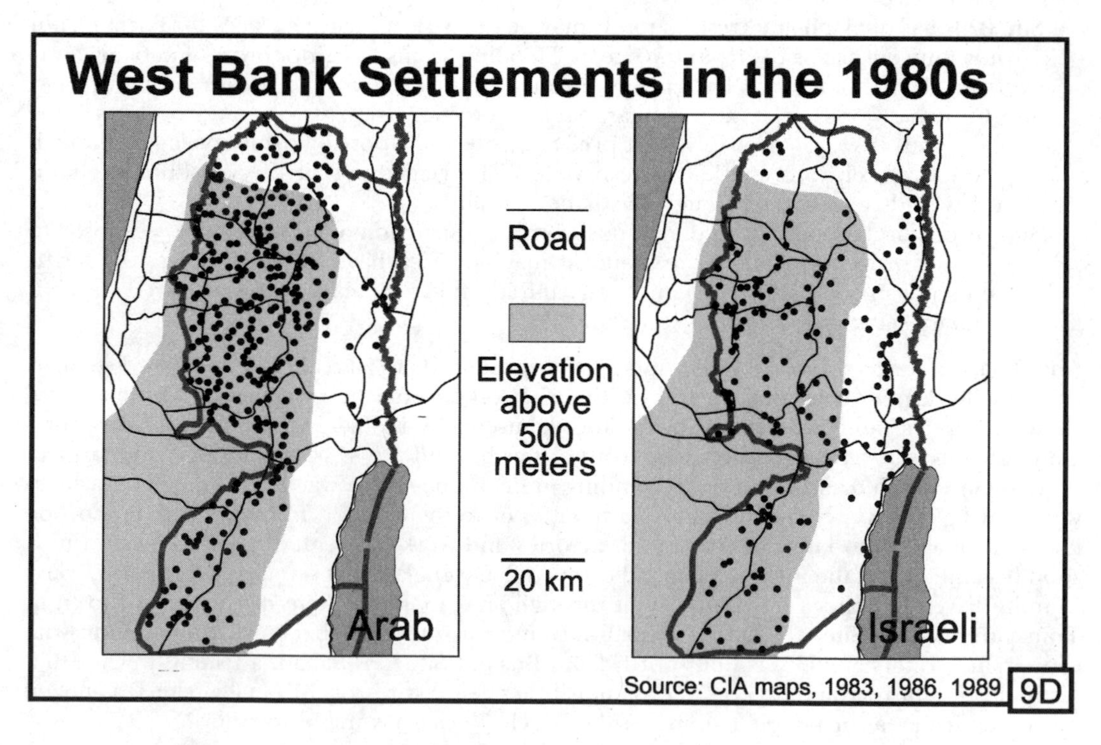
West Bank Settlements in the 1980s
Road
Elevation above 500 meters
20 km
Arab
Israeli
Source: CIA maps, 1983, 1986, 1989
9D

Teacher's Guide for Transparency 9C

These two maps show the geographic pattern of Arab and Israeli settlements in the hilly region known as the West Bank (the dry hills on the west side of the Jordan River). Shortly after the Israelis captured this area in 1967, the government began a program of settlement.

The maps on Transparency 9C are examples of simple thematic maps. They clearly show that the Arab and Israeli settlements have quite different geographic patterns. The older Arab settlements are "unbalanced." Their geographic pattern has a pronounced "bias" toward the west side of the area. The newer Israeli towns are more even, but at a more detailed scale you might describe many of them as arranged in "strings."

Activity: Have students draw a line that separates the area into a half with only a few dots and a half that has most of the dots. Then, they can count the dots in each half, add them together to get a total for the entire area, and calculate percentages in each half. When that is done, the students can make a generalization about the pattern - is it balanced, obviously biased, or somewhere in between? You can run this activity with this map or any other dot map with a distinctive pattern.

For example, one might draw a diagonal line through Charlotte, North Carolina, and count shopping centers on each side of the line. The conclusion? "about 85 percent of the major shopping centers are in the southeast half of the urban area." Similar analysis of a map of African-Americans in Charlotte leads to the conclusion that "more than 70 percent of the black people in Charlotte live in the northwest half of the city." Together, these two statements could be cited as proof that different groups within the population of Charlotte do not have equal access to large shopping centers. Before judging too quickly, however, one should study other maps, in order to see if factors such as industrial areas, parks, or road patterns would explain the imbalance.

This is one way in which map pattern analysis can contribute to a public discussion of issues of fairness and efficiency.

Teacher's Guide for Transparency 9D

The reasons for geographic patterns are usually easier to see if a thematic map includes at least a little bit of reference information. These two maps have exactly the same thematic information as the ones on Transparency 9C, but the map maker added a few lines to show roads and a pale gray shading to show areas of high elevation. With this background, the map reader can see that the majority of the Arab settlements are at higher elevations. This makes sense, because the weather gets cooler and rainier as you go up higher in the mountains. Both of those climate trends are advantageous for farming and grazing in a hot and dry region such as the Middle East.

The Israeli settlements, by contrast, appear to have been located for defensive or military reasons rather than to take advantage of favorable climate. Most of them are along major roads, with a sizeable number in the very dry area close to the border with Jordan.

Activity: Find some interesting data that occur at discrete points - for example, burglaries, street festivals, car accidents, video rental stores, or schools with winning football records. Have students make a bare thematic map of the data by putting dots or other point symbols in appropriate places on a blank outline map of the area. Then have groups of students add different kinds of background information to their maps - streets, rivers, political borders, landmarks, and so on. Finally, ask students to compare the maps and decide what kind of background helps a map communicate its main message best.

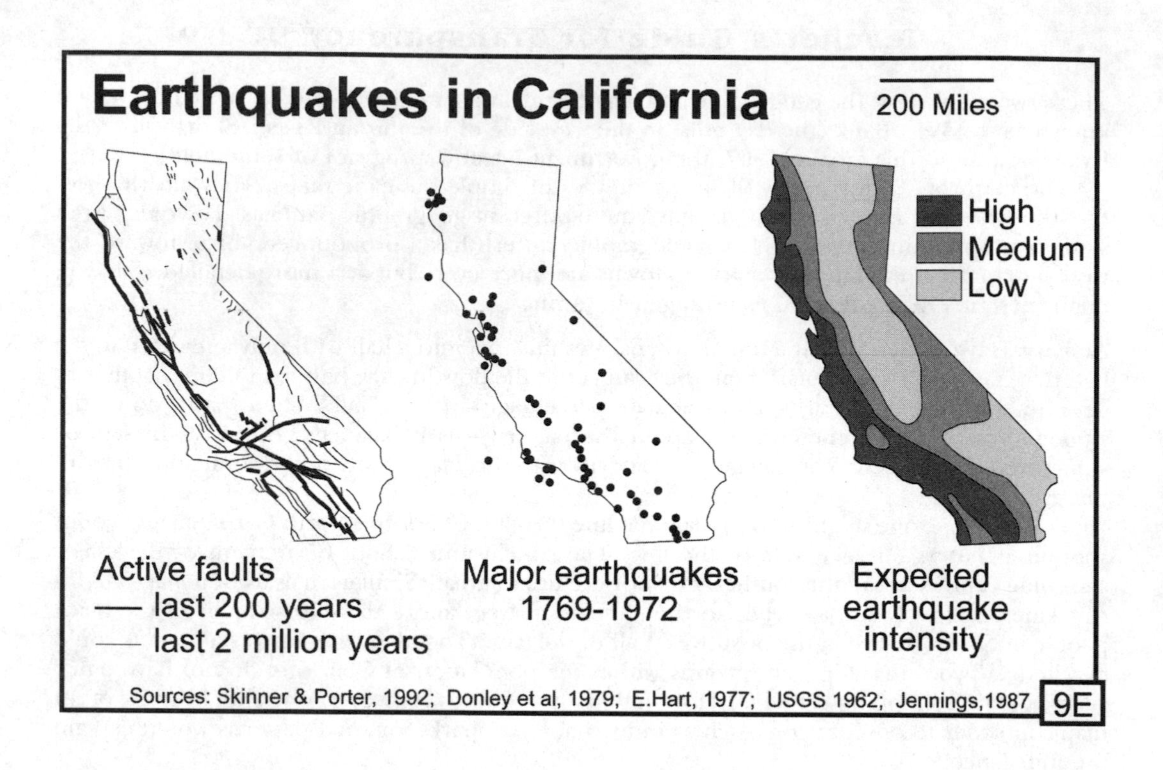

Alquist-Priolo Special Studies Zones Act

Potentially active faults under the Alquist-Priolo Act

This 1972 state law requires the state geologist to:

"Delineate . . . appropriately wide Special Studies Zones to encompass all potentially and recently active traces of the San Andreas, Calaveras, Hayward, and San Jacinto Faults, and other such faults . . . that . . . constitute a potential hazard . . ."

Source: E.Hart, 1977

9F

Teacher's Guide for Transparencies 9E and 9F

Environmental risks, to buildings and people, are greater in California than in other parts of the country.

Some of the risks are due to California's unique climate. California is located at the northern edge of the northern hemisphere's subsidence region (Transparencies 4H and 4I). The zone of sinking air (the subsidence) moves northward in summer, bringing hot and dry weather to most of the West Coast from Mexico to Oregon. In winter, the dry subsidence shifts to the south. As a result, San Francisco has about 6 months of rainy "winter"; Los Angeles about 3; and San Diego has only a few rainy weeks.

This seasonal shift in the subsidence (and the resulting pattern of dry summers and rainy winters) brings with it four distinctive climatic hazards to much of the state of California:

- In summer, smog accumulates in the hot and dry air.
- In autumn, fires burn in vegetation that has been dry all summer.
- In winter, floods and mudflows occur when rain falls on hilly areas that burned in autumn; the fires often leave bare ground that is especially vulnerable to erosion.
- In spring, coastal cliffs collapse when waves beat against shores that have lost their sandy beaches during winter storms

These seasonal hazards occur in various parts of the state nearly every year. As a result, many Californians have what some geographers call a "disaster culture." This says that people expect climatic hazards every year, and they plan for them. Such preparation is exhausting, and therefore the people don't have enough political energy left to plan well for longer-term hazards, such as earthquakes.

Part of their complacency about earthquakes stems from the way the earthquake hazard is perceived. For example, after an especially damaging earthquake in 1970, the state legislature passed a tough law, the Alquist-Priolo Special Studies Zones Act. This law requires sellers to notify buyers if a house for sale is located in an earthquake-prone area. Realtors and homeowners realized that this law would hurt house prices and sales. They therefore lobbied the legislature to adopt fairly lenient criteria for what constitutes an earthquake-prone area. These political pressures are evident on maps. The first map on Transparency 9E shows the location of known earthquake faults; the middle map shows historic earthquakes; the third map shows a measure of earthquake risk. Meanwhile, Transparency 9F shows the official map of earthquake-prone areas. Clearly, you can get a different impression of the earthquake hazard, depending on the criteria and symbols that were used in making a map. For example, try counting the major earthquakes (on the dot map) that occurred in places that are *not* shown as risky on the map of special studies zones.

This is a major goal of liberal education, to give people enough background to evaluate the maps and data that are presented to them in the course of their everyday lives as citizens.

Activity: Ask students why real estate brokers might prefer to use Transparency 9F than the other maps in following the rule about notifying buyers about earthquake risk. Ask about the financial implications of declaring a house to be in a quake-prone area if it is not really in danger. "Wouldn't that unfairly deprive a house owner of profit?" The issue is too complex to accept either extreme, and careful comparison of these maps can help students move toward a more reasonable middle ground. Have students try to draft a fairly worded warning that could be given to potential homebuyers.

Map-evaluation skills are a major goal of geographic education. To make sure that students acquire the skills needed to make reasonable comparisons and interpretations of different kinds of maps, a textbook author or course designer should clearly show the skills that are taught within each major unit of the course. Authors who do not provide this kind of information for textbook buyers are not doing their job. Teachers deserve better!

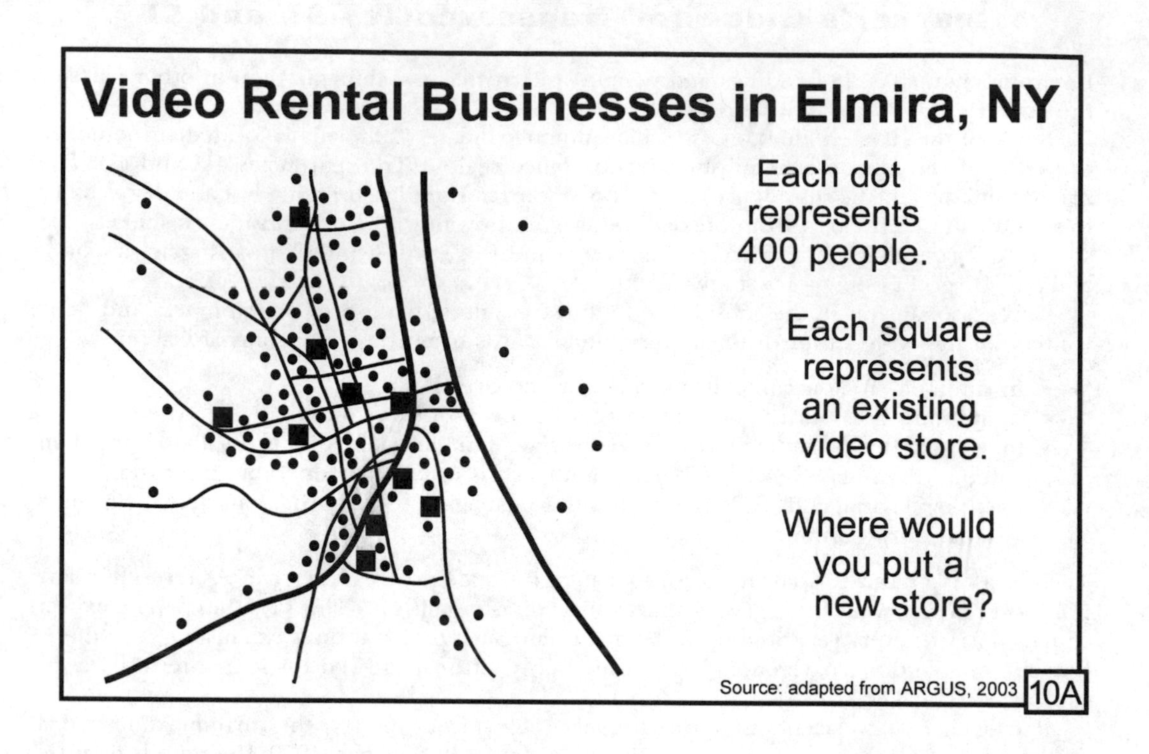
Video Rental Businesses in Elmira, NY
Each dot represents 400 people.
Each square represents an existing video store.
Where would you put a new store?
Source: adapted from ARGUS, 2003
10A

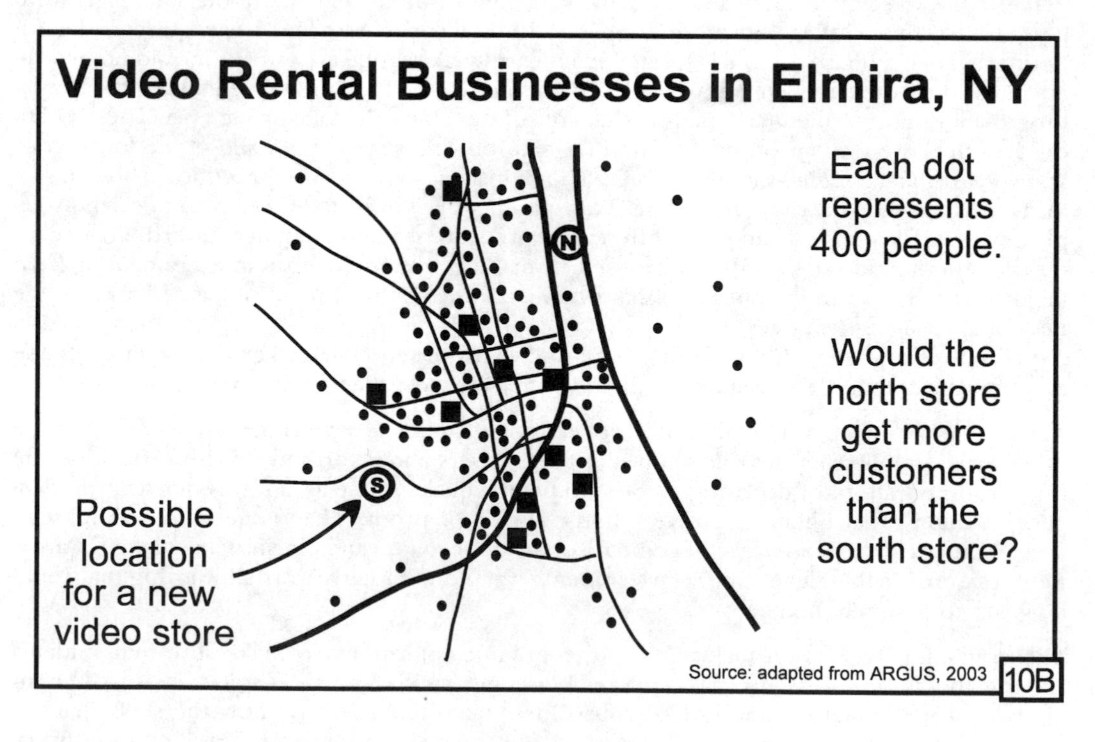
Video Rental Businesses in Elmira, NY
N
S
Each dot represents 400 people.
Would the north store get more customers than the south store?
Possible location for a new video store
Source: adapted from ARGUS, 2003
10B

Teacher's Guide for Transparencies 10A and 10B

This map depicts Elmira, New York, a city that used to be an important producer of picture tubes and other TV components. The decline of U.S. television manufacturing hit Elmira hard. People lost jobs. Residents no longer had enough money for vacations or fashionable clothes. Ironically, conditions like these that hurt travel agents and clothing stores can help some other businesses. For example, video-rental stores provide an inexpensive and useful service.

Activity: Imagine that a company has asked for a permit to build a new video store. The yellow pages of the phone book show the locations of existing video stores (triangles on the map). Ask students to decide which star is in a better location for a new store. There are several steps in the process of using maps to analyze the potential market for something. The first two steps are done for this example.

1. Find a basemap (a dot map of population is exceptionally good, if available).
2. Mark the locations of existing businesses that are potentially competing for the same customers.
3. Use a pencil and draw faint marks halfway between each pair of competitors. (If two are located close to each other, you might treat them as one location and give each one half of the customers in that area.)
4. Adjust the marks to reflect other considerations (if desired):
   a. If products are of different quality, move the marks to give a larger share of the area to the business that appears to produce the better product.
   b. If transportation is not equal, adjust the marks so that the seller who has better access gets more area.
   c. If political or economic borders have an influence, adjust the marks to give more area to the seller who gains from that influence.
   d. If personal preferences or loyalties have an influence, adjust marks to give more area to the seller that customers seem to prefer.
5. Use those marks as starting points for dividing lines. These lines separate the people who are likely to go to one seller from those who are more likely to go to another. The goal is to make a map that shows the "territory" of each seller.
6. Count dots within the territory of each seller. Multiply the number of dots by the population represented by each dot. The result is an estimate of total population in the territory.
7. If you need a precise estimate, you could conduct a phone or mail survey to find out what proportion of the people in an area rent videotapes. The result is an estimate of the number of customers a seller is likely to have at a given location.

These maps and this Activity are adapted from ARGUS Activity N, which also has an extended simulation of the process of deciding where to locate expansion teams in major-league sports. Note that the exercise teaches theory (location allocation) and a skill (market estimation) as well as some facts about a specific place (Elmira).

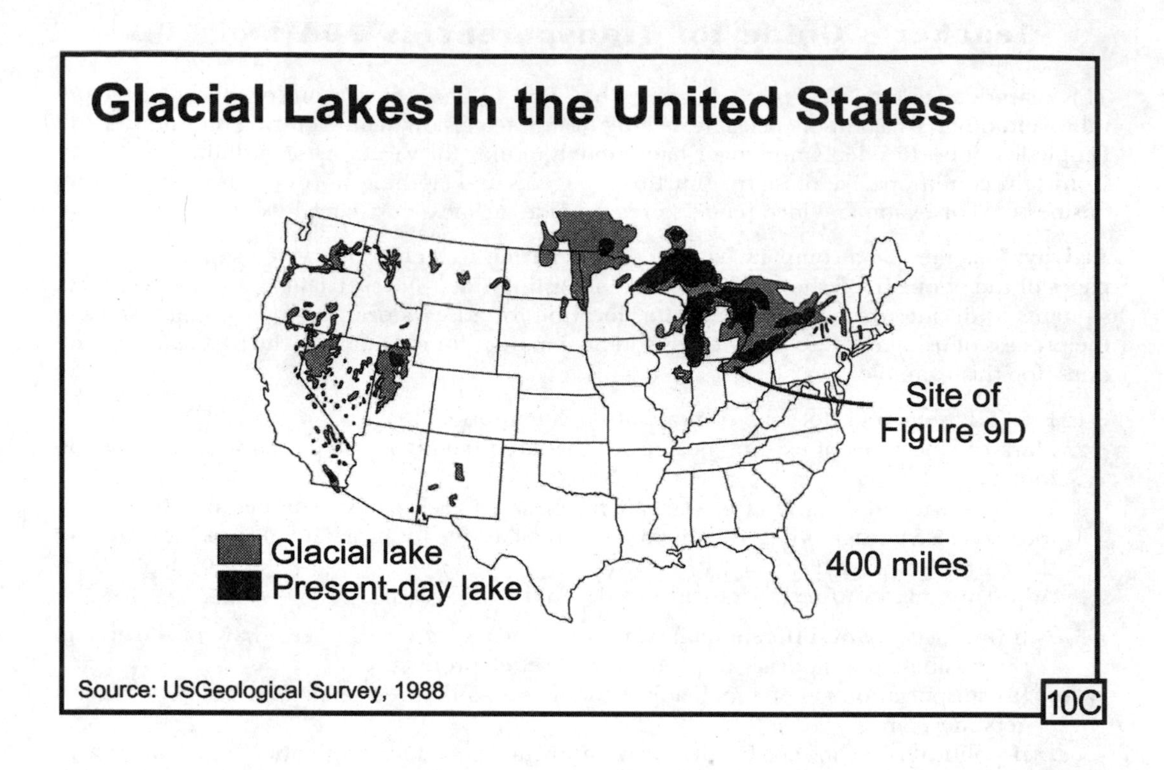
Glacial Lakes in the United States
Site of
Figure 9D
Glacial lake
Present-day lake
400 miles
Source: USGeological Survey, 1988
10C

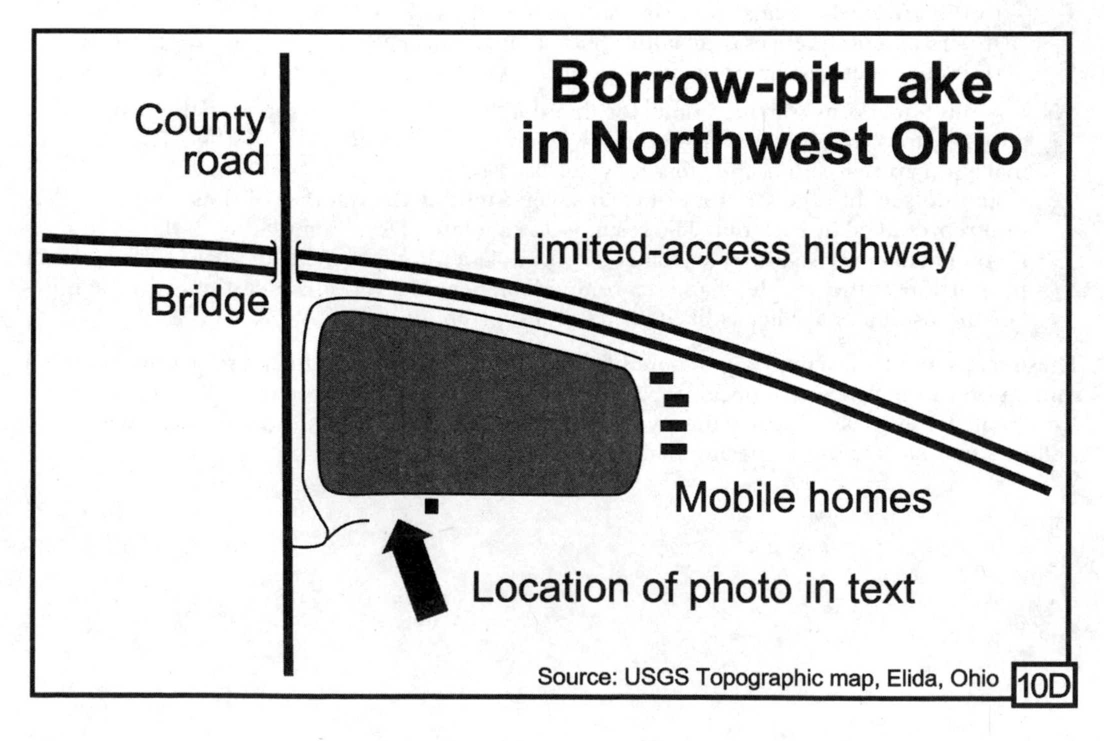
Borrow-pit Lake
in Northwest Ohio
County
road
Limited-access highway
Bridge
Mobile homes
Location of photo in text
Source: USGS Topographic map, Elida, Ohio
10D

Teacher's Guide for Transparency 10C

Transparency 3L shows the land that was covered by ice sheets during the Pleistocene period in geologic history. During that time of colder and wetter climate, lakes formed in many mountain and desert areas that are now dry and barren. As the ice melted, water often was trapped between the wall of ice and hills beyond the farthest extent of the ice sheet. (The hills either were already there and may have stopped the ice from spreading further or they may have been created during the advance of the ice as it pushed rocks and dirt ahead of it.)

Ice Age lakes of this type have an important common feature. They tended to trap sediment. Here's how: winds and streams brought dust and sediment into the lakes, which settled as mud at the bottom of the lake and filled in the deeper areas. When these lakes drained and disappeared, the result was often a nearly flat plain.

Activity: Put a few small objects into a bottle. Fill the bottle with muddy water and allow the mud to settle. Ask students why the mud tended to settle into the low areas around the objects. Then ask what they would see if the water were all drained away.

Activity: If you are lucky, a competition to set a new land speed record will take place in Utah at the test track on the Bonneville Salt Flats during your class. Ask students why the test track is there. The Bonneville Salt Flats are an exceptionally flat area that was covered by the ancient glacial lake, Lake Bonneville. The Great Salt Lake is a remnant of this Ice Age lake (see the Transparency). Occasionally, heavy rains in the surrounding mountains will raise the level of the Great Salt Lake a few feet. This does not sound like much, but because of the flatness of the terrain, it can flood many square miles of land around the present lake (including the test track).

Teacher's Guide for Transparency 10D

Activity: The mostly straight-sided lake on this map is not an isolated case. Find aerial photographs or topographic maps for any of the former glacial lakes shown in Transparency 10C (a good source is *www.terraserver.com*). Among the features you are likely to see on these maps are straight ditches, rectangular ponds, and other obviously artificial water features that were made as people tried to figure out how to live on an extremely flat plain. This kind of map-aided exploration of "ordinary" landscapes would be good counterpart for a host of published map interpretation workbooks that focus on volcanoes, canyons, ski slopes, and other spectacular terrain features.

In this book I have said more than once that the ability to apply the information on a broad-scale thematic map to a narrow-scale reference map (and vice versa) is one of the key strategies for geographical analysis. This is the skill that makes information on thematic maps both accessible and useful to those searching for the background info they need to solve a local problem. Learning to do this by using a faraway, exotic landscape is a way to focus on the skill without the emotional overtones and rhetoric that accompany a controversial local issue. Then, students can use the same skill to find information that can help shed light on a local landscape puzzle or controversy.

If they learned to do that, teaching geographical skills would really pay off!

Glossary and Index

Note: Transparency number in parentheses following page number.

capital - political center of an area, 2
capital - also, tools and money used to make production more efficient, 47
cartogram - map that shows the amount or value of something in an area by distorting the size of the area, 205 (*3O*)
cartographic convention - generally accepted way to use symbols on a map, 44
cartography - the art, science, and craft of making maps, 5
categorilla - term coined for the largest feature in any category: the highest mountain, longest bridge, biggest ball of twine, and so on, 63, 119
causality - the idea that one event may be the cause of another event, 56, 58
census - official count of population, including age, income, occupation, and so on. In the United States, the government has to do a census every ten years, 131
central place theory - theory to explain the size and locations of *service centers*, ranging from many scattered small places providing general services that are needed every day, up to a few large central places providing specialized services needed only occasionally, 41, 116, 187 (*2G*), 187 (*2H*)
Central Valley - important farming area in central California between the Coast Ranges and the Sierra Nevada, 5, 105
Chamber of Commerce - local business organization that often produces community profiles to give to interested individuals, 70, 132, 133, 191 (*3B*)
chaos theory - idea that big effects may have tiny, unpredictable causes, 57
checklist - data form with labeled blanks to be filled in by the observer, 68
China-bashing - blaming Chinese competition for the loss of jobs in the United States, 47
chokepoint - place where traffic is funneled through a narrow strait, pass, or other feature that constricts traffic, 78, 79, 102, 225 (*4Q*)
choropleth map - map in which political units or census areas are colored or shaded according to the amount of something in them; best used for calculated ratios such as population density; the arbitrary borders of the map units can sometimes be misleading, 5, 102, (*1A*), (*1B*), 233 (*4Y*), 233 (*4Z*), 249 (*8D*)
Civil War (1861-1865) - war that occurred when people in southeastern States tried to secede and form their own country, 43, 108, 203 (*3N*)
classification - putting things into categories of similar features, 65
climate - "normal" weather in a place, both the average conditions and the extremes one is reasonably likely to experience over a period of time, 217 (*4I*), 217 (*4J*)
climatic analog - place that has a similar location on another continent and therefore has similar climatic conditions, 106, 116, 143
cognitive psychology - the study of how people learn, 24, 123
community profile - concise description of the population, economy, and other relevant facts about a local community, 70, 122, 191 (*3B*), 243 (*7D*), 243 (*7E*)
computer programs in geography - programs that store data, assist with simulations, make maps, and do other tasks to help someone do or teach geography, 154
condition - characteristic or feature in a given *location*, 100
Confederacy - the states that seceded from the Union in 1860-1865, 43
connection - natural or human-made *link* with another *location*, 101
container port - place where people efficiently can move sealed containers of cargo from ships to trucks or railroads and vice versa, 77
contour line - *isoline* separating higher from lower elevation, 237 (*7A*)
Coriolis "Force" - tendency for objects in motion on the surface of the earth to be deflected from an apparent straight path; the deflection is to the right in the Northern Hemisphere and to the left in the Southern Hemisphere, 215 (*4G*)
Corn Belt - region of grain/livestock farms, from western Ohio to eastern Nebraska, 225 (*4R*)
correlation - statistical measure of the relationship between variables; a high correlation means that a place with a high value on one measure will probably also have a high value on the other measure, 82, 108
credibility - likelihood of being believed, 13
creed - a statement of basically unprovable faith, 56

cultural definition of resources - idea that the people in an area decide what things can or will be used as resources; compare *determinism*, 71, 141
cultural legacy - persistent effect of the use of land by a particular *culture*, 71
culture - shared mental rules about appropriate behavior, 71
curricular position - time set for a subject in a weekly class schedule, 17, 127, 247 (*8A*)

deductive logic - moving "downward" from a general principle to a specific application, 8
deforestation - removal of trees from an area, 34
Delta - flat floodplain of the Mississippi in Arkansas, Louisiana, and Mississippi, 231 (*4X*)
density - number of objects per unit of area, 6, 157
design - using material efficiently to do something worthwhile (as opposed to *art*), 31
determinism - idea that nature "determines" what people do in an area, 71, 142
dietary deficiency - lack of an essential nutrient in what a person eats, 152
diffusion - the spread of something through an area, 36, 109
direction - orientation of a path or sight-line with respect to north, 213 (*4E*), 255 (*8I*), 237 (*7A*)
directions - instructions about how to go to a specified place, 101
distance - amount of space between two locations, a *spatial primitive*, 211 (*4D*)
dogma - belief that "this is absolutely true, period" (see *relativism*), 48
dot map - uses dots (or other repetitive symbols) to show the general pattern of something in an area, 116, 221 (*4M*), 225 (*4R*), 231 (*4X*), 239 (*7B*), 257 (*9A*), 259 (*9C*), 263 (*10A*), 263 (*10B*)

earthquake - ground vibrations, often destructive, that occur when rocks slip along a geologic fault (a break in the rock), 149, 261 (*9E*), 261 (*9F*)
ecological fallacy - presumption that all persons within an area have all of the traits that are shown on maps of the area; for example, a given *census tract* may have high average income and many domestic servants, but it is an ecological fallacy to conclude that domestic servants there have high incomes, 171 (*1A*), 171 (*1B*)
El Niño - exceptionally warm ocean water off the west coast of South America, 110
empire - large territory with many different kinds of people under the control of one political group, 44
engine of an activity - the basic thing that a student does to make a classroom or homework activity run, 4, 54
environmental conditions - the mix of natural features such as climate, soil, and wildlife in a specific place, 39
environmental impact statement - mandatory planning document that describes the likely environmental effects of a proposed activity, 132
epidemiology - study of the causes of diseases, 81
Erie Canal - waterway that made it easier to travel from the frontier to New York than to any other city on the East Coast, 76
essay test - set of questions that ask students to write answers in their own words, 122
ethics - issues of right and wrong, a concern of the humanist *perspective*, 7
etiquette - rules about how to behave, 1
Eurocentrism - focus on European people and their actions, 63
European explorers and settlers - people who came to America from Europe during what European historians call the Age of Exploration, 241 (*7C*)
evaluation, student - assessment of performance on a specific task, 112-125
evaluation, geographic - opinion about a place, 29, 45-49, 54, 113
exception - place with conditions that do not follow a perceived "rule," 108

fan - "apron" of eroded rock around a mountain in a desert region, 67
federal land ownership - land owned by the federal government rather than by private individuals or other government agencies, 205 (*3O*)
field trip - excursion to make observations and test hypotheses in the world, 26, 122-123, 245 (*7F*)
flowline map - map that varies the size of a line to show the number of people or amount of something moving along a route, 143, 227 (*4S*)

fluid dynamics - principles that govern the motion of fluids such as air or water, 66
formal region - group of places with similar features, 80
frontier - area where people are moving in, building homes, and starting farms and other businesses, 43, 203 (*3M*)
frost-free season - part of the year that has no significant risk of frost, 43, 199 (*3J*), 201 (*3K*)
functional region - group of places linked together by a flow of something, 80

Gardner, Howard - author of *Frames of Mind*, about cognitive orientations, popularly called learning styles, 123
GENIP - Geographical Education National Implementation Project, an organization formed to help coordinate educational efforts of several geographical professional associations, 86
geo-diary - record of landscapes and objects a person sees in a typical day, 26
Geographic Information System (GIS) - computer file of locations and their traits (e.g., ownership, climate, yield, family income, and so on), 231 (*4X*)
geographical irony - realization that the very things that make a place good for some purposes can also pose a threat for others, 15, 84
geographical skill - analytical procedure that can help us understand spatial patterns and connections, 97
geography - scholarly discipline with focus on spatial patterns and links, 2, 7
ghetto - area in which a particular race or class of people is more or less forced to live (like "tree," it is a word with different meanings in different places), 34
Gibraltar - strait that separates the Mediterranean Sea from the Atlantic Ocean, 78, 225 (*4Q*)
GIGI - Geographic Inquiry into Global Issues, an NSF-funded project to develop and test classroom activities for secondary schools, 3
glacier - moving mass of ice that forms in a cold, snowy environment, 43, 201 (*3L*), 265 (*10D*)
GPS - global positioning system, device that determines location by triangulation from orbiting satellites, 100
grade - letter that indicates level of performance on a task or test, 123
graffiti - messages painted on walls, bridges, and other "public" places, 1, 162, 169 (*IntroA*), 169 (*IntroB*)
Grand Canyon - large gully (!) formed by the Colorado River, 30
gravity model - method of predicting the amount of traffic, phone calls, and other *spatial interaction* between places, 35, 195 (*3E*), 195 (*3F*)
growing season - length of time between the last killing frost of spring and the first one in autumn, 43, 199 (*3J*), 201 (*3K*)
guild - medieval organization in which members train new apprentices to take their places when they end their careers, 126

Harpers Ferry - strategically important town where the Shenandoah and Potomac Rivers join and flow through a gap in the Blue Ridge, 80, 173 (*1C*), 173 (*1D*), 251 (*8F*)
hemisphere - half of the globe; also, half of the brain that appears to "focus" on specific kinds of mental tasks, 24
hermeneutics - attempt to understand how people get meaning from a text, 28
history - scholarly discipline with a focus on the influence of time, 2, 7
Homestead Act of 1862 - law granting land to settlers, 72, 107
HOTS - high-order thinking skills, such as synthesis, evaluation, and other cognitive activities that are "higher" than mere memorization, 48, 50, 120
human-environment interaction - the geographical theme that deals with the mutual influences between humans and their environment, 37, 89, 90, 157
hypothesis - statement about a possible relationship between phenomena, 68, 122

image - scene, sound, or other sensory impression, 29-32, 114
imaginary places - places that do not exist; often used in instruction, though their effect often is a confusing message, at best, 62, 121
in-service training - summer courses, workshops, and other educational activities for teachers who are currently employed in classrooms, 126, 133

inductive logic - thinking "upward" from specific *observations* to derive a general principle, 8, 92
industrial orientation theory - idea that people tend to locate factories near the input or output that costs the most to transport, 175 (*1E*), 175 (*1F*), 177 (*1G*), 177 (*1H*)
infrastructure - canals, roads, powerlines, and other built features that support human activity in a place (some analysts include institutions such as welfare, schools, and so on), 40, 73, 75, 78, 160
international rationale for geography - knowledge about other places is useful in learning how to live in an interconnected world, 3
inventions - new ideas that can result in new products and economic growth, 82, 233 (*4Z*)
Islam - one of the world's major religions, 16, 183 (*2D*)
isoline map - map that uses lines to separate areas of higher value from areas of lower value, 189 (*2I*), 199 (*3I*), 199 (*3J*), 201 (*3K*), 203 (*3M*), 217 (*4J*), 229 (*4U*)

land division - method of marking land that is owned or controlled by specific individuals or groups, 40, 72
land use - changes people make (and structures they build) in order to use land, 39-43
language - using words and other symbols to communicate images, theories, and value judgments, 27, 61, 113
latitude - angular distance north or south of the Equator, 66, 116, 209 (*4A*)
layout devices - boxes, tinted pages, special type fonts, and other mechanical devices to help a reader stay oriented in a book, 19
lead - chemical element used in paint and as a gasoline additive until its adverse health effects were discovered, 152
learner outcomes - statements of desired student knowledge, skills, and attitudes in a subject such as geography, 14, 92, 96, 113, 115, 117, 132
learning style - preferred method of learning, reading, experimenting, and so on, 124
legacies (relics) - landscape features left by previous inhabitants of a place, 71
levee - long dam to protect an area from flooding, 141
link - something that connects two places together, 73
local interactions - interplay of various factors in a specific place, 16, 19, 100
local rationale for geography - knowledge about local places is useful in making them more safe, fair, beautiful, and so on, 3
location - position in space, a cornerstone geographic idea, 3, 8, 59, 100, 159, 209 (*4B*), 211 (*4C*)
location rent theory - tries to explain the pattern of land value in an area in terms of locations with respect to markets, resources, and so on, 22, 40
location-allocation analysis - deciding how many of something (such as convenience stores) to put in an area and where to put them, 41, 157
locational advantage - having better site conditions and/or situational connections, 140

malaria - disease carried by mosquitoes in hot and wet places, 81, 108
map pattern analysis - study of the arrangement of features in order to get ideas about what might cause them, 42, 107
map pattern correlation - mathematical method of comparing map patterns to measure the degree of similarity, 44, 82, 108
matrix - x-y grid of cells with labels on the columns and rows, 20, 143, 150, 189 (*2J*)
Maumee Lake Plain - flat bottom of a former glacial lake in northwest Ohio, 159
mental map - conceptual image of the spatial arrangement of something, 43, 122
migration - movement of people from one place to another; migration theory tries to identify the pushes, pulls, or both that persuaded people to move, 116
moisture balance - balance between precipitation (moisture income) and evaporation (moisture outgo) in a particular place; a moisture surplus can make rivers, leach soils, cause floods, and so on; a moisture deficit can cause deserts, 43
multiple-choice test - set of questions that ask students to identify which of several alternative responses is correct, 118
musical theme - sequence of notes that recur in a musical composition and help tie it together, 87

National Geography Standards - list of concepts and skills that students should

the quantity of something at a point or in an area (also called graduated-symbol or scaled-symbol map), 217 (*4I*), 223 (*4P*), 227 (*4T*), 249 (*8C*)

Public Land Survey – system of using straight lines to divide land into square Townships and Sections, 213 (*4F*)

pure geographer – uses geographical methods to understand the world, 56, 144

random – not organized in any way, 107

range – distance a typical person is willing to travel to get a particular good or service, 72

rationale – reason for doing a particular thing, such as studying a subject such as geography or refusing to grant a request, 3, 8, 127

raw data – population counts, production amounts, rainfall measurements, and other absolute numbers (see *processed data*), 7

reasons, reasonable and unreasonable – explanations for locations of things, 57

redlining – drawing lines on a map to delimit areas where a bank might refuse to grant loans for home purchase or improvement, 46

reference map – map that shows the locations of a variety of things in an area that is usually fairly small (contrast *thematic map*), 16, 70, 147, 161, 183 (*2C*)

region – sizeable area with generally similar conditions (formal region) or internal connections that tie it together (functional region), 80, 104, 131

regional geography – study of interaction of factors at a local scale, 16, 19, 146, 189 (*2I*)

relative location – location expressed in terms of relations with a known location, 59

relativism – belief that "nothing is absolutely true" (see *dogma*), 48

resource – something that people in an area have learned how to use, 71, 141

resume – list of qualifications and experience, used in seeking a new job, 131

role-playing simulation – classroom activity in which different students are assigned different roles and asked to interact, 49, 55

rubric – statement of criteria and expected performance levels, used in evaluating written answers, projects, or presentations, 120

safety net – system of insurance and/or public subsidy to help people that are affected by accidents, unemployment, or other catastrophes, 83, 233 (*4Y*)

Sahara – desert in northern Africa, 67, 79, 217 (*4J*)

satellite image – photograph or electronic image taken from a satellite in orbit, 38

scale, map – mathematical relationship between the size of a map and the part of the real world it shows, 116, 173 (*1C*), 173 (*1D*)

scientific method – term for the idea that scientists proceed in an orderly manner to gather observations in order to test hypotheses, 50

scissors – tool with two opposing blades that cut better when working together than they could working separately, 15, 93, 132, 141, 146, 157, 181 (*2A*)

sequent occupance – sequence of ways in which people occupied land at various times in the past; many present-day landscape features are left from previous occupance, 72

sickle-cell anemia – disease associated with malaria in tropical regions, 81

site – all local conditions (terrain, climate, soil, vegetation, energy, mineral resources, and so on) that affect what people can do in a place, 140, 157, 235 (*5B*)

situation – connections (transportation routes, corporation ties, political associations) between a place and other places, 140, 157, 235 (*5B*)

skill – ability to perform a specific task, such as locating a place or measuring distance between places, 97, 120-121

slaves – people who are "owned" by and forced to work for other people, 231 (*4W*)

slide – small transparent photograph that can be projected on a classroom wall or screen, 21, 31, 52, 133

soil survey – book with detailed maps and tables about the soil, climate, wildlife, and other resources of a county, 131, 253 (*8G*)

Soil Taxonomy (USDA) – set of rules to classify soils into groups, 39

spatial – refers to distances, directions, areas, and other aspects of space, 7

spatial cognition, spatial thinking – thinking about spatial relationships appears to take place in a distinctive part of the brain, 98

spatial imbalance (or bias) – tendency for features to be found in one part of a map rather than throughout the area, 42

spatial interaction – traffic, phone calls, and

other flows of people, goods, or information between places (see *gravity model*), 35, 74, 195 (*3E*), 195 (*3F*)
spatial model - theory about how conditions in one place can affect conditions in another place, often quite far away, 110
spatial pattern - distinctive arrangement of features or objects in space, 107
spatial primitive - basic spatial concept, such as location, distance, direction, or enclosure, 61
spreadsheet - computer program that puts data in rows and columns for analysis, 69
standardized test - test that is given to a large number of people and can be used to compare schools, states, or even countries, 123, 125
standards, educational - statements of desired learner outcomes in a subject such as geography, 14, 88, 96, 136
strand - one part of a rope or other woven object, 28, 112, 124, 132
street pattern - arrangement of roads in an area, 197 (*3H*)
string - spatial pattern of features arranged in a narrow line, 42, 107
subregion - smaller part of a region, 82
subsidence - zone of sinking air about 28 degrees of latitude north or south of the zone of instability near the equator, 66, 215 (*4H*)

tax deduction - reduction of taxable income for business expenses, 133
teachable moment - time when students are primed to learn a specific idea, 150
territorial marker - landscape feature that marks the edge of someone's space, 1
test - method of evaluating mastery of a knowledge or skill, 22, 112, 118
thematic map - map that shows the spatial arrangement of a few things in an area (contrast *reference map*), 16, 71, 116, 147, 160, 183 (*2D*)
theme - idea that permeates a discussion and holds it together by defining what ideas are important, 87, 91, 147, 160, 235 (*5A*), 235 (*5B*)
theory - statement to explain something (or at least allow us to predict something that may happen under similar circumstances), 35, 112, 114, 122
Tokyo transit region - area connected by subways and bus lines in Tokyo, Japan, 80
tolerance - willingness to "put up with" other ideas or people, 48
topical geography - study of the spatial patterns of various phenomena, usually at a broad scale (see *thematic map*), 16, 146, 189 (*2I*)
topographic map - a detailed *reference map* that shows terrain, water bodies, forest cover, roads, houses, other buildings, and many other features; these maps are published by the United States Geological Survey, 116, 131, 173 (*1C*), 237 (*7A*)
tragedy of the commons - idea that what works for a few in an area may not work if there are many there, 157
transect - line along which an observer records data, 69
transition - mixture or blend of features in a border zone between regions, 104
tree - an example of a word with different meanings in different places, 33
trivia - isolated and therefore meaningless facts, 1, 24, 33, 93, 113, 119, 154

unemployment benefits - payments to unemployed people, 83, 233 (*4Y*)
URISA - the Urban and Regional Information Systems Association, which publishes a *Proceedings* of each national meeting, 145

value judgment - opinion about the beauty, truth, morality, or importance of something, 29, 46, 54, 112, 122, 160
vernacular theory - explanatory idea that is believed by the people in an area, 38
volcano - place where molten rock from the earth's interior erupts onto the surface, 108

West Bank - disputed area between Israel and Jordan, 42, 259 (*9C*), 259 (*9D*)
where - a little word with big meanings, 10
world-class geography - geographic knowledge and skills that are equal to the average for peer countries around the world, 14
Yellow pages - listing of business addresses and numbers in a telephone book (or a CD-ROM), useful in making a community profile, 131, 247 (*8B*)

Index of Geographic Locations

About the Author

Phil Gersmehl, PhD, is Professor/Director of the New York Center for Geographic Learning, Department of Geography, Hunter College, City University of New York. He has worked with geographic alliances in 23 states, as well as with similar groups in Canada, England, Japan, Korea, and Russia. He was Director of the ARGUS Project and Codirector of the ARGWorld Project, both funded by the National Science Foundation and other sources, and both administered by the Association of American Geographers.

Prior to moving to Hunter, he worked on Distance Learning courses for the Continuing Education and Extension program at the University of Minnesota, and on a number of instructional computing projects for IBM and for several university programs. A highlight of his work at that time was a pilot TV episode for the Corporation for Public Broadcasting.

He is author of two textbooks, *Physical Geography* (1980) and the *Language of Maps* (1996), a regular contributor to the journals of several professional societies, and the recipient of awards for research and geographic education.

About the CD

The facing CD on the inner back cover of this book includes the following:

- A multimedia presentation of extended activities on a number of useful topics for geography students (**Mac users** will have better results by using Virtual PC version 6.0 or higher. Copy the files from the disk into the Virtual PC desktop, then execute the files and programs from there.)
- .pdf files of the tranparencies in the Illustrations (Transparency Masters) chapter of this book, pages 168–265
- The geography standards for the U.S. states that have them
- Sources of Local Geographic Information (©1995 Darrell Napton, South Dakota State University)
- Suggestions/guidelines for how to write course objectives effectively